Praise for the Book

A true professional, Mike Fagel arrived at FDNY WTC Incident Command Post on Duane Street, a short distance from Ground Zero, as chaos was still not contained. He organized, directed, and cajoled until order again appeared in our health and safety efforts for the thousands of personnel struggling at rescuing and recovering the victims of 9/11. Many of the Ground Zero workers have their health still intact because of Mike's courage and efforts. The Fire Department was well served by his knowledge and expertise.

Charles R. Blaich
Deputy Chief of Department, FDNY (Ret.)
Colonel, USMC (Ret.)

. . . a must read for emergency managers, planners, first-line responders plus faculty, and students involved in the study of emergency response, homeland security, and public health. Mike Fagel has a rare combination of both superb academic and hands-on, first-responder credentials.

Colonel Randall J. Larsen, USAF (Ret.)
Director, Institute for Homeland Security

Mike Fagel demonstrates in his third textbook his on-the-job expertise as an emergency manager; as someone who has known Mike for many years, I highly recommend his approach and his concepts. He continues to pursue the professional development of the field of emergency management and this is demonstrated in his most recent work. Dr. Fagel is committed to using his real world "on-the-job" approach to making the rest of us safer.

Edward Plaugher, Fire Chief (Ret.)
Arlington County Fire Department
Arlington, Virginia

If you have ever had an emergency management situation, Mike's classroom teachings and publications are a must for your agency. We have the good fortune of not only having Mike's high level of expertise, but also to have him as one of our local residents with a longstanding tie to the Kane County Sheriff's Office. Homeland security and first-responder safety are of paramount importance in today's world, and Mike's teachings give a highly detailed guideline to help first responders make order out of chaos. Mike's real-world experience, most recently involving many events we see in the news and his willingness to educate our first responders, is an opportunity that should be utilized by all agencies.

Patrick B. Perez, Kane County Sheriff
Sheriff of Kane County Illinois
Chicago, Illinois

Dealing with an emergency is like a battle or even a war. The powers are but the weapons. They are useless without a plan. A plan must address any scenario that may occur. It is the plan that will guide emergency leaders to bring resources to bear quickly as the catastrophe develops. Further, it is in the process of developing the plan that opportunities for prior mitigation emerge through hard engineering, procedures, or training. To achieve plans, one needs adequate resources, good people, wide involvement from all parties, and good morale and esteem in the planning team.

Finally, creating contingency plans is a lonely and unsung profession. No member of the public ever knows what one does. Everybody, including the planners themselves hope that their plans are never used. It is essential, therefore, that national and local leaders go out of their way to recognize the work of the contingency planners so as to raise their morale and esteem. One day, the future of the nation may depend on them.

Regardless of the comprehensiveness or process of planning, NO plan survives first contact with the enemy. Having alternate solutions is imperative.

Jonathan Best LP, EMSI, CHS-III
Director, Public Health Preparedness and Response
State of Connecticut Department of Public Health

Effective emergency and crisis management requires vigilance across a panoply of evolving threats, hazards, technologies, and operational capabilities. No matter how experienced one is, as an emergency responder or emergency manager, there are new lessons to be learned everyday. This book complements earlier treatments of EOC design and operations by Dr. Fagel, and offers the practitioner new confidence-building measures for confronting a range of public health, agroterrorism, and active shooter incidents that can impact a community and shake the confidence of the populace to return to normalcy. His focus on the best use of social media and other communication modalities is timely and important in shaping contemporary planning and community resilience. Maintaining the trust and confidence of the general population from incident onset through long-term recovery is an essential element of effective emergency management and this book is a toolkit for best practices in citizen-centric preparedness.

Robert J. Coullahan, CEM, CPP, CBCP
President, Readiness Resource Group

Emergency management planning must consider myriad elements to successfully prepare for any incident. When you are dealing with agriculture and food elements, the challenges can become much more complex. Dr. Fagel has experience in both traditional emergency management and agriculture operations that provide a unique understanding required for successful crisis management and emergency planning.

Jeff M. Witte, Director/Secretary
New Mexico Department of Agriculture

Dr. Michael J. Fagel has assembled a group of experts in a variety of areas of emergency management and has edited a highly usable book that belongs to the desks of EM professionals. Most emergency operation plans have appendices relating to specific critical events. The organization of Fagel's book around hazard-specific issues makes it easy to find useful guidance when planning for a wide range of critical incidents from agroterrorism to pandemics to active shooters to large-scale public events. The coverage is very up to date, as evidenced by references in 2013 and coverage of such modern topics as the impact of social media on emergency management. Having taught with Dr. Fagel, I see in this book the effective classroom style that I associate with his work, but translated into a very practical and useful manual. In conclusion, it is a book that is easy to recommend.

Frank K. Cartledge
Alumni Professor of Chemistry Emeritus
Louisiana State University

I have worked beside Dr. Mike Fagel for more than 5 years. He is a professional in every sense, a committed emergency responder/manager, an agribusiness expert, an educator, and a good friend. The diversity of thinking, working, and experience Dr. Fagel offers on the subject matter of agroterrorism can only be matched by a selected group of experts. Dr. Fagel has meticulously detailed all the important aspects associated with preventing, responding, and recovering from an attack on agribusiness and the food supply. Mike introduces the subject by showing the immense scope and size of the number-one industry in the United States, agriculture, and the allied industries of food production. He outlines the complexity of the farm-to-table continuum making a special effort to point out where security should be improved. He goes on to point out the recognition of agriculture as critical infrastructure formally recognized by the federal government. Several presidential directives and the Department of Homeland Security place an updated emphasis on the importance of agricultural infrastructure. And finally, Dr. Fagel effectively emphasizes the devastating psychological and economic consequences of an agroterrorism event.

Stan W. Casteel, DVM, PhD
Professor of Veterinary Pathobiology
Veterinary Medical Diagnostic Laboratory
College of Veterinary Medicine
University of Missouri

Effective emergency and crisis management requires vigilance across a panoply of evolving threats, hazards, technologies and operational capabilities. No matter how experienced one is as an emergency responder or emergency manager there are new lessons to be learned every day. This book complements earlier treatments of EOC design and operations by Dr. Fagel, and offers the practitioner new confidence building measures for confronting a range of public health, Agroterrorism and active shooter incidents that can impact a community and shake the confidence of the populace to return to normalcy. His focus on best use of social media and other communications modalities is timely and important in shaping contemporary planning and community resilience. Maintaining the trust and confidence of the general population from incident onset through long-term recovery is an essential element of effective emergency management and this book is a toolkit for best practices in citizen-centric preparedness.

Robert J. Coullahan, CEM, CPP, CBCP
President, Readiness Resource Group

Crisis Management and Emergency Planning

Preparing for Today's Challenges

Crisis Management and Emergency Planning

Preparing for Today's Challenges

Michael J. Fagel, PhD, CEM

CRC Press
Taylor & Francis Group
Boca Raton London New York

CRC Press is an imprint of the
Taylor & Francis Group, an **informa** business

CRC Press
Taylor & Francis Group
6000 Broken Sound Parkway NW, Suite 300
Boca Raton, FL 33487-2742

Version Date: 20131009

International Standard Book Number-13: 978-1-4665-5505-1 (Hardback)

Visit the Taylor & Francis Web site at
http://www.taylorandfrancis.com

and the CRC Press Web site at
http://www.crcpress.com

I dedicate this book to the victims of recent events, the responders, and their families.

And, I must thank my parents for their support of my endeavors, along with my children and my wife, Patricia. They have all had their plans changed when the pager goes off (yes, there are still pagers) and multiple cell phones! Their support has been crucial to my success and I sincerely thank you all! So, please read, learn, and share some of these hard-fought lessons. Reach out to our authors, and always keep learning and communicating. Be prepared, stay safe, and take care of yourself and your family!

Michael J. Fagel, PhD, CEM, CHS-IV

Contents

SECTION II RESPONSE PLANNING AND PREPAREDNESS

SECTION V EXERCISE DESIGN AND DEVELOPMENT

SECTION VI VULNERABILITY ASSESSMENTS AND CRITICAL INFRASTRUCTURE

Foreword

Dr. Mike Fagel is an internationally recognized expert in the emergency management profession and is an extraordinary conveyor of his profound knowledge to those he teaches and mentors. I have had the honor and privilege of working with Dr. Fagel on numerous local-, national-, and international-level security, and emergency management programs. This experience includes codeveloping and coinstructing several courses, and codeveloping a national crisis and emergency management agency for a Middle Eastern ally of the United States. During my extensive work with Mike, I have found him to be a consummate professional, setting or upholding the highest levels of academic and professional, standards when conducting operations and delivering products. This book is among those extraordinary products that every emergency manager should possess and refer to routinely in their important work.

In this book *Crisis Management and Emergency Planning: Preparing for Today's Challenges,* Dr. Fagel and an elite team of professionals provide an outstanding tool that captures and communicates essential information of value to anyone involved in disaster science and emergency management. It is not a secret that the challenges emergency management professionals face are complex and dynamic, and this edition addresses both the critical components and the emerging trends and systems within emergency management.

The contents in this book deliver state-of-the-art strategies and procedures useful for identifying potential or actual hazards, preparing for such hazards, mitigating the cascading system failures during an incident, and facilitating a community bouncing back economically and culturally from disasters. In essence, this book is essential for anyone focused on the art and science of community resiliency and the whole of community emergency management approach—focused on saving communities—the primary role of an emergency manager. Recent incidents such as "Superstorm" Sandy, Hurricane Katrina, the Haitian 2010 earthquake, and the potential of an incident such as an earthquake in the New Madrid Seismic Zone require emergency managers to look beyond a government-centric approach and determine what other resources can assist within the whole of the community. Emergency managers benefit from remembering that in the processes associated with emergency management, the focus should remain on the communities served—it is not about the process, but the products and it is not about the incident, but the individual (www.continuitycentral.com).

A "landscape-scale disaster" (Leonard and Howitt, 2010) incident response and recovery effort (e.g., Hurricane Katrina, Haitian 2010 earthquake) would benefit from this book's guidance and frameworks for resource and information coordination that facilitates effective emergency management. Craig Fugate, the Federal Emergency Management Agency (FEMA) administrator, asserts that historically, within the emergency management community, practitioners of emergency management typically planned for what their respective capabilities can handle and do to respond to the government (www.continuitycentral.com). Administrator Fugate further asserts that emergency management practitioners really need to be planning for disasters that force them to look beyond a government-centric approach and determine what outside (or other) resources can assist. He challenges emergency management practitioners to plan for the worst-case scenarios or metascenarios, which go beyond the capabilities of government solutions. He refers to these scenarios as the "Maximum of Maximums" (MOM).

Administrator Fugate's statements concerning emergency managers' roles within local, tribal, state, and federal jurisdictions within the United States are also applicable to the international emergency management community. Evidence of the necessity for a whole of community perspective within the international disaster/emergency management construct, along with the need for a meta-leadership (leadership of leaders) approach, emerges during studies of major disasters, including the 2010 Haiti Earthquake, the 2011 Japan Earthquake/Tsunami/Nuclear Emergency, and other "predictable surprises" (Bazerman and Watkins, 2004) the world knows in terms of major disasters and catastrophes. As Max Bazerman and Michael Watkins provide in their book titled *Predictable Surprises*, the predictable surprise is a problem that will get worse when left unattended, eventually creating a far bigger problem; yet, the organization ignores the problem (Bazerman and Watkins, 2004). The risks posed by major disasters are projected to increase due to environmental changes, urbanization, mass population movement, and continuing technological innovation (or the lack thereof in many communities).

Owing to the interdependency of communities economically, environmentally, socially, and politically, whether considering a developed national-level community or the international community of nations, there appears to be clear realities associated with emergency management. Today's realities include the fact that a major disaster's impacts are more rapid, farther reaching, and interconnected (Aniskoff, 2011). These disasters require, in most cases, more than most state-level and/or national-level government-centric systems can handle, thus inferring that they are not government problems as much as they are societal problems requiring societal solutions. This is a changing perspective pertaining to stakeholders. Emergency managers will need to stop perceiving the public as a liability and more as a resource. Success or failure of a disaster response in the first few hours and days rests with the local emergency responders, and, the true first responders are the bystanders and survivors of the incident. Therefore, the public should be engaged as part of the solution (Fugate, 2010). Emergency management, therefore, is more about communications and networking than about tools and processes.

The best practice to build community resilience involves developing a broad-based national and/or international dialogue to explore and define new, community-oriented models for the practice of emergency management at all levels (United Nations, 2009). Concurrently, emergency managers at all levels (local, tribal, regional/state, federal/national, and international) may benefit from expanding partnerships with "outside" actors, working directly with advocacy groups, the private sector, community organizations, state/local/tribal/state governments, and other nations in collaborative relationships addressing all phases of emergency management. And, working with these partners, the stakeholders should operationalize these community-based emergency management concepts through a catastrophic planning initiative referred to earlier as the Maximum of Maximums (MOM), building and testing response and recovery capabilities against the basic tenants of a whole community approach to emergency management (Webster, 2010; FEMA, 2010).

These are lofty endeavors, not for the faint of heart and weak of mind. These are the endeavors of the prudent but courageous who seek to continuously grow within the disaster science and emergency management profession. *Crisis Management and Emergency Planning: Preparing for Today's Challenges* is a "must have" for those professionals choosing to engage in such noble endeavors.

J. Howard Murphy, MBA, MSS, FAcEM, CEM
Senior Homeland Security Program Manager and
Former Commander of the U.S. Army's first CBRNE Incident Response Force

References

Aniskoff, P. *Toward "Whole Community" Emergency Management* (PowerPoint). Washington, DC: FEMA, 2011.

Bazerman, M.H., Watkins, M.D. Predictable Surprises: The Disasters You Should Have Seen Coming and How to Prevent Them. Boston, MA: Harvard Business School Press, 2004.

Continuity Central Plan for the "Maximum of Maximums" Fugate tells state emergency managers. www.continuitycentral.com/news05436.html, 2010.

Federal Emergency Management Agency (FEMA). A "Whole Community Approach" to *Emergency Management.* Washington, DC: FEMA, 2010.

Fugate, C. Keynote Address at 2010 IAEM Conference. San Antonio, TX: International Association of Emergency Managers, 2010.

Leonard, D., Howitt, A. Rethinking the Management of Large-Scale National Risk (Lecture). Cambridge, MA: National Preparedness Leadership Initiative, Harvard University John F. Kennedy School of Government, 2010.

United Nations International Strategy for Disaster Reduction. Disaster Risk Reduction in the United Nations: 2009 Roles, mandates, and areas of work of key United Nations entities. United Nations, 2009.

Webster, W.R. *NEDRIX Mid-Year Update: Better EM Involvement by Private Sector* (PowerPoint). Boston, MA: FEMA Region I, 2010.

Preface

As we worked collectively to bring you our latest book for publication, we developed a very fitting name for the chapters we are presenting to you. The landscape of crisis planning and emergency planning has evolved greatly since our last two books were published in 2011 and 2012. The challenges of today have changed our mindset and many of our thought processes, but they are still met with the same resolve and professionalism that we use to help prepare our leaders of TODAY for the challenges of tomorrow.

I have been in the disaster response and public safety planning arena for nearly 40 years, and while much has changed, many things can still best be solved by communication. And I am not talking about two-way radios, or the latest handheld or pocket device; I am talking about personal communication. What we really need to do to continue being successful is to incorporate face-to-face planning, open discussion, and honest communication into our daily lives. Not technology, but REAL understanding and COMMUNICATION.

In these four decades of planning in law enforcement, fire, EMS, emergency management, as well as public positions in government, the common thread to operational success that I have seen is effective COMMUNICATION! All of us must be OPEN to communication at every level and at all times.

This book is the third in our series, with chapters offered to you on a variety of subjects. We have pushed the envelope on some of our chapters, and explored new ideas and thought processes. While putting this book together and getting it ready for publishing, the news flow of horrific events was constant. From multiple public shootings, bombings, and massive weather systems, it has been very difficult for our communities and our infrastructure to bounce back from these events! The Aurora Colorado shooting, Hurricane Sandy, the Newtown, Connecticut school tragedy, the Boston Marathon Bombing, and the deadly May 2013 Oklahoma Tornado outbreaks, to name a few, will continue to tax the response and planning communities.

My trusted colleagues and friends have endeavored to help compile some important ideas and lessons learned for you to understand their thoughts and insights better. As I have told my students in their master's classes at Northwestern University, Illinois Institute of Technology, Benedictine University, Eastern

Kentucky University, and Northern Illinois University, my goal and mission as their instructor, teacher, and mentor ARE to prepare people to take over behind us when we are gone. The greatest gift we get as educators is to learn that OUR students are practicing in the field, and have honed their skills to GIVE BACK to the community and organizations that they serve.

Authors

Michael J. Fagel, PhD, CEM, CHS-IV, has been involved in many phases of public service. His professional career spans nearly four decades in fire, rescue, emergency medical services, law enforcement, public health, emergency management, as well as corporate safety and security. Since 2003, he has supported many phases of Homeland Security operations in numerous capacities.

Currently, he is an instructor at the Illinois Institute of Technology-Stuart School of Business, Masters in Public Affairs Program, as well as at Northwestern University in the Masters of Public Policy and Administration Program, delivering master level courses in biodefense, terrorism, and homeland security. He also teaches homeland security at Northern Illinois University, Benedictine University's Masters in Public Health Program, as well as an instructor at Eastern Kentucky University, Safety Security Emergency Management Masters Program. Also, he supported the U.S. Army's SBCCOM at Aberdeen Proving Grounds in their WMD facility support operations for 48 months. He spent 32 months standing up the National Guard Bureau's CERIAC Fusion Center operations. He is a senior instructor at Louisiana State University's National Center for Biomedical Research and Training (NCBRT). He serves as an Subject Matter Expert (SME) for the National Center for Security and Preparedness, based in Albany, supporting the New York State Division of Homeland Security and Emergency Services. He has been involved in the training of Fusion Center and intelligence officials in numerous training classes for United States Department of Homeland Security (DHS).

Dr. Fagel has delivered several hundred lectures across the nation and written over 200 articles on safety and disaster planning. Also, he served the National Domestic Preparedness Office SLAG team (NDPO) at the FBI in Washington.

Fagel spent 10 years at FEMA in their Occupational Safety and Health Cadre in Washington, responding to incidents and disasters such as the Oklahoma City Bombing where he worked as a safety officer and CISD debriefer.

He spent over 100 days at the World Trade Center for FDNY at Ground Zero after the 9/11 attacks.

He was involved in numerous NLE efforts as well as Salt Lake City EOC operations in 2002. He has been an exercise developer and lead for several regional operations as well as for specific federal partners.

Dr. Fagel has spent several deployments in the Middle East helping to create a national response plan and a new FEMA-type organization. He was a delegate to the European Conference on Emergency Management held in Budapest in 2007.

Along with other assignments, Fagel is a homeland security analyst at the Argonne National Laboratories engaged in the protection of critical infrastructure. He has served on numerous OSHA VPP inspection teams as an SGE, with a background in safety, security, and disaster preparedness.

He is a member of the Northern Illinois Critical Incident Stress Debriefing team, the International Association of Fire Chiefs Committee on Safety and Health, and served on their Terrorism Committee. He served on the Illinois Terrorism Task Force and was the Region V President for the International Association of Emergency Managers; he was a Certified Emergency Manager Commissioner (CEM) for IAEM as well. He spent 28 years at North Aurora Fire as EMS coordinator and emergency management planner. Currently, he is a member of the board of trustees for the Sugar Grove (Illinois) Township Fire Protection District; he was a sheriff's deputy for 10 years, and has returned to the Kane County Sheriffs office in various training and support roles.

Dr. Fagel has published four textbooks on emergency planning, emergency operations and food safety law and is an editor for numerous trade textbooks, among them the following:

Fagel, M.J., *Principles of Emergency Management: Hazard Specific Issues and Mitigation Strategies)*, Boston, MA, Taylor & Francis, 2011.

Fagel, M.J., *Principles of Emergency Management and Emergency Operations Centers (EOC)*, Boston, MA, Taylor & Francis, 2010.

Fagel, M.J., *Emergency Operations: EOC Design,* Louisville, KY, Chicago Spectrum Press, 2008.

Fagel, S.S., *Food Safety Law*, New York, Van Nostrand Reinhold, 1997.

He serves as a columnist for several national trade publications and has appeared on FOX, NBC, CBS, NPR, NY1, and local media outlets.

Andrew M. Bramnik is a health physics and emergency response professional who has worked for the U.S. Nuclear Regulatory Commission (NRC) and the United Nations International Atomic Energy Agency. As an NRC inspector, he conducted health, safety, and security inspections at licensed facilities such as hospitals, blood banks, universities, construction sites, research laboratories, and production factories. As a member of the NRC's regional emergency response staff, he designed a new emergency operations center, coordinated emergency exercises with nuclear power plants, and responded to declared nuclear power plant emergencies. He also developed emergency procedures for incident response, flu pandemic, and continuity of operations. He is currently pursuing an MA in public policy and administration at Northwestern University and holds a BS in electrical and computer engineering from The Ohio State University.

Mark Chambers, NREMT-P, CHEP, is a leading international consultant for disaster preparedness management. He specializes in healthcare system and terrorism preparedness but retains a broad and extensive background in several major classifications of emergency services and disaster preparedness. He has held senior management and Emergency Support Function (ESF) leadership positions for both the states of Mississippi and North Carolina. Within that realm, he was responsible for leading the hospital preparedness programs for 8 years until late 2009 whereupon he moved from the public sector into private practice with several national and international consulting firms. He is intimately familiar with many best practices regarding emergency management and emergency operations center conduct. He has served on various national and state advisory committees helping to set policy on everything from National Incident Management System (NIMS) and the National Response Framework (NRF) to local healthcare preparedness policy. He has also collaborated with disaster healthcare advisors to both Bush and Obama administrations and continues to offer support as needed. He has led statewide exercises testing state and regional capability to manage large-scale disasters with a specific focus on the need to provide medical surge capacity and capability as well as evacuation and shelter-in-place strategies. Most recently, he deployed to Libya and completed a healthcare system and disaster preparedness analysis under combat conditions to help facilitate an increased survival rate for wounded Libyans during the revolutionary conflict in 2011.

Ernest P. Chiodo, MD, JD, MPH, MS, MBA, CIH, is a physician, attorney, industrial hygienist, industrial toxicologist, and biomedical engineer licensed to practice medicine in New York, Michigan, and Illinois as well as law in Michigan and Illinois. Dr. Chiodo earned his medical degree (MD) from Wayne State University School of Medicine, his juris doctor (JD) from Wayne State University Law School, his master of public health (MP.H.) from Harvard University School of Public Health, his master of science in biomedical engineering (MS) from Wayne State University, his master of science in threat response management (biological, chemical, and radiological defense) from the University of Chicago, his master of science in occupational and environmental health sciences with specialization in industrial toxicology from Wayne State University, and his master of business administration with a concentration in economics from the University of Chicago. He is board certified in the medical specialties of internal medicine, occupational medicine, and public health and general preventive medicine. Public health and general preventive medicine is the medical specialty focused on epidemiology. He is also certified in the engineering and public health discipline of industrial hygiene by the American Board of Industrial Hygiene as a Certified Industrial Hygienist (CIH) in the comprehensive practice of industrial hygiene. He has served as the president of the Michigan Industrial Hygiene Society. Dr. Chiodo is an assistant clinical professor of family medicine and public health at Wayne State University School of Medicine and has clinical privileges in the Henry Ford Health System. He is also an adjunct professor of industrial hygiene and industrial toxicology at Wayne State University. He has served as the medical director of the City of Detroit and of the Detroit Health Department and was the chief physician responsible for measures designed to protect the public health of over 1 million persons living or working in the City of Detroit. He has also served as the medical director of the pension boards of the City of Lansing, Michigan, which is the capital city of the State of Michigan. In addition, he has served as the medical advisor to the final appeals committee of Jefferson Pilot Financial disability insurance carrier.

Dr. Chiodo serves as the chairman of the Environmental Litigation and Administrative Practice Committee of the Environmental Law Section of the State Bar of Michigan. He also serves as an adjunct professor of law at John Marshall Law School and Loyola University Law School in Chicago.

Ruth Cover has had a noteworthy career in healthcare administration and public health with additional experience in emergency management, health security intelligence, healthcare quality, and adult education. Twenty years of her career were devoted to staff and management positions with urban, rural, and Department of Veteran Affairs hospitals and clinics. Positions in public health included emergency response coordinator for a county health department and a hospital preparedness program coordinator at state level. Selected as the first health security member of the Interagency Threat Assessment and Coordination Group at the National Counterterrorism Center in 2010, she worked with Department of Homeland Security and other state, local, and tribal representatives on terrorism-related products applicable to first responders. These full-time endeavors were interlaced for 15 years as a county emergency management agency volunteer, adjunct facility assignments for three universities, and a 21-year U.S. Naval Reserve career that culminated for 18 months in Asia and the Balkans with U.S. and NATO military commands. She holds a master's degree in health care administration and certification in homeland security (CHS-III).

J. Lawrence (Larry) Cunningham retired from the U.S. Secret Service in 1994. As the agent in charge of the Secret Service field office in San Jose, California, he coordinated and supervised the advance security arrangements for visiting dignitaries, who included Pope John Paul II, Mikhail Gorbachev, U.S. presidents, and other foreign heads of state. His oversight of law enforcement personnel, motorcade routes, sites, and communications included the development of emergency operations plans. His office developed precedent-setting cases resulting in updated federal criminal statutes.

He served on the White House Presidential Protective Division for 5 years from 1982 to 1987 and conducted lead security advances for presidential visits to numerous domestic and international cities. He supervised detail agents while traveling with the president. He received seven performance awards during his Secret Service career.

Cunningham established Essential Security Strategies, LLC in 1995. He conducts security surveys, risk assessments, executive protection seminars, security training, and investigations for public and private organizations worldwide. Assessments include the evaluation and formulation of integrated crisis plans and the implementation of a wide range of security protocols. Clients include the Defense Threat Reduction Agency, Exxon/Mobil, Harvard University, The International Monetary Fund, NASCAR, The Bush/Cheney Presidential Campaign, Harpo Studios, RLJ Companies, The American Battle Monuments Commission, The George Washington University, U.S. Airways, Gate Gourmet, The United Arab Emirates Royal Family, The Jordanian Royal Family, and other international companies.

Currently, Cunningham serves as the Department of Homeland Security (DHS)-certified adjunct instructor, subject-matter expert, and course developer in 10 law enforcement and emergency response disciplines for the National Center for Biomedical Research and Training Academy of Counter-Terrorist Education at Louisiana State University in Baton Rouge, LA. These courses are delivered to domestic and international responder organizations.

Elizabeth Dawson, MSc and Jacob J. Dickman are the founding partners of the Emergency Management Research Group Inc. Both partners are graduates of the University of Chicago's Threat and Research Management Program. Elizabeth Dawson has several years of experience with the Chicago Police Department in patrol, gang investigation, crime scene investigation and is currently assigned as the Crime Lab's liaison officer to the Cook County Medical Examiner's Office.

Jacob Dickman, MSc, is a principal and founding partner of Emergency Management Research Group Inc., a small research firm specializing in small business recovery models. Jacob Dickman has many years with the Chicago Fire Department where he has developed extensive experience in teaching and writing training drills for large groups of adults. He also worked at the Overall NATO-G8 Summit on training and course development for the Chicago Fire Departments participation in the event. Jacob is also an active member of the International Association of Emergency Managers (IAEM) and is active on the public–private partnership caucus. Dickman is also a member of the Chicago Fire Department currently detailed as a field training officer. He is currently a PhD student at Capella University's School of Public Safety Leadership, with a concentration in emergency management. He holds a master of science in threat and response management with an administrative concentration from the University of Chicago and a bachelor of arts in management from Benedictine University. Dickman earned his bachelor's with summa cum laude honors. He is a member of the Sigma Beta Delta international honor society in business, management, and administration. He was also the student chapter vice president of the University of Chicago IAEM student chapter.

Randall C. Duncan, MS, is a local government emergency manager living in Kansas. He has been a local government emergency manager since 1986 in both Kansas and Oklahoma, and has administered a "baker's dozen" of presidential declarations of major disaster and emergency. In 2000, Duncan was the first local government emergency manager to accompany FEMA officials to the Republic of Turkey to assist Istanbul Technical University in setting up a center of excellence for emergency management. In 2001, Duncan provided support to FDNY at the World Trade Center in conjunction with the USDOJ. In 2012, Duncan provided assistance to the State of New York Department of Homeland Security and Emergency Services (DHSES) in their response to Hurricane Sandy. Duncan holds a master's degree in public affairs with an emphasis on disasters and emergency management from Park University and a bachelor of arts in political science from Southwestern College.

Jeremiah W. Dunlap, MS, BA, has worked for Northwestern University's Center for Comparative Medicine (CCM) since 2005. After graduating from Washington University in St. Louis with his BA in biology, Dunlap joined CCM as an assistant manager of quality and training. As such, he is responsible for the training of laboratory research personnel, maintaining the security of department facilities on both campuses, and occupational health and safety issues. During his 7 years as an employee of Northwestern University, he has earned his certification as a laboratory animal technologist through the American Association of Laboratory Animal Science (AALAS), as well as a master's in public policy and administration through Northwestern University. Specializing in public safety and security, Dunlap has taken public policy courses on national security, terrorism, and biosecurity.

Ronald E. Fisher, PhD, is the U.S. Department of Homeland Security manager at Idaho National Laboratory. He brings more than 25 years of experience from Argonne National Laboratory, most recently as deputy director for the lab's Infrastructure Assurance Center. He served as a senior consultant to the National Petroleum Council on oil and natural gas infrastructure vulnerabilities and for the President's Commission on Critical Infrastructure Protection. His research activities include developing vulnerability assessment methodology, risk, and resiliency analyses. The methodologies he helped to develop have been conducted at thousands of critical infrastructure facilities throughout the United States. Dr. Fisher has more than 150 publications including contributions to multiple books on homeland security. He also has contributed to multiple copyrights and trademarks. His dissertation was on homeland security and is titled, *Taking a Normative Approach to Organizational Culture Change on Critical Infrastructure Protection.*

Stacey Hall, PhD, is the associate director of the National Center for Spectator Sports Safety and Security (NCS4) and an associate professor of sport management at the University of Southern Mississippi (USM). Dr. Hall has been published in international sport management, homeland security, and emergency management journals. She has coauthored two textbooks—*Global Sport Facility Operations Management* and *Security Management for Sports and Special Events.* Additionally, she has been invited to publish in national magazines such as *Athletic Management, Athletic Administration*, and *Security Magazine.* Dr. Hall has been referred to as one of the nation's leading experts in sport security with interviews in *USA Today, ESPN the Magazine, CBS New York*, and *ESPN Outside the Lines.* Dr. Hall has presented papers at international and national conferences, and conducted invited presentations for U.S. federal and state agencies, college athletic conferences, and professional sport leagues. Dr. Hall has been the principal investigator on external grant awards in excess of $4M from the U.S. Department of Homeland Security to develop sport event risk management curriculum, conduct risk assessments at college sport stadia, and develop training programs for sport venue staff.

Kelly Hamilton, MPA, is biosecurity director and ESF #11 coordinator for the New Mexico Department of Agriculture and codirector of the Southwest Border Food Safety and Defense Center. Prior to his work in New Mexico, Kelly served the citizens of Wyoming for over 20 years as a certified law enforcement officer with a number of those years leading the investigative, enforcement, and regulatory programs at the state veterinarian's office.

Douglas E. Himberger, PhD, is a retired senior vice president and director for NORC at the University of Chicago, where he led the Security, Energy, and Environment Department, leading independent, public interest–based research in those domains. Prior to NORC, Dr. Himberger was a partner and vice president at Booz Allen Hamilton, where he led defense, security, and resilience teams on converged science and technology service offerings ranging from basic research and development planning to technology transition as a knowledge management subject-matter expert. His team supported the Department of Homeland Security (DHS), Department of Health and Human Services (HHS), Federal law enforcement, and counterterrorism agencies, where he fostered broad collaboration ("Megacommunities") with companies and academia to build strong information-sharing networks and advanced technology solutions. Additionally, he created and led a Global Risks Initiative Taskforce for analyzing complex all-hazard threats (e.g., pandemics, natural/man-made disasters), and provided multidisciplinary solutions for both government and industry. Dr. Himberger serves on the boards of multiple nonprofit organizations (such as the Safe America Foundation and the Center for Excellence in Education), has published broadly, and earned an MS and

PhD in physics from Georgetown University and a BS in physics from Nebraska Wesleyan University.

Stephen Krill is a senior vice president for business development at Guardian Centers LLC, with 24 years of emergency management experience. He started his career in the Incident Response Branch at the U.S. Nuclear Regulatory Commission and the Emergency Preparedness Section at Oyster Creek Nuclear Power Plant, and supported government, nongovernment, and commercial preparedness and response as a senior associate at Booz Allen Hamilton and an assistant vice president/division manager at SAIC. He received numerous awards, including the prestigious U.S. Secretary of Transportation's Team Award for his leadership role during the 2005 hurricane season.

Krill holds professional certifications in emergency management, business continuity management, and project management. He coauthored a textbook on emergency management and wireless security, published multiple journal and magazine articles, lectured on terrorism preparedness, and holds two U.S. Patents. Krill earned a BS in nuclear engineering from the University of Cincinnati and a master of environmental engineering from the Johns Hopkins University. He is pursuing a PhD in engineering management with a focus in crisis, disaster, and risk management at the George Washington University.

Kathryn R. Juzwin, PsycD, is a clinical psychologist and associate professor of clinical psychology at Argosy University–Schaumburg. She specializes in first-responder mental health, high-risk personnel assessment, critical incident response, disaster mental health, and trauma. Dr. Juzwin is the regional coordinator for mental health response for the Illinois Medical Emergency Response Team (IMERT), a responder on the Illinois Disaster Assistance Team (D-MAT IL-2), and is the current education coordinator and has also been a coordinator for the Northern Illinois Critical Incident Stress Management Team (NICISM). For the IMERT team, she has written the training curriculum for mental health responders and trains personnel related to mental health issues in crisis and disaster response. Clinically, Dr. Juzwin focuses on trauma, forensic trauma, self-injury, eating disorders, and high-risk patients. She is active in education, consultation, writing, and research in these areas. She is the director of Self-Injury Recovery Services at Alexian Brothers Behavioral Health Hospital in Hoffman Estates, Illinois, which is the only JCAHO-DSC–certified program for self-injury in the country. Dr. Juzwin is the chief psychologist for COPS and FIRE Testing Service and conducts outcome research, assessment, protocol development and training, and supervision in the area of testing and assessment for high-risk hiring for law enforcement, fire/EMS, and emergency dispatch personnel. At Argosy University, Dr. Juzwin teaches testing and assessment, ethics, professional development, police psychology, trauma, forensic trauma, eating disorders, self-injury, and suicide assessment courses.

Matthew Lawrence is an accomplished emergency management professional. His principal areas of expertise include exercise design, development and coordination, and emergency response planning for man-made incidents, agroterrorism, and natural disasters. Lawrence developed the first international agroterrorism functional exercise for the Border Governors' Agriculture Worktable and provided exercise design support for the first full-scale agriculture movement control exercise in the United States. He also designed, developed, and delivered multiple emergency management training programs for local and state agencies. Lawrence currently serves as the chair for the International Association of Emergency Managers' Food and Agriculture Ad Hoc Committee.

Joseph Lombardo, CEM, is a training and exercise specialist for Ada City–County Emergency Management, overseeing emergency management training and exercise activities for Ada County, Ada County Highway District, and the cities of Boise, Eagle, Garden City, Kuna, Meridian, and Star, Idaho. Previously, Lombardo worked as a program specialist with the Federal Emergency Management Agency contributing to the development of national preparedness policy. He also supported multiple homeland security and emergency management programs as an associate with the consulting firm of Booz Allen Hamilton, focusing on infrastructure protection issues across multiple sectors, including food and agriculture, communications, information technology, and transportation. He has supported the response to multiple disasters across levels of government, ranging from catastrophic events to local emergencies and was deployed by the Red Cross in response to Hurricanes Katrina and Rita.

Raymond McPartland is the founder and chief executive officer of Tier One Associates LLC. Tier One consults and develops training on matters of life safety, emergency preparedness, and homeland security for both the private and public sector and specializes in creating unique, client-specific, realistic training, and preparedness programs. McPartland is a subject-matter expert with the National Center for Biomedical Research and Training (NCBRT) at Louisiana State University, and the Chemical, Biological, Radiological, and Nuclear Defense Information Analysis Center (CBRNIAC), as well as an active trainer and detective with the New York City Police Department.

Currently assigned to the NYPD's Counterterrorism Division's Training Section as a lead instructor and curriculum development specialist, Detective McPartland is a leading instructor with his regional training team responsible for instructing patrol, specialty personnel, and regional partners in various aspects of terrorism and counterterrorism. He is currently the Division's subject-matter expert on active shooter events and the primary author of the New York City Police Department's published research work—*Active Shooter: Recommendations and Analysis for Risk Mitigation*. His other topics of instruction include active shooter preparedness and response, critical infrastructure protection, maritime terrorism, WMD and radiological awareness, hostile surveillance detection, and behavioral analysis, and observation.

McPartland has attended numerous schools and training through the Department of Homeland Security's National Preparedness Consortium and FEMA's Emergency Management Institute.

Concurrent to his numerous duties as an NYPD detective and CEO of Tier One Associates LLC, McPartland works part time as an adjunct professor at Metropolitan College of New York in their MPA Program in Emergency and Disaster Management, and at Mercy College in their undergraduate program for Corporate and Homeland Security.

James A. McGee holds a bachelor of science (BS) in natural resource management from California Polytechnic State University and a master of science (MS) in criminal justice from Virginia Commonwealth University. He has 25 combined years of law enforcement experience, 21 years as a special agent with the Federal Bureau of Investigation (FBI). His experience includes 35 years addressing international security issues, counterterrorism investigations, crisis management, critical infrastructure protection, risk assessments, tactical operations, and homeland security initiatives. During his FBI career, McGee received numerous awards including the FBI Shield of Bravery, the FBI Medal of Merit, the FBI Medal of Valor, the Federal Law Enforcement Officer's Association Medal of Valor, the U.S. Attorney General's Award for Exceptional Heroism, the Department of Justice Certificate of Merit, and the Department of Justice Certificate of Special Recognition. McGee has been employed by The University of Southern Mississippi as a faculty member and as an adjunct professor with William Carey University Department of Criminal Justice and Tulane University Department of Homeland Security Studies where he teaches undergraduate courses in terrorism, counterterrorism, critical infrastructure protection, sport security management, and homeland security. He also frequently teaches critical incident management, major event security management, major case management, hostage negotiation, and interviewing terrorist suspects for the United States Department of State Anti-Terrorism Assistance Program. McGee is designated as an expert witness in security issues regarding critical infrastructure protection, specifically venues of mass gatherings and has testified in this capacity in U.S. Federal District Court. McGee is a member of the Global Security Team for ESPN and a senior consultant for the Soufan Group. He is an award-winning author, having written and published the book *Phase Line Green—The FCI Talladega Hostage Rescue* (2007). He also coauthored the textbook *Security Management of Sports and Special Events: A Team Approach to Creating Safe Facilities* (2011).

Rick C. Mathews is the director of the National Center for Security and Preparedness (NCSP) at the State University of New York as well as a public service professor at the Rockefeller College of Public Affairs and Policy at the University at Albany, State University of New York. Prior to his July 2007 appointment at SUNY, he served as the assistant director for research and development at the National Center for Biomedical Research and Training at Louisiana State University where he led the center's efforts to develop national level training in support of the U.S. Departments of Homeland Security, Justice, and Health and Human Services among others. The NCSP team led by Mathews develops and delivers training in the areas of counterterrorism, active shooter response, mass casualty incident management, emergency management, and homeland security. Mathews is a frequent public speaker, press and media consultant on counterterrorism, and technical advisor to various agencies.

Robert Muller, MD, has six degrees including an MD from Louisiana State University and MBA/MHA from Auburn University. He has qualified for the CEM degree for the past 15 years. Dr. Muller has taught in various law enforcement academies on local and federal levels. He has served as the medical director of numerous fire departments, the New Orleans Police Department, St. Tammany Parish Sheriff's Department and Southeast Louisiana Search and Rescue, as well as the New Orleans Airport. He served as a deputy chief of the New Orleans Police Department in command of Search and Rescue, Hurricane Preparedness, and Special Events. He has served as the deputy director of the St. Tammany Parish Office of Emergency Preparedness and as a deputy coroner in both Orleans and St. Tammany Parishes. He has over 45 years of educational and real-life situational experiences.

Jacob Neufeld is the emergency planning associate at Memorial Sloan-Kettering Cancer Center, in New York City. In addition to supporting the director with the hospital's preparedness program, Neufeld is directly responsible for enterprise-wide clinical and business continuity planning for the hospital, 3 research facilities, 10 business locations, and 17 regional care sites. He is a graduate of Metropolitan College of New York where he earned an MPA in emergency and disaster management. He also attended the Disaster Recovery Institute International where he achieved certification as Associate Business Continuity Planner. He also holds a certificate from the Israeli Military Industries School for Security and Anti-Terror Training. Neufeld is an active volunteer with his local American Red Cross chapter, serving as a coordinator of continuity of operations planner and disaster responder.

James Peerenboom, PhD, is responsible for leading multidisciplinary teams of scientists and engineers in developing innovative solutions for infrastructure assurance, systems analysis, decision and risk analysis, and advanced modeling and simulation problems. For the last 15 years, he has focused on critical infrastructure protection and resilience issues, providing technical support to the Departments of Energy and Homeland Security, President's Commission on Critical Infrastructure Protection, and White House Office of Science and Technology Policy. He has participated in developing a critical infrastructure assurance R&D roadmap, and led infrastructure interdependencies and vulnerability assessment programs for various government agencies. Dr. Peerenboom is the author of more than 70 publications, including journal articles, book contributions, reports, and conference papers.

Dr. Peerenboom also served as the director of the Argonne Infrastructure Assurance Center. The mission of the Center is to leverage Argonne's expertise, knowledge, technologies, and specialized research capabilities and facilities to meet infrastructure assurance needs at the local, regional, and national levels. This includes applying innovative and cost-effective protection, mitigation, response, and recovery tools and technologies and analyzing the data collected. Dr. Peerenboom earned a PhD in energy and environmental systems from the Institute for Environmental Studies and an MS and a BS in nuclear engineering from the University of Wisconsin–Madison.

Sheriff Patrick Perez* is serving his second term as the sheriff in Kane County, comprising a population of 515,000, approximately 40 miles west of Chicago. The Kane County Sheriff's Office is actively involved in ILEAS (Illinois Law Enforcement Alarm System) and in the WMD Team. His agency also provides the only bomb squad in Kane County. The blizzard of 2011 and the NATO Summit of 2012 are two of the most recent incidents in which his agency was tasked with response. Sheriff Perez is a second-generation law enforcement officer and has made preparation for critical incidents of all types a priority for his agency.

Linda Reissman has over 30 years of experience in the municipal, private, and governmental sectors of the emergency management, public health, and EMS disciplines and holds a master of science in emergency and disaster preparedness. Linda is currently the director of emergency management for Memorial Sloan-Kettering Cancer Center (MSKCC) and Network. She is responsible for emergency preparedness and business continuity for the 450-bed hospital, including 30+ clinical, research, and business sites in NYC and regionally. From 1996 to 2010, Reissman served as training officer for HHS/NDMS Disaster Medical Assistance Team (DMAT), NY-4, and is currently assigned to NY-5, a developmental DMAT team. As a member of DMAT, Linda has participated in the development and execution of integrated federal regional exercises and has deployed to disasters for team activations, including 9/11.

Prior to her position as MSKCC, Reissman served 8 years as the senior EMS representative with the NYS Bureau of EMS where she oversaw EMS agency compliance and coordinated hospital and EMS disaster-response plans in cooperation with local, state, and federal emergency management authorities. She has also served with the NYC Office of Emergency Management as a planner during its early inception and assisted in the development of the NYC Bio-Terrorism Plan and Family Assistance Plan, and spearheaded a $500,000 Federal Terrorism Grant in 2010, in cooperation with Mount Sinai Medical Center. Reissman coauthored a $500,000 ASPR/NYCDOH Consortium grant that will address disaster planning for patients with special healthcare needs.

Reissman is also a consultant/emergency management planner, with Strategic Emergency Group (SEG), specializing in eight ESF project aspects, including

* Reviewer, not contributor.

comprehensive emergency management planning, HSEEP exercises, response and recovery, and critical infrastructure protection and resilience. She also serves as a core team member on a number of SEG projects including the Statewide HSEEP exercise program for the State of Missouri Emergency Management (SEMA) and the Missouri Hospital Association (MHA) where she supports exercise design, operations, and analysis as an exercise planner, controller, and lead evaluator.

Reissman has been a national speaker on resiliency planning for healthcare and most recently presented at the 2011 International Association of Emergency Managers (IAEM) Annual Conference and the Baylor Emergency Management Symposium.

Derek Rowan has been involved in adult education for 30 years, teaching a variety of technical tasks, high-stress skills, emergency services training, and exercise design. He has designed, developed, conducted, and evaluated hundreds of preparedness exercises spanning all seven types of HSEEP for all levels of the government and private sectors—including international. He was the chief of control for our nation's largest and most complex exercises to date—National Level Exercise 2011 and the two operational exercise series of our nation's first national Cyber Exercise—National Level Exercise 2012 and the largest national continuity exercise ever—Eagle Horizon 2012.

S. Shane Stovall, CEM, serves as the coordinator of Emergency Management Programs for True North Emergency Management. Before coming to True North in January 2012, Stovall served as the director of emergency management for the City of Plano, Texas for 5 years. Before coming to Plano, he served for 2 years as a project manager with the General Physics Corporation's Homeland Security and Emergency Management Unit and 8 years with the Charlotte County Office of Emergency Management in Punta Gorda, Florida. Stovall graduated in December 1995 from the University of North Texas with a bachelor's degree in emergency administration and planning. In addition to his education, Stovall also received his certification as a certified emergency manager (CEM) from the International Association of Emergency Managers in 2004. Stovall is an active member of the emergency management community and has served many national roles promoting emergency management programs.

Wade Townsend has more than 24 years of federal experience of increasing responsibility in emergency management and security program management areas. He is known as a dynamic, result-oriented professional with a successful record developing, implementing, and administering new programs in Homeland Security and emergency management operations. Wade serves in the Office of Infrastructure Protection, National Protection and Programs Directorate, United States Department of Homeland Security (DHS) as a technical authority on physical security relating to policies and proposed legislation affecting the security pertaining to the chemical infrastructure, and is currently the inaugural DHS Academic

Chair at Marine Corps University. Townsend, a plank owner in the initial startup of DHS, began his DHS career by coordinating with private, local, state, and federal agencies in identifying critical infrastructures and systems, identifying, mitigating and correcting vulnerabilities, and preparing and responding to national emergencies. Prior to his 8 years with DHS, Townsend started his career as an intern with the Department of Energy (DOE). He was with DOE for more than 14 years working on a broad range of energy-related issues including energy emergency planning, critical infrastructure protection, nuclear safeguards and security, and national security policy and planning. He is a native of the Commonwealth of Virginia and earned a bachelor of science from George Mason University in public administration in 1988 and earned a master of military studies from the Marine Corps University in 2012.

Introduction

"Disasters and incidents: we read about them all the time, and many people continue to say... it won't happen here."

Edmund Burke once said "Those who don't know history are doomed to repeat it."

We can list the numerous events that have occurred worldwide dealing with natural and man-made disasters. Weather, terrorism, chemical accidents, fire, and violence are just a few of the elements that come to mind.

As emergency planners, WE research and do our best to plan for the events, as it is not a matter of IF, but a matter of WHEN again.

As history often repeats itself, it did happen again with Hurricane Irene and then most recently, again with Super Storm Sandy. And for all planning and preparedness that had been put into place, especially from lessons learned with Hurricane Katrina, people were still largely unprepared. With the advent of new technology, people were WARNED a week in advance of the storm's potential influences in the Northeast. Multiple states felt the impact of the super storm BEFORE it ever impacted New York, New Jersey, Connecticut, and other areas.

For years, we have become more focused on improved methods for communications and we have some of the best devices in the world available to us. BUT, communications are NOT JUST TECHNOLOGY; it is the act of receiving and sending information, data, and information so that it can be processed. As the event was occurring, many federal and state assets were on heightened alert. The governors of the impacted states made impassioned pleas to the federal government at the highest levels. Aid was promised, and then it struck. The wind, water, waves, and rising tides were NO match for the sandbags and natural barriers. Power failed, critical infrastructure was severely impacted, and the affected public was left cold, powerless, and hungry in the ensuing days. The government entities had tried in the week beforehand to inform the public of the impending event. BUT, not even one community was prepared for the massive force of the storm. Gasoline shortages, food supplies, and many other basic needs were impacted, and, even to this day, are still impacted, a year after the event.

"It won't happen here" are words that I often hear as an emergency planner. So, what occurred in New York CAN happen anywhere? I was deployed to New York City several times for Hurricane Sandy relief efforts and will restate the obvious:

> "The CITIZENS must be prepared for a minimum of 72 hours without any of the normal items that we rely on."

Agencies need to really understand the INTERDEPENDENCIES of all the critical infrastructures that work together. There are 16 critical infrastructure sectors as defined by Presidential Policy Directive (PPD)-21:

1. Chemical
2. Commercial facilities
3. Communications
4. Critical manufacturing
5. Dams
6. Defense industrial base
7. Emergency services
8. Energy
9. Financial services
10. Food and agriculture
11. Government facilities
12. Healthcare and public health
13. Information technology
14. Nuclear reactors, materials, and waste
15. Transportation systems
16. Water and wastewater systems

The PUBLIC needs to understand that preparedness is a culture that they must embrace 365 days a year, and that is for their own resilience and planning. Make a kit, have a plan, AND practice it! Be prepared to leave your home, office, business, and organization for an EXTENDED period and NOT be able to come back to it for some time, be it days, weeks, or months.

In my 10 years with FEMA, and my years as a responder and emergency manager, several lessons that we keep on learning again and again come to mind:

- Planning
- Preparation
- Practice

AND IT CAN HAPPEN TO ME!

Preparedness is a way of life, a daily routine that cannot be stressed enough. Citizen preparedness, resiliency, and planning are things that must be thought about 365 days a year, not just a week before the forecasted storm is about to impact!

The same holds true for the municipalities where we live. ALL disasters are local and that fact has been reinforced many times. Local governments WILL BE THE FIRST RESPONDERS and the help must come from within the community, the region locally as the responders know THEIR community better than anyone else. They know where the population densities are, the special needs of population concentrations (now known as functional needs), as well as other high-risk population areas.

As the response to Hurricane Sandy is looked at over the next few years, many things will come to light. One area that I saw with my deployment was that, as we have always said, COMMUNICATION issues come to the forefront.

If we go back just a bit to look at lessons learned from other events, it may be a simple failure of IMAGINATION. One has to look at the 9/11 Commission report, the after actions from Hurricane Katrina in 2005, and many other events.

It is about CULTURE and the culture of PREPAREDNESS. We cannot implement a plan in 4 hours that has NOT been practiced!

As you read through some of the chapters in this book, my colleagues have shared with you some simple solutions that, if implemented effectively, may help you in your preparedness and planning culture.

Regardless of the event, whether it is Sandy, Katrina, World Trade Center and Pentagon attacks, active shooter situations, or the next crisis, all of us can learn from the chapters in this book.

Quoting Edmund Burke once again, "The only thing necessary for the triumph of evil is for good men to do nothing." We, as men and women in today's emergency planning and response community, cannot stand idle but we must strive to learn from history and be PREPARED.

The emergency planners of today, the leaders of tomorrow, the responders, the planners, and the elected and appointed officials MUST share one common theme: PLANNING, PREPARATION, and PRACTICE that may help in the next event!

Be safe, and remember the 3 Ps.

Michael J. Fagel, PhD, CEM, CHS-IV

I

POLICY CONSIDERATIONS

Chapter 1

Policy and Laws Relating to Emergency Management Planning

Michael J. Fagel, Stephen Krill, and Matthew Lawrence

Background

Throughout its history, the United States has faced many disasters—from natural disasters to hazardous materials releases to terrorist attacks. While all disasters are local, the Federal Government will provide support—personnel, equipment, resources, and funding—when local capabilities and capacities are exceeded.

Before the 1970s, various federal agencies and programs provided disaster relief services. At one point, more than 100 federal agencies handled disaster and emergencies. However, until four devastating hurricanes struck in the 1970s, the need for improved disaster coordination at the federal level did not happen.

- *Hurricane Agnes (1972):* This hurricane caused significant East Coast flooding and $2 billion in damages.
- *Hurricane Eloise (1975):* This hurricane caused $200 million in damages and 76 fatalities.
- *Hurricanes David and Frederick (1979):* These hurricanes were among the deadliest ever seen in the Caribbean, with Frederick causing $2.2 billion in damages.

These disasters and the need for disaster preparedness nationwide pushed the Carter Administration to establish the Federal Emergency Management Agency (FEMA) to coordinate all disaster relief efforts at the federal level.

Authorities and Directives

Robert T. Stafford Act

To bring a more orderly and systemic means of federal natural disaster assistance for State and local governments in carrying out their responsibilities to aid citizens, the Robert T. Stafford Disaster Relief and Emergency Assistance Act (Public Law 93-288) was passed in 1988. This Act describes the programs and processes by which the Federal Government provides coordination and support to State and local governments, tribal nations, eligible nongovernment organizations (NGOs), and individuals affected by a declared major disaster or emergency. The Stafford Act covers all hazards, including natural disasters and terrorist events.

The Stafford Act was amended several times, most recently with the Post-Katrina Emergency Management Reform Act (PKEMRA) in 2006 (Public Law 109–295) to address the issues and challenges that arose during the response to Hurricane Katrina.

Presidential Decision Directives

In 1995, the Clinton Administration issued Presidential Decision Directive (PDD-39), "U.S. Policy on Counterterrorism" in response to the worst terrorist act on U.S. soil—the bombing of the Alfred P. Murrah Federal Building in Oklahoma City. PDD-39 built on prior directives and outlined three key elements of a national counterterrorism strategy:

1. Reduce vulnerabilities to terrorist attacks and prevent and deter terrorist acts before they occur (threat/vulnerability management).
2. Respond to terrorist acts that occur, end the crisis or deny terrorists their objectives, and apprehend and punish terrorists (crisis management).
3. Manage the consequences of terrorist acts, including restoring essential government services and providing emergency relief, to protect public health and safety (consequence management).

The Directive elaborated on specific roles and responsibilities for several federal agencies with respect to each element of the strategy. For example, PDD-39 gave the Federal Bureau of Investigation (FBI) lead agency responsibility for crisis management, and FEMA similar responsibility for consequence management. Reflecting the need for greater interagency coordination, PDD-39 also directed the National Security Council to coordinate interagency terrorism policy issues and to ensure implementation of federal counterterrorism policy and strategy.

Homeland Security Act of 2002

Title I of the Homeland Security Act of 2002 established the Department of Homeland Security, with the mission of

1. Preventing terrorist attacks within the United States
2. Reducing the vulnerability of the United States to terrorism
3. Minimizing the damage, and assisting in the recovery from terrorist attacks that do occur within the United States
4. Carrying out all functions of entities transferred to the Department, including by acting as a focal point regarding natural and manmade crises and emergency planning

The act organized not only FEMA into DHS, with a direct line of report between the administrator and the secretary, but it also kept it separate from the new Preparedness Directorate. The Preparedness Directorate consolidated preparedness assets from across the Department. It facilitated grants and oversaw nationwide preparedness efforts supporting first responder training, citizen awareness, public health, infrastructure, and cyber security.

Post-Katrina Emergency Management Reform Act

The Post-Katrina Emergency Management Reform Act of 2006 clarified and modified the Homeland Security Act of 2002 with respect to the organizational structure, authorities, and responsibilities of FEMA and the FEMA Administrator. Among other changes, PKEMRA transferred a significant portion of the DHS Preparedness Directorate into FEMA, notably

- The United States Fire Administration
- The Office of Grants and Training
- The Chemical Stockpile Emergency Preparedness Division
- The Radiological Emergency Preparedness Program
- The Office of National Capital Region Coordination

Homeland Security Presidential Directives

Following the terrorist attacks of September 11, 2001, the Bush Administration issued several Homeland Security Presidential Directives (HSPDs), either augmenting or replacing the PDDs established by the Clinton Administration. Among several, the two most notable HSPDs around preparedness and response include

- **HSPD-5, Management of Domestic Incidents**, which the White House issued February 28, 2003, establishes a single, comprehensive national incident management system. It also designates the Secretary of Homeland

Security as the principal Federal official for domestic incident management and recognizes the statutory authorities of the Attorney General, Secretary of Defense, and Secretary of State. In addition, HSPD-5 directs the heads of all Federal departments and agencies to provide their full and prompt cooperation, resources, and support, as appropriate and consistent with their own responsibilities for protecting national security, to the Secretary of Homeland Security, Attorney General, Secretary of Defense, and Secretary of State in the exercise of leadership responsibilities and missions assigned.

- **HSPD-8, National Preparedness**, issued by the White House December 17, 2003, establishes policies to strengthen the preparedness of the United States to prevent and respond to threatened or actual domestic terrorist attacks, major disasters, and other emergencies by requiring a national domestic all-hazards preparedness goal, establishing mechanisms for improved delivery of Federal preparedness assistance to State, local, and tribal governments, and outlining actions to strengthen the preparedness capabilities of Federal, State, local, and tribal entities. Annex 1, Integrated Planning System, published in January 2009, establishes a standard and comprehensive approach to national planning.

Other References

HSPD-5 and HSPD-8 also help establish several important policy references around homeland security and emergency management, including

- **National Incident Management System (NIMS)**, December 2008, provides a systematic, proactive approach to guide departments and agencies at all levels of government, nongovernmental organizations (NGOs), and the private sector to work seamlessly to prevent, protect against, respond to, recover from, and mitigate the effects of incidents, regardless of cause, size, location, or complexity, in order to reduce the loss of life and property and harm to the environment.
- **National Infrastructure Protection Plan (NIPP)**, January 2009, establishes a risk management framework for the Nation's unified national approach to critical infrastructure/key resource protection.
- **National Preparedness Guidelines**, September 2007, finalizes development of the National Preparedness Goal and its related preparedness tools as mandated in HSPD-8. The Guidelines consist of four elements: the National Preparedness Vision, the National Planning Scenarios, the Target Capabilities List, and the Universal Task List.

Presidential Policy Directives

In 2009, the Obama Administration issued Presidential Policy Directive (PPD) 1, calling for the organization of a National Security Council System. PPD-8 (http://

www.dhs.gov/presidential-policy-directive-8-national-preparedness) was issued 2 years later on March 30, 2011 and focused on National Preparedness, introducing the Whole Community Preparedness initiatives that have restructured emergency managers' thought processes by focusing on core capabilities and the National Preparedness Goal. At its core, PPD-8 requires the involvement of everyone—not just the government—in a systematic effort to keep the nation safe from harm and resilient when struck by hazards, such as natural disasters, acts of terrorism, and pandemics. PPD-8 is organized around six elements:

- **The National Preparedness Goal.** The National Preparedness Goal was introduced to define the core capabilities necessary to enhance preparedness relating to the specific types of incidents that pose the greatest threat to the security of the United States. It emphasizes actions aimed at achieving an integrated preparedness approach optimizing the best use of available resources.
- **The National Preparedness System.** The National Preparedness System was designed to help build and sustain the capabilities outlined in the National Preparedness Goal. The focus on this system is to engage all levels of government, including the private and nonprofit sectors to achieve the Whole Community Preparedness initiatives. The National Preparedness System includes guidance for planning, organization, equipment, training, and exercises to build and maintain domestic capabilities.
- **National Planning Frameworks** and Federal Interagency Operational Plans.
- An annual National Preparedness Report.
- **Building and Sustaining Preparedness.** To build and sustain national preparedness, this element of PPD-8 includes public outreach initiatives and community-based and private-sector programs to enhance national resilience.
- In addition, a number of new guidance documents will help the general public, businesses and nonprofit organizations, and all levels of government make the most of their preparedness activities.

To date, there have been 21 Presidential Policy Directives, with PPD-21 being issued on February 12, 2013, focusing on critical infrastructure security and resilience, a common risk management theme.

Chapter 16, Presidential Policy Directive 8: An Overview, provides additional information about PPD-8.

Shift from Target Capabilities to Core Capabilities

The National Preparedness Guidelines, released in September 2007, defined the need for Target Capabilities and the Universal Task List. Target Capabilities provided metrics for measuring response times and capabilities. These metrics provided emergency managers with benchmarks for improving their response capabilities and allowed an opportunity to build discussion-based and operations-based exercises that were easily evaluated based on these predefined metrics. The shift to Core Capabilities has removed those metrics and focused more on preparedness through the whole community initiatives that PPD-8 brought along with its new focus. This has caused some concern in several mission spaces, including the food and agriculture sector where defining the need to properly respond and act to a foreign animal disease emergency has been moved to ensuring "supply chain integrity." The International Association of Emergency Managers (IAEM) USA committee on Food and Agriculture has been working to reevaluate some of these changes and submitted a position paper to IAEM that was approved on December 13, 2012. The position paper outlines three primary areas where the committee believes DHS needs to focus:

1. Food, agriculture, and animal disease emergency preparedness involves capabilities with unique planning, organizing, equipping, training, exercising, and assessing needs; these are not adequately addressed by the Core Capabilities in the National Preparedness Goal.
2. The move to Core Capabilities from Target Capabilities has presented challenges to many organizations as the shift from a metric-based system has left many groups uncertain about how capabilities are exercised and evaluated.
3. Programs, guidance, and initiatives related to the food and agriculture sector are generally set in motion prior to sufficient sector engagement and include ambitious deadlines for the implementation of various provisions.

The committee is currently working with other organizations to solicit support on these issues.

Federal Disaster Assistance Nonprofit Fairness Act of 2013

On February 8, 2013, Congress passed HR 592, amending the Robert T. Stafford Disaster Relief and Emergency Assistance Act to clarify that houses of worship are eligible for certain disaster relief and emergency assistance. This was in direct response to the catastrophic damage caused by Hurricane Sandy in the Northeastern U.S. While this is not the first time that Congress has enacted legislation providing financial assistance to religious nonprofit institutions, this legislation is important because it amends the Stafford Act to include these groups.

Response Plans

For nearly a decade, a progression of response plans—Federal Response Plan (FRP), National Response Plan (NRP), and National Response Framework (NRF)—was written to address "lessons learned" from actual and potential disasters, as well as changes in statutory and policy directives, with the intended aim of improving the Nation's preparedness and response coordination.

The creation of the FRP was driven by the PDD-39, which itself was developed following the Oklahoma City bombing of the Murrah Federal Building. The NRP superseded the FRP, following the terrorist attacks of September 11, 2001. Finally, the NRF superseded the NRP after the devastation of Hurricane Katrina. Creation of both the NRP and the NRF was driven by HSPD-5.

These Federal plans are supported by emergency support functions (ESFs) annexes that describe the missions, policies, structures, and responsibilities of Federal agencies for coordinating resource and programmatic support. The ESFs provide the structure for coordinating Federal interagency support for a Federal response to an incident. They are mechanisms for grouping functions most frequently used to provide Federal support.

The plans are guidelines, not requirements for the States in the beginning, but States are supposed to follow Federal plans to get funding. Some States follow the plan, others do not.

The development of these plans was based on several premises:

- A basic premise of all the plans is that the State is FEMA's primary client. Response doctrine is rooted in America's Federal system and the Constitution's division of responsibilities between Federal and State governments.
- Another premise is that incidents are handled at the lowest jurisdictional level possible. In the vast majority of incidents, State and local resources and interstate mutual aid will provide the first line of emergency response and incident management support. When State resources and capabilities are overwhelmed, governors may request Federal assistance. That strategy provides the framework for Federal interaction with local, tribal, State, territory, commonwealth, and private-sector and nongovernmental entities in the context of domestic incident management.
- A third premise is that mass care is traditionally a community response. NGOS, such as the American Red Cross and the Salvation Army, and other faith-based and community-based organizations have traditionally provided mass care services to communities during disasters.

All of these plans focus upon all-hazard emergencies: natural disasters, technological emergencies (such as hazardous material releases), and acts of terrorism.

Federal Response Plan

The FRP established a new process and structure for the systematic, coordinated, and effective delivery of Federal assistance to address the consequences of any major disaster or emergency declared under the Stafford Act. The plan organized the types of Federal response assistance that a State is most likely to need under 12

ESFs. It also described the process and methodology for implementing and managing Federal recovery and mitigation programs and support/technical services.

The FRP provided a focus for interagency and intergovernmental emergency preparedness, planning, training, exercising, coordination, and information exchange, serving as the foundation for the development of detailed supplemental plans and procedures to implement Federal response and recovery activities rapidly and efficiently.

The FRP applied to a major disaster or emergency as defined under the Stafford Act for which the President determines that Federal assistance is needed to supplement State and local efforts and capabilities. The FRP covered the full range of complex and constantly changing requirements following a disaster: saving lives, protecting property, and meeting basic human needs (response); restoring the disaster-affected area (recovery); and reducing vulnerability to future disasters (mitigation). The FRP did not specifically address long-term reconstruction and redevelopment.

The FRP engaged 22 Federal agencies, plus the American Red Cross, to assist states with disaster preparedness and response. The FRP augmented other responses such as the National Contingency Plan (NCP) for oil and hazardous materials spills and the Federal Radiological Emergency Response Plan (FRERP).

Under the FRP, the Federal Government and the American Red Cross shared responsibilities for sheltering victims, organizing feeding operations, providing emergency first aid at designated sites, collecting and providing information on victims to family members, and coordinating bulk distribution of emergency relief items. As the primary agency for mass care under ESF 6, the American Red Cross coordinated Federal mass care assistance in support of State and local mass care efforts. The American Red Cross was the only NGO signatory to the FRP. Other NGOs became formally involved in later plans.

While the FRP was revised in 2003 in response to the terrorist attacks of September 11, 2001, it was acknowledged that further revision was needed to include long-term recovery activities (e.g., housing) as well as response services.

National Response Plan

In 2003, the President George W. Bush directed DHS, through HSPD-5, to develop a new NRP to align Federal coordination structures, capabilities, and resources to form a unified, all-discipline, and all-hazards approach to domestic incident management. This approach eliminated critical seams and tied together a complete spectrum of incident management activities to include the prevention of, preparedness for, response to, and recovery from terrorism, major natural disasters, and other major emergencies.

In December 2004, Secretary Tom Ridge issued the NRP which described how to improve coordination among Federal, State, local, and tribal organizations to help save lives and protect America's communities by increasing the speed, effectiveness, and efficiency of incident management.

As noted by the preface, Secretary Ridge acknowledged:

> [I]mplementation of the plan and its supporting protocols requires extensive cooperation, collaboration, and information-sharing across jurisdictions, as well as between the government and the private sector at all levels.

This approach was far reaching in that it, for the first time, eliminated critical seams and tied together a complete spectrum of incident management activities. These included prevention of, preparedness for, response to, and recovery from terrorism, major natural disasters, and other major emergencies. The plan described improved processes for coordination among local, tribal, State, and Federal organizations by increasing the speed, effectiveness, and efficiency of incident management.

The NRP was built on the template of NIMS, which provides a consistent doctrinal framework for incident management at all jurisdictional levels, regardless of the cause, size, or complexity of the incident. It superseded other response plans, namely

- FRP
- United States Government Interagency Domestic Terrorism Concept of Operations Plan
- Federal Radiological Emergency Response Plan

The NRP and its coordinating structures and protocols provided mechanisms for coordination and implementation of a wide variety of incident management and emergency response activities. The NRP was an integration of the State, local, and Federal assets. Included in these coordinating activities were Federal support to local, tribal, and State authorities; interaction with nongovernmental, private-donor, and private-sector organizations; and the coordinated, direct exercise of Federal authorities, when appropriate.

While the FRP addressed response activities only, the NRP included response and recovery, as well as a need for long-term recovery activities that was recognized following Hurricane Katrina. The NRP was considered by the emergency management community not to be a plan but rather to set boundaries for hierarchical framework for planning.

The NRP also established a new term of reference for disasters—*incidents of national significance*—in order to address potential acts of terrorism. Under the authority of the Secretary of Homeland Security, the Federal response to incident of national significance could include

- A federal department or agency, responding under its own authorities, requests DHS assistance
- Resources of State and local authorities are overwhelmed:
  - Stafford Act major disasters or emergencies
  - Other catastrophic incidents

- More than one federal department or agency is involved:
  - Credible threats or indications of imminent terrorist attack
  - Threats/incidents related to high-profile, large-scale events

Given its sweeping changes, the NRP included its own implementation schedule.

The plan addressed the full spectrum of activities related to domestic incident management, including prevention, preparedness, response, and recovery actions. The NRP focused on those activities that are directly related to an evolving incident or potential incident rather than steady-state preparedness or readiness activities conducted in the absence of a specific threat or hazard.

Additionally, since incidents of national significance typically resulted in impacts far beyond the immediate or initial incident area, the NRP provided a framework to enable the management of cascading impacts and multiple incidents as well as the prevention of and preparation for subsequent events. Examples of incident management actions from a national perspective include

- Increasing nationwide public awareness
- Assessing trends that point to potential terrorist activity
- Elevating the national Homeland Security Advisory System alert condition and coordinating protective measures across jurisdictions
- Increasing countermeasures such as inspections, surveillance, security, counterintelligence, and infrastructure protection
- Conducting public health surveillance and assessment processes and, where appropriate, conducting a wide range of prevention measures to include, but not be limited to, immunizations
- Providing immediate and long-term public health and medical response assets
- Coordinating federal support to State, local, and tribal authorities in the aftermath of an incident
- Providing strategies for coordination of Federal resources required to handle subsequent events
- Restoring public confidence after a terrorist attack
- Enabling immediate recovery activities, as well as addressing long-term consequences in the impacted area

On August 30, 2005, Secretary Michael Chertoff invoked the NRP the day after Hurricane Katrina hit the Gulf Coast. By so doing, Secretary Chertoff assumed the leadership role triggered by the law to bear primary responsibility to manage the said crisis.

Almost a month later, in advance of the landfall of Hurricane Rita, Secretary Chertoff declared the storm an incident of national significance and put preparations in place in the gulf region of Texas.

Because of the lengthy implementation schedule, the increased level of coordination did not sufficiently materialize. This situation became severely problematic when Hurricane Katrina roared into the Gulf of Mexico, then made landfall in Louisiana. Hurricane Katrina caused severe destruction along the Gulf coast. The most severe loss of life and property damage occurred in New Orleans, Louisiana, which flooded as the levee system catastrophically failed, in many cases hours after the storm had moved inland.

Following Hurricane Katrina, the plan was updated on May 25, 2006. The notice of change stated the update "emerged from organizational changes within DHS, as well as the experience of responding to Hurricanes Katrina, Wilma, and Rita in 2005."

National Response Framework

Published in January 2008, the NRF was developed to address the requirements of PKEMRA. It is framework that guides local, State, and Federal entities enabling all response partners to prepare for and provide a unified national response to disasters and emergencies. This framework establishes a comprehensive, national, all-hazards approach to domestic incident approach.

As identified by DHS, the NRF

> [P]resents the guiding principles that enable all response partners to prepare for and provide a unified national response to disasters and emergencies—from the smallest incident to the largest catastrophe. This important document establishes a comprehensive, national, all-hazards approach to domestic incident response. The Framework defines the key principles, roles, and structures that organize the way we respond as a Nation. It describes how communities, tribes, States, the Federal Government, and private-sector and nongovernmental partners apply these principles for a coordinated, effective national response. It also identifies special circumstances where the Federal Government exercises a larger role, including incidents where Federal interests are involved and catastrophic incidents where a State would require significant support. The Framework enables first responders, decision-makers, and supporting entities to provide a unified national response.

An underlying basis of the NRF is a set of key principles:

- **Engaged Partnership**. Leaders at all levels must communicate and actively support engaged partnerships by developing shared goals and aligning capabilities so that no one is overwhelmed in times of crisis.
- **Tiered Response**. Incidents must be managed at the lowest possible jurisdictional level and supported by additional capabilities when needed.
- **Scalable, Flexible, and Adaptable Operational Capabilities**. As incidents change in size, scope, and complexity, the response must adapt to meet requirements.

- **Unity of Effort through Unified Command.** Effective unified command is indispensable to response activities and requires a clear understanding of the roles and responsibilities of each participating organization.
- **Readiness to Act.** Effective response requires readiness to act balanced with an understanding of risk. From individuals, households, and communities to local, tribal, State, and Federal governments, national response depends on the instinct and ability to act.

Owing to the confusion brought about by the term of reference—incident of national significance—in the NRP, the NRF eliminated this term.

An important concept presented in the NRF included the preparedness life cycle, which represents a systemic approach to build the right capabilities for the Nation in response to all hazards. The preparedness life cycle (see Figure 1.1)

- Introduces National Planning System
- Defines response organization
- Requires training
- Advocates interoperability and typing of equipment
- Emphasizes exercising with broad-based participation
- Describes process for continuous evaluation and improvement

The NRF establishes 15 ESFs:

- ESF #1—Transportation
- ESF #2—Communications

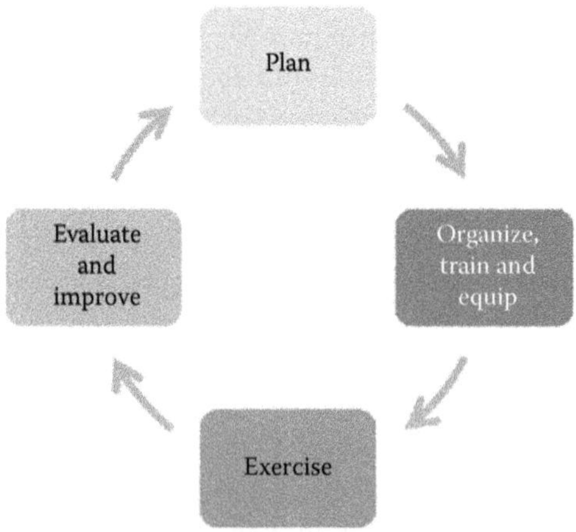

Figure 1.1 Preparedness life cycle of the NRF.

- ESF #3—Public Works and Engineering
- ESF #4— Firefighting
- ESF #5—Emergency Management
- ESF #6—Mass Care, Emergency Assistance, Housing and Human Services
- ESF #7—Logistics Management and Resource Support
- ESF #8—Public Health and Medical Services
- ESF #9—Search and Rescue
- ESF #10—Oil and Hazardous Materials Response
- ESF #11—Agriculture and Natural Resources
- ESF #12—Energy
- ESF #13—Public Safety and Security
- ESF #14—Long-Term Community Recovery
- ESF #15—External Affairs

It also includes several support annexes and incident annexes to help improve coordination:

Support Annexes

- Critical Infrastructure and Key Resources
- Financial Management
- International Coordination
- Private Sector Coordination
- Public Affairs
- Tribal Relations
- Volunteer and Donations Management
- Worker Safety and Health

Incident Annexes

- Biological Incident
- Catastrophic Incident
- Cyber Incident
- Food and Agriculture Incident
- Mass Evacuation Incident
- Nuclear/Radiological Incident
- Terrorism Incident Law Enforcement and Investigation

To promote awareness and education of the NRF, FEMA developed an independent study training course, IS-800, *An Introduction to the NRF,* which is available free of charge. FEMA continues to develop other general orientation courses for ESFs and the Support and Incident Annexes though its online study program at the Emergency Management Institute.

Emergency Support Functions

All three plans use ESFs as a means to provide the interagency staff to support federal response operations of the National Response Coordination Center (NRCC), the Regional Response Coordination Center (RRCC), and the Joint Field Office (JFO). Depending on the incident, deployed assets of the ESFs may also participate in the staffing of the Incident Command Post. Under the NRF, each ESF is structured to provide optimal support for evolving incident management requirements.

ESFs may be activated for Stafford Act and non-Stafford Act implementation of the NRF (although some Incidents of National Significance may not require ESF activations). ESF funding for non-Stafford Act situations will be accomplished using NRF Federal-to-Federal support mechanisms and will vary based on the incident.

Within the NRF, each ESF Annex identifies the ESF coordinator and the primary and support agencies pertinent to the ESF. Several ESFs incorporate multiple components, with primary agencies designated for each component to ensure seamless integration of and transition between preparedness, prevention, response, recovery, and mitigation activities. ESFs with multiple primary agencies designate an ESF coordinator for the purposes of preincident planning and coordination.

A Federal agency designated as an ESF primary agency serves as a Federal executive agent under the Federal Coordinating Officer (FCO) (or Federal Resource Coordinator for non-Stafford Act incidents) to accomplish the ESF mission. When an ESF is activated in response to an Incident of National Significance, the primary agency is responsible for

- Orchestrating Federal support within their functional area for an affected State
- Providing staff for the operations functions at fixed and field facilities
- Notifying and requesting assistance from support agencies
- Managing mission assignments and coordinating with support agencies, as well as appropriate State agencies
- Working with appropriate private-sector organizations to maximize the use of all available resources
- Supporting and keeping other ESFs and organizational elements informed of ESF operational priorities and activities
- Planning for short- and long-term incident management and recovery operations

ESF Support Agencies

When an ESF is activated in response to an Incident of National Significance, support agencies are responsible for

- Conducting operations, when requested by DHS or the designated ESF primary agency, using their own authorities, subject-matter experts, capabilities, or resources

- Participating in planning for short- and long-term incident management and recovery operations and the development of supporting operational plans, procedures, checklists, or other job aids, in concert with the existing first-responder standards
- Assisting in the conduct of situational assessments
- Furnishing available personnel, equipment, or other resource support as requested by DHS or the ESF primary agency
- Providing input to periodic readiness assessments
- Identifying new equipment or capabilities required to prevent or respond to new or emerging threats and hazards, or to improve the ability to address the existing threats
- Nominating new technologies to DHS for review and evaluation that have the potential to improve performance within or across functional areas
- Providing information or intelligence regarding their agency's area of expertise

Reference

http://www.dhs.gov/presidential-policy-directive-8-national-preparedness (accessed September 13, 2013).

II

RESPONSE PLANNING AND PREPAREDNESS

Chapter 2

Emergency Operations Center Readiness Continuum

Derek Rowan

Introduction

Emergency Operations Centers (EOCs) and emergency management agencies across the country have a strong commitment to prepare for, mitigate effects of, respond to, and recover from large-scale incidents requiring the activation of their EOCs, Multiagency Coordination (MAC) Systems, Incident Command Systems, and requisite staff.

Some of the more common issues that arise within EOCs during real-world incidents are challenges revolving around

1. Roles and responsibilities of the EOC staff
2. Coordination with outside jurisdictions or agencies
3. Collaboration with different agencies within the jurisdiction
4. Unfamiliarity on position requirements and duties
5. Unfamiliarity on how to share situational awareness and common operating picture
6. Inconsistent planning procedures for future operations or needs
7. Communication challenges within the EOC elements
8. Effective and clear public messaging
9. Managing the incident and field responders
10. Unfamiliarity for proper resource ordering

Not every EOC will have all of these challenges with every incident, but many will have more than one. Each of these challenges generally is caused by inadequate or unclear

1. Plans, policies, and procedures for both the EOC and the agencies providing staffing
2. Organizational structure for the EOC and how it will be expanded, contracted, and organized for a specific incident
3. Clearly defined roles and responsibilities for each position complete with job action sheets, job aids, or checklists
4. Equipment for the EOC and positions to assist in situational awareness
5. Training for EOC staff on all of the above

While it is easy to point the finger at lack of training, it is a much more complex challenge than just simply providing training classes. In the book *Principles of Emergency Management*, Dr. Michael J. Fagel quotes consultant Art Botterell as saying "no matter who you train, someone else will show up." Training is difficult to schedule in our complex and busy environments. Everyone is expected to do more with less. We are all required to function on multiple levels in our professional lives while operating in an environment that is not allowed to take a "coffee break." Organizations must continue to operate and cannot be stopped because there is a training class scheduled.

Creating an effective emergency operations center organization varies based on the needs of the jurisdiction, staffing levels, hazards, and more. Thus, there is no "right way" of organizing an EOC. The Federal Emergency Management Agency (FEMA) "Multiagency Coordination Systems" course IS-701, provides for four common organization structures:

1. Major Management Activities
2. Incident Command System (ICS)
3. Emergency Support Function (ESF)
4. Multiagency Coordination (MAC) Group

The actual design of an EOC will often be combinations of these, based on the needs, statutes, agencies, and staff of that jurisdiction. Regardless, the need for position-specific training tends to be the major element that is missing from many EOC preparedness programs. In the many exercises that we have conducted for EOCs around the country, probably the most common occurrence in after action reports (AARs) is the need for position-specific training and recurrent training for those that have had training in the past. Because most staff in an EOC (including full-time staff) do not utilize their EOC-specific incident training often, there is a gradual decay in the knowledge, skills, and relationships that have been built at previous incidents.

This inability to continually train on skills and knowledge that we use infrequently translates into a lower level of readiness in our emergency response and management capabilities. Since emergency response, emergency management, incident command, and other similar elements are not something that we practice often or even really perform in our regular duties, it is imperative that these skills be constantly kept up to date as they atrophy over time without use. The professional emergency services such as the fire department train on basic skills such as incident command constantly. This is noteworthy because they also perform under the incident command system daily at emergencies. Why do they constantly practice? They recognize that skills have an inherent weakening over time. This seemingly obvious problem is identified again and again in the AARs of major national incidents.

The report "*Lessons we don't learn*" by the *Homeland Security Affairs Journal* explores the reasons for these same issues of incident and emergency management on major incidents. The results identified referencing the challenges of using a planning system such as ICS during an incident:

1. Agencies and departments within agencies lack the commitment to coordinate with each other. Especially if they compete for resources.
2. ICS is in common use, but it is not understood and implemented in a consistent manner. Different departments, locations, or disciplines do their training in isolation and do not exercise together.

The report continues to say that "responders cannot be expected to learn the functions of incident management during the heat of an event." The report maintains "if they haven't already been training in logistics for example, it will take them a long time to figure it out. Yet, absent sound training, this is exactly what happens, with the needless result that recognize and well-developed incident management functions are carried out poorly."

For an EOC, it is even more critical to constantly refresh your knowledge skills and abilities, due to the changing nature of emergency threats, the constant influx of new staff, and updates to various emergency operating plans from within the organization and from external partners.

Initial training must be relevant, solid, effective, memorable, and based on the specific context of the roles of each staff person. You must train as you would respond. Too often, this training is generic training without specific activities based on how it is actually performed within that entity. Part of the problem is inadequate position-specific job action sheets or ineffectively defined roles.

Recurrent training must be relevant, easy to take, scalable, low cost, and perhaps, most importantly, fast. Staff do not have the time to constantly attend training classes.

Validation events and exercises need to be realistic, plausible, and provide a mechanism for a true evaluation of readiness and capabilities. This evaluation needs to provide a direct link for lessons learned and a methodology to create identified areas for improvement that can be tracked over time.

Recurring training that requires little time and commitment, but still provides meaningful training and exercises can be accomplished with a blend of web-based seminars and activities, instructor-led in-person learning, and both virtual and real exercises.

While many EOCs have training schedules, we have seen that a large number of those do not have a comprehensive training strategy, plan, and schedule. This plan and schedule should be based on lessons learned, previous exercise corrective actions, vulnerabilities, and upcoming events. This approach is an effective readiness enhancement.

An EOC needs a unique continuum of preparedness. Rather than providing single classes on specific topics as is typical in the traditional training plan, they need an interwoven learning environment that builds upon previous classes, lessons learned, and new themes that are being practiced in the EOC or region.

This constant approach to curricula development allows moving into the next theme based on the audience, input from previous classes, and client direction and input. The result is a cycle of learning that is easy to attend, short, and recurring to accommodate all schedules. We have five types of training delivery methodologies:

1. Instructor-led classroom training
2. Online independent study training
3. Online instructor-led distance learning training
4. Online facilitator-led discussion-based exercises
5. In-person exercises

The Training and Exercise Plan

Developing a training and exercise plan is the first step to defining the needs that training can address. The creation of a training and exercise plan will allow your EOC to have a clear roadmap of future needs, training, and validation efforts going forward. This will assist the organization with the creation of budgets, replacement of resources, retraining of personnel due to staff turnover, along with the ability to create a self-sustaining program. The training and exercise plan will outline the steps necessary to get from where you are to an effective emergency operations center with integrated evaluation.

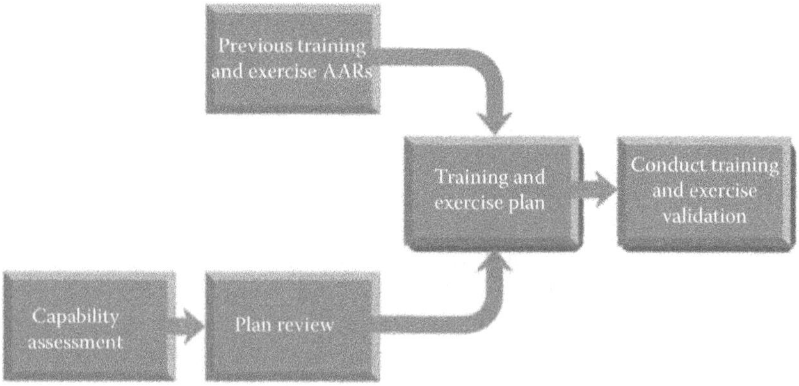

Each class in the training plan must work through a standard instructional systems design process. Most utilize the Analyze, Design, Development, Implementation, and Evaluation (ADDIE) model. This methodology looks at the goals, outcomes, objectives, subject matter, plans, audience, timeframes, and budget, to determine the most efficient and effective course curricula and delivery. As the courses and exercises are produced, the best practice is to employ a real-time feedback mechanism and verification of relevance. Feedback includes not only course results, but also a review of doctrine, real-world incidents, and student-provided feedback to produce improvements and updates to courseware and to relevant plans and checklists.

Instructor-Led Classroom Training

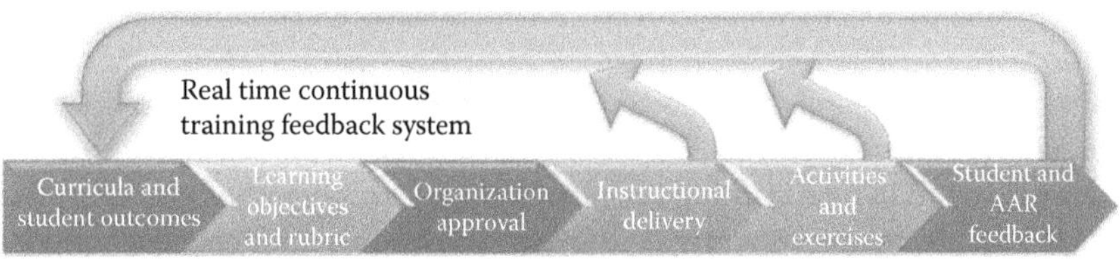

Traditional classroom training led by an experienced instructor is familiar to all of us. Improving this training can be done in several ways. First, making it relevant is an important addition that is so often left out, especially from training that is "off the shelf." For example, when providing a standardized course, changing the photographs to local, relevant photos makes a tremendous difference in the student's mind. This simple change provides additional buy-in, student attentiveness, and interactivity. Second, make sure the instructor examples and activities are relevant to the student audience. While this is "Instructor 101," it is very common for this not to occur, especially when using existing course materials. Third, using good, experienced instructors that abide by four simple rules will greatly improve the course. Those rules:

1. Prepare for the instruction.
2. Care about the students.
3. Do not read the slides.
4. Be on time.

Traditional classroom training is great for interactivity, group activities, and student relationship building. It is also a good method for "initial" training for those that are new to the subject matter.

Online Independent Study Training

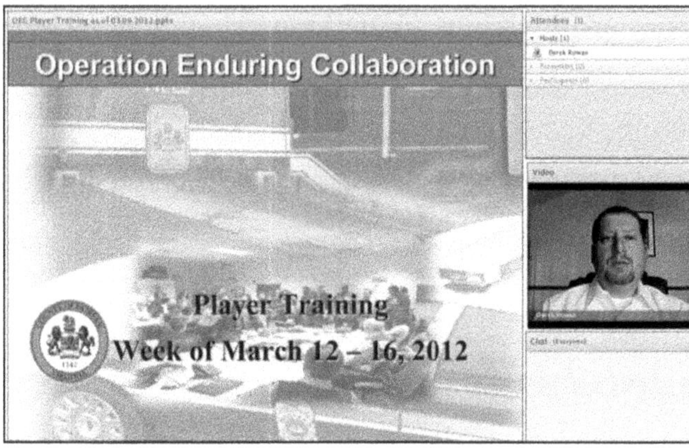

Another popular training methodology is online independent study training. Almost all of us have received our ICS 100 and 200 certificates via independent study. But there are many additional training sessions that can be accomplished via this method.

It is a simple matter to create courseware and create an independent and even interactive course utilizing easily purchased tools. These tools provide for video, slide presentations, student interactive modules, quizzes, and more.

Online independent courses can be a fantastic adjunct for new policies or procedures, or reviews of existing materials. They are also good for longer instructor-led courses to provide a module that is easily learned through an online method, thereby shortening the classroom time.

Online Instructor-Led Distance Learning Training

Using the Internet to provide a classroom for instructor-led training has been around for some time. However, having it be taken seriously for credentialed classes is relatively new. The complexity of these can vary from simple online "webinars" to complex multiple-week courses that are credentialed to provide accredited certificates from FEMA.

There are several FEMA courses that are being developed into an online instructor-led delivery model. These will have the same content as the classroom versions, but are being delivered through a web-based system with a live instructor. Students must be online at the same time so that they can participate in interactive student-to-student activities—similar to classroom style groups.

This format is most effective for "short chunk" training—training that is less than 30 minutes in duration and focused on specific job performance enhancement tasks or procedures. These will typically be set up into different audience tracks and provide specific training for that audience.

Online Facilitator-Led Discussion-Based Exercises

Online discussion-based exercises have been attempted throughout the last few years. They have not caught on due in part to complex systems, time requirements, and difficult interaction. With the increase in technology web conferencing system and pervasive web cams, interactive discussion-based exercises that are conducted online are a convenient method for small quick exercise conduct. These should be viewed not as a replacement for traditional table-top exercises, but rather for quick validation of focused planning elements.

In-Person Exercises

Traditional exercises, both discussion and operations-based, are still the king of preparedness activities for EOCs. There is no better or more realistic method for validating and increasing readiness for an EOC. The key to effective exercises is agency buy-in and commitment, and realistic design and simulation. We have found that exercise dissatisfaction can be traced to choosing the wrong exercise type, not having adequate agency buy-in, or inadequate design and simulation. When these items are correctly aligned, exercises can provide the best way to see strengths, areas for improvement, and truly validate all capability elements. (See Chapter 19 for additional information on exercise design.)

Creating the Continuum

When you merge each of the preceding elements into an organized plan, with structure and a feedback loop, you will achieve unsurpassed training for your staff. To accomplish this, your training and exercise plan should note which levels of which elements should be employed to develop this level for each audience.

One example could include a cycle and mix of online webinars conducted monthly, several instructor-led classroom courses, an in-person tabletop exercise, and an in-person functional EOC exercise. Each element builds on the previous and allows incorporating rapid evaluation of one into the next.

Because it is important to track progress, the continuum must include a tracking database that will record each person that received training, the subject and objectives for that training, capabilities met during exercises, and corrective actions identified. This system will provide an objective review and instant reports on the readiness of the entire system.

Benefits

The benefits of this training continuum are

1. Recurring training: Continual training in the use of incident management protocols in the EOC is critical to being successful during an incident.

2. Getting the staff's mind thinking about readiness: Because EOCs typically utilize a different operational model, then most staff are used to during their day-to-day jobs, keeping it on the mind promotes greater awareness and it allows for the natural usage of procedures during high-stress incidents. The more people remember a tool, the more they will use the tool.
3. Immediate recall of critical emergency operation plan procedures: Because of the nature of incidents, many plan and procedure components require instant knowledge of and application during an incident. It is said the first 5 minutes of a critical incident is worth the next 5 hours due to this heavy reliance on an initial strong command and control system to direct coordination when things are at their most chaotic.
4. Less repeating of the same mistakes: Lack of coordination and collaboration continue to show up in numerous AARs for both exercises and large-emergency incidents because we continue to not heed the advice of so many. EOC operations only work when everyone is trained properly, knows their specific role and responsibility, and understands the mission of the EOC.
5. Easier and faster to show compliance with standards: Because of the many requirements that an EOC must follow, having a detailed system of training and exercise showing compliance, and results by tracking all training is critical.
6. Less expensive training: Web-based training is inherently less expensive due to the decreased time to develop, deliver, and maintain. There is no travel for instructors or staff and there is increased value due to the efficiencies of having multiple training and exercises under one contract.
7. More concentrated and useful—designed for true needs: Each training class will be designed to be built upon the previous one. Using the results of the feedback received and the student performance witnessed, you can deliver specific concentrated learning objectives in a short amount of time targeting the true needs of the staff.

Conclusion

Training and exercises are a critical part of an EOC preparedness and readiness strategy. Combining them into a consistent mix and cycle of events that continually raises the bar for participants will, over time, result in an increased culture of preparedness for an organization.

Chapter 3

Stress Management and Responders

Kathryn R. Juzwin

Why Stress Management Is Important to Emergency Managers

In crisis situations, community management has two populations to attend to and manage: their civilian citizens and their own employees. This chapter offers emergency planners and community management professionals a framework of important components to build into their emergency response plan in a pre-disaster situation. There are a number of benefits for building this structure. The first is obvious, having a solid plan ready to put into place in a disaster or emergent situation helps optimize success in response and recovery. However, if the organization is focused on long term, consideration of specific elements has been shown to predict organization commitment. The organization commitment has been shown to directly relate to the perceived leadership style, especially where the leadership defines roles of personnel toward defined goals through assigned tasks, using specific procedures, timelines, and communication (Dale and Fox, 2008). Further, research demonstrates that planning targeting reduction of perceived role stress and increasing or solidifying interpersonal factors have benefits to the organization (Dale and Fox, 2008).

Ever evolving, emergency managers have begun to anticipate events that meet and exceed the scope of what they can manage within their own municipality, region, or state. The mechanics of the response structure appears well dictated, as evidenced by the trend to use the National Incident Management System (NIMS; FEMA.gov) approach, and the protocol of various agencies mandating response within their scope of service. In the case of disasters, what happens? "On rare occasions, emergencies occur that are so large in scale and so sever that local responders may not have the resources—people, equipment, expertise, funds—to effectively and safely respond. Even in such cases, local responders do not hesitate to do what they have been trained to do—go to the site prepared to save lives, protect property, and remove the threat" (Jackson, Baker, Ridgely, Bartis, Linn, Science and Technology Policy Institute, RAND Corporation and the National Institute for Occupational Safety and Health [NIOSH], 2004, p. iii).

Law enforcement has built into its pre-incident strategic planning interventions to add to their incident command approaches. Five components have been identified as necessary to build into preincident planning (Sheehan et al., 2004). This includes assessment and triage, specific crisis intervention with individuals, small and large group crisis intervention, and strategic planning. Within the context of an emergency situation, the community manager will want to optimize the chance for retention and ability to return to function in a typical daily work routine. Strategic planning should take into account elements of the design of tasks, management style, interpersonal relationships, work roles, career concerns, and environmental concerns (Goetzel et al., 1998 in Colligan et al., DHHS-NIOSH). Specifically, each of these is important to consider in their contribution to stress for the employee:

- The design of tasks involves heavy workload, infrequent breaks, long work hours, shifts, tasks with little personal meaning, underutilizing of skills, and lack of control.
- Management style where employees do not contribute in decision making, poor communication, and the absence of family-friendly policies.
- Poor social environment and interpersonal relationships among coworkers and supervisors.
- Work roles that have conflicting or uncertain job expectations, or too much responsibility.
- Job insecurity, lack of potential for advancement, or rapid changes for which workers perceive they are unprepared contribute to career concerns.
- Unpleasant or dangerous physical conditions contribute to environmental concerns.

Minimal attention in emergency planning has focused on the potential secondary survivors, specifically the responders and emergency service workers who attend to all of those demands. One element often not accounted for in planning is the nonemergent and nonimmediate human impact element for the *responders*. It is important to consider the well-being of the employees because when a disaster strikes their community they become victim responders. Their commitment will be to do what they have to do for their job and community. Their minds and hearts will be preoccupied with their own families, homes, and neighborhoods. It is important to note that there are findings (Galea, 2007) suggesting that the serious long-term consequences of disasters and mass trauma are most frequent in specific groups of survivors: in the order of severity of consequences, the people injured in the incident, rescuers, people who have lost their personal belongings and homes, families of those injured, and then the general population (Galea, 2007, Sheehan et al., 2004).

Stress-Related Disorders

It is estimated that stress-related disorders will be the second leading cause of disability by the year 2020, and is the primary strategic goal for the World Health Organization's Global Burden of Disease of the NIOSH Work Organization and Stress-Related Disorders Program (Ray and Sauter, 2008). In the United States, work-related stress is estimated to cost $171 billion annually, which is the same as cancer, cardiovascular disease, and greater than Alzheimer's or HIV or AIDS (NIOSH). There is an estimated $300 billion annual cost due to lost hours from absenteeism, decreased work productivity, and cost of health expenditures (APA, 2004).

Organizations such as the World Health Organization and the National Institute of Mental Health suggest that emotional health-related problems, suicide, and mental illness account for over 15% of the burden of disease costs to an economy. The number one cause, cardiovascular conditions, contributes to more than 18% of the disabling conditions. However, when alcohol and drug-use problems are factored into the emotional problem-based disabling conditions, this increases disability to more than 20%.

In a disaster situation, the rate of impact is generally greater. Research demonstrates that individuals without the benefit of prior disaster training or experience are at greater risk for PTSD (post-traumatic stress disorder) (Perrin et al., 2007). In a study of PTSD post-WTC, the rate was 21.2% for construction, engineering, sanitation, and unaffiliated workers, compared to the rate of 12.4% for rescue/recovery workers (Perrin et al., 2007). The National Mental Health Information Center (DeWolfe, 2000) makes important considerations regarding people who have any exposure to disaster situations. While all of these are important, one is absolutely the most critical point. These are important to highlight so as to consider the mental health needs of not only primary survivors but also the secondary survivors and personnel: No one who sees a disaster is left untouched by it.

Understanding Stress along the Continuum

Stress occurs as a function of living. Stress reactions are normal, but can be adaptive or maladaptive. Stress can be motivating and positive or negative and destructive. In extreme situations, reactions can be viewed as "normal responses to abnormal events rather than signs of psychopathology" (Rogers, 2007, p. 3). Stress can be short term and in the form of both small and large demands. Typical stress symptoms can impact the body's ability to be healthy and can interfere with the ability to fight illness or infection. Stress, at its extreme, is connected to heart disease, cancer, respiratory problems, accidents, suicide, depression, anxiety disorders, and alcohol-related problems. As discussed above, these contribute to the largest debilitating illnesses and economic impact. This impact has direct and indirect costs to the individual, employer, and society. These costs include days out sick, paid/unpaid time off, shift or work load coverage, treatment, recovery, and lost productivity to the employer.

When something happens that is out of the norm, the potential for problematic stress occurs. An unanticipated event with serious and unexpected impact, such as a tornado with damage to the community, a significant fire or flood, has far-reaching impact with regard to the length of time, extent of response needed to meet the demands of the extent of and recovery from the event. Exposure to these situations causes reactions that may include terror, horror, fear for one's safety and well-being. Reactions may be anywhere from mild to severe, and impact people in a number of different ways over a long time. Most reactions are temporary, although can be quite unnerving to the individual because they are so out of the range of typical reactions that usually happen for the person. Symptoms are patterns of behaviors, reactions, feelings, thoughts, or problems that impact how one manages in their life.

First responders and support personnel in emergency situations performing rescue and recovery work are exposed to stressors beyond their normal daily emergency-task demands. Their "normal" emergency response is generally above and beyond the situations that most of the general population ever faces. However, when there is a situation outside of that "normal" or "typical" scope, exposure to these events increases the risk of emotional and physical trauma, and PTSD. While the general population is considered to be at risk for PTSD at a prevalence rate of about 4% (Perrin et al., 2007), this increases upwards of 32% for rescue and recovery service personnel in the WTC study, 25% for search and rescue, and for firefighters (21%) (Perrin et al., 2007).

Stress Reactions

There are a number of symptoms that interfere with emotions or feelings, physical health or with thinking. Because these problems affect our thinking, feeling and physical functioning, our behaviors and the way we manage in our daily lives can become problematic and influenced in many ways.

Most of us have a typical baseline for recovering from stress and stressful events in our lives. We may be thrown off for a period of time, for example, a couple of days. However, when these problems do not resolve, then stress may be moving to distress or even cumulative stress. At the worse, these cause negative impact on physical health as well.

It is important to note that everyone handles stress differently. Not everyone responds the same way to stress. Some people have immediate signs of stress, others would not demonstrate any. Sometimes it is very obvious, as some of the symptoms show dramatically in physical complaints. Other people show only subtle signs that can often be attributed to something else entirely. When these last for more than a few days, it can be a signal that the problems may be more than the typical stressor, and that some additional support might be needed.

Educating people on stress in terms of physical, emotional or cognitive signs gives them an opportunity to watch for changes in themselves and in each other. Creating a culture of positive self-care and encouraging people to practice stress management daily is a solid foundation for preventative care for future situations.

These symptoms may include

Physical Symptoms	*Emotional Symptoms*	*Cognitive Problem*
• Constant fatigue or lack of energy • Inability to relax • Sexual problems • Weight changes • Cold sweaty palms • Changes in sleep patterns, trouble falling asleep, poor sleep or lack of restorative sleep • Difficulties with attention or poor concentration • Aches, pains, and muscle tension • Headaches • Grinding teeth, clenched jaw • Stomach and gastrointestinal problems such as upset stomach, nausea, diarrhea, and indigestion • Loss of appetite • Racing heart beat, shallow breathing, tightness in the chest	• Feeling irritability or short tempered • Sad, blue, or down • Worry, apprehension • Isolating self • Feeling isolated • Crying, tearfulness • Lessened ability to relax or to enjoy things • Feeling abandoned or targeted • Feeling out of it, overwhelmed, ineffective • Thinking of death as an option • Hopelessness • Anxious, fearful • Panic, terror, hypervigilance • Compulsions, checking, putting things in order • Anger/rage • Apathy, lethargy	• Feeling overwhelmed or unable to manage demands in daily life • Problems with remembering things and new learning • Forgetfulness • Problems with problem-solving • Difficulties concentrating • Difficulties making sense of things • Slowed thinking • Rigid or obsessive thinking • Problems with word finding • Reliving an event over and over • Denial or blame • Difficulties dreams • Denial/blaming • Suspiciousness • Doubting faith • Disorientation • Confusion • Easily overwhelmed

Acute Stress

Acute stress reactions are generally very distressing to the individual, and often the individual does not recognize these reactions as connected to the event. And in emergency situations, everyone can be so consumed with responding to the crisis, people may not realize that either they themselves or their peers are experiencing acute stress response. In the heightened focus of the moment, people may not be aware of their own reactions. There may be panic, shock, disbelief, disorganization, and/or distress. Many people experience a sense of being out of control.

One experience is described as feeling "disconnected" or often results in feelings of disorientation or that the individual is "going crazy." Unfortunately, the person often also feels as though they are losing control or are unable to manage, which can be compounded when, in fact, their ability to manage is somewhat compromised. It is important to make certain to connect the problems as reactions to the event, and that when managed, they tend to be lessened.

People who experience this acute stress report a variety of problems and changes in their behavior. These changes or problems occur in feelings, thinking, physical functioning, and behaviors. These problems include those listed above, and can be very intense and disturbing to the individual. *This is especially problematic if they are in the form of recurrent images, smells, sounds or thoughts about the stressful events.*

Manager's Responsibilities in a Critical Event

There are two levels of attention that a manager must take into account during a critical event. First is their responsibility in organizing and implementing their crisis response plan. The second is their responsibility to their people responding to and implementing the crisis response plan. *Without the second piece being managed effectively, the first piece may be compromised.* Each person involved in responding to your community plan will have people who have families, loved ones, friends, property, and worries about each of those. While they may be your emergency response team, they are also human beings with human connections.

Understanding stress responses helps insulate people and normalize stress. It also helps them recognize that when stress symptoms become too much, that it is important they are encouraged and supported to get the help they need before it becomes traumatic or critical incident stress.

Planning: Helping Take Care of Your Responders in Advance

The cornerstone of public health includes accessing resources from local to international to assure health of the people. Preplanning and having all of these elements in place may help alleviate some aspects of stress for the employee, knowing they have provided the best for their family as they were able, even in their absence from the family.

Our policy dictates and guides our eventual practice, like a blueprint dictates the building of a house. Many administrators understand and accept the need for moving beyond the reactive. Traumatic stress experts advocate that there should be representation at the policy making level, to offer guidance, consultation, and ongoing collaboration (Fairbank and Gerrity, 2007). Using an approach that includes emphasis on psychological health and early identification of psychological problems, followed by early intervention, inclusive of cultural considerations, and family and community support, is strongly recommended in public health policy (Fairbank and Gerrity, 2007).

There is at least one way to help mitigate the stress incurred to responders in a disaster situation: *assure their families are as safe as possible because of anticipated preparedness. The family is prepared to respond, and therefore in a more secure position to react and be self-sufficient. Controlling those elements that can be foreseen and pre-managed is the foundation to any good strategic plan.*

In the development of the community response plan, it is necessary to emphasize the importance that all employees have emergency plans in their own homes for their families that would include all the necessary general elements that are recommended for all citizens. It is a general rule of thumb that citizens should plan to be able to be self-sufficient for up to 3–4 days (72–96 h). If the disaster is of severe enough scope, many elements will not be able to be supplied for a longer period of time.

What elements go into a family emergency plan? In these times, almost every village or city has a website that directs its citizen's to an emergency plan. One of the most thorough is presented by the Federal Emergency Management Agency (FEMA) and is located on their website.

A general family plan should at least have elements including:

- Emergency planning for potential risks and specific hazards
- Emergency financial planning
- Emergency and alternative communication plan and contact numbers, especially if in the event of mass communication disruption
- Designated contact and/or meeting places and contact schedule
- Disaster supplies kit for each family member
- Disaster supplies kit for each pet
- Alternative shelter arrangements
- Food and water
- Sanitation needs
- Cash monies and credit cards
- Plan for communication, heat, light, power, and so on
- Anticipating the environmental risks and the elements, including bugs/wildlife
- Protective clothing, gloves, and hats

Among the components listed above, the agency can provide suggestions to encourage employees to have their crisis response plan in place, so if they are not there to assure the safety of their families, their family members know what to do related to:

- Where to meet in the house, community, or out of the danger area
- Whom to contact about status, especially if separated
- Important documents and money, access to it, if necessary
- Emergency supplies, resources, documentation, and identification (emergency preparedness lists)
- Resources within the community for contact
- Household information (gas, electric, etc.)
- Identification procedures and documentation
- What the rules are, and that everyone knows the plan
- Who is responsible for whom and what to do during an evacuation
- Who takes what with them and what gets left behind
- When to evacuate
- How to communicate the plan, with written information of contacts, numbers, back up plan, identified people whom everyone will contact
- Who to go to for help, if necessary
  - A call-tree for resources and supports
- When to leave and where to regroup and other back-up plans
- Identification, insurance, and vital records, inventory of home possessions
- Information related to utility shut-off and safety

Here are some ideas to build into the emergency plan the following elements that focus specifically on anticipating the needs of employees for their families and homes.

- Educate all employees about the specific agency policies and expectations for their responses during a major incident or disaster. By informing them in advance that the scope of their jobs has demands of community response, they can anticipate the conflicts that arise between the professional and personal demands. If the agency has a specific plan, put in succinct relevant information in bullet point formats, date it (so there is always a time reference), and distribute it. The goal is that this information can be put in the employee's home disaster kit.
  - Encourage your employees to have a disaster or emergency plan at home and to talk about it in advance of any situation.
- Include an avenue for communication and contact for employees and their families. This can include an identified go-between who has the responsibility of being the contact person and relaying of information in a planned and anticipated manner.
  - Assure that the agency has the current updated emergency contact information and an identified person for each employee. This should be

portable and taken up with the communication officer. It should be in electronic as well as hard copy.
 - Identify a point of contact for the organization and how that contact should be made for the employee and for the family who may be trying to contact the organization.
 - Develop a system for communication of status that involves a back-up plan for this information to be conveyed in the event of local communication failure.

- Have a system in place to assure that your responders can have communication with their family members and support persons. Build in a mechanism, time, and a place for employees to have status contact with their family contact person. If they have a sense they know their families are safe, they can focus on what their job is with more clarity and less stress. There may be less walking off the job or outright quitting if your responders are able to have information about the status of their first priority, their families.
- Encourage employees to have as part of their own disaster responses strategy an ability to reach out and connect with nondeployed employees or responders to be available to support families of deployed responders to assist them to assure the families' welfare and safety in their absence. If this is not possible, then have the employees identify other neighbors or friends who can help out. Everyone should have someone they are attached to and cooperating with, to get them through the immediate crisis and danger phases and into recovery phase. This is particularly important for prolonged service and absence.
  - Encourage employees to develop a network of spouses/significant others and families who can provide support, relief, and help to the family during a deployment. Include a call tree or a flow chart of who can provide what resource to others. Many spouses of military have regular support group meetings and functions to offer each other support and resources during the deployment of their spouses.
- Identify resources within the community that would be available to assist families in the event of an emergency. This may include church organizations, other coworkers, service agencies, schools, and so on. Your plan might include developing support agreements among your community agency and other community resources so that they know to extend their support to these families. Additionally, encourage your responders to develop relationships within their community so that there are established connections before something happens. Identify who is available within your community to act as a center for various supports (family reunification, child services, or homeless).
- Anticipate a centralized and accessible area for responders only, place where information can be posted. This can include a schedule of informational briefings, centralized posting of information, and post schedules of when updates are going to occur.

Suggestions for Supporting Your Responders

There are several strategies to assist your responders to organize themselves and attain their goals while lessening their stress levels during a critical event.

- Communication is key. Have regular planned briefings with your personnel. This helps people anticipate when information is going to be given. They should occur at intervals that are known and anticipated. Make it easy for your people to know who is who, and who to go to for what.
- When appropriate, post information in a centralized location for those who are not able to be at a briefing, but need the information. Make certain that there is nothing in a posted brief that you would fear a reporter getting a hold of, and put posted briefs in secure locations.
- Make certain that during the briefings your lines of communication are never cut, put on hold, or sent to phone mail. Always be able to have your command center accessible by a live person who can interact with the caller.
- You may want to have a meeting at the end of a shift or when people go off duty, to thank them for their efforts. When possible, use their names, shake hands, pat them on the back, and make eye contact. Ask about their family's well-being during this time. Tell them their efforts are appreciated and noted.
- Allow for clear lines of communication and responsibility. People forget to communicate the information about the why's and the specifics of why things are and are not happening. Your plan might want to post this information, to demonstrate that requests for resources have been acknowledged and that there is legitimate reason for why a supply may not be here. Status reports are vital, they help people anticipate and gain a sense of control. Include anticipated times, dates, and so on when you know them. When you do not have that information, you can indicate that information is as yet unavailable, but will be provided when you get it.
- Have a place for your responders and workers to go to escape the situation, and rest and refuel. This may mean assigning someone to make sure that adequate nourishment and hydration is available. This can also be a role for a mental health clinician or chaplain who can be available for status check-ins for the workers.
  - For example, as a devastating crisis was unfolding, one hospital put out sandwiches, cookies, coffee, water, and cold drinks. People would gravitate there just to rest. Workers gravitated there, people who worked there but were off shift, showed up, just to be there with their coworkers. The administration blocked the media from the grounds, and escorted people to and from their cars. Hardly anyone spoke, a number of them sat there crying, but the unspoken support was there. The responding crisis responders walked around the cafeteria, offering support, contact, and when asked, a prayer. As the days post-incident passed, their hospital

chaplain and crisis workers continued to be present and prayed with some people and sat and talked about anything with people. The loyalty to the agency increased, and the organizations' morale pulled them through this difficult time with few people choosing to leave their jobs because of the incident.

- Your command staff should watch for signs of stress as discussed above. Irritability, loss of temper, anger, withdrawal, and hostility are big cues that your staff are stressed and potentially struggling. Remind them to give people time down for a quick break, and if they would not take it, ask them to take it anyway. Acknowledge their desire to stay focused, busy, and involved, but not at the expense of their health or burn out. When someone loses their temper or their cool, give them a time down to regroup. They have hit their wall; they need a minute or two to get it back together.
- It is important to monitor if any of your responders are becoming angry, hostile, irritable, or acting out of their normal range of typical responses. For managers, it is important to see this as an important information about their stress tolerance, fatigue, or overload. Often emergency workers cannot or would not say that they are scared, overwhelmed, and so on. Many times, out of fatigue and exhaustion, they cannot recognize it in themselves. If they had to repeat their experiences to others of what they see or do, they would not be able to be out there doing what they do. So, they have developed ways of managing the awful stuff that overwhelms the everyday person. It is important not to personalize their behavior or reactions, but to recognize it as this person is stretched even beyond their own limits. Ask them what they might need or want. It may be reasonable, it may not be possible. Talk in private, give them some time to gather themselves and regroup. Do not retaliate or respond in anger. Give them a couple of minutes to regroup, have some quiet time, and then regroup. Check in with them later. They may say nothing; they may have a lot to say.
- Watch for your responders' reactions to one another. Make sure you do not have a scapegoat situation developing, where one person is the identified target for the groups' feelings of anger, helplessness, or overwhelm. Encourage their teamwork and effort as a team is what makes them the strongest they can be. If there is a personal overlap with any of them in this situation, watch for that. Your young team members may be very vulnerable. This is a time where you want to be aware of the "group think" and the teams bond around the stronger members, and cull out the one that is emotionally the one not like the rest. This group thinks that behavior can cause the best of teams to crash and not be able to do what they have always relied on each other as a team to do.
- Make certain your people have access to a routine and schedule. A routine establishes some element of predictability during times of uncertainty. Establish active and down routines.

- Have a plan for removing someone if necessary and get them the assistance they need.
- Integrate support into maintaining their ability to keep doing their jobs, as a preventative and as a way for them to refuel and recharge, and reconnect with their teams.
  - Make sure everyone has known time in and time out of service.
  - When they are out of service, they need to be out of proximity of the situation as much as possible. Additionally, when setting up for their space, taking physical characteristics of the location is necessary to consider before the tents get pitched.
  - Build a place for quiet and escape. Avoid constant access to media coverage of the event.
  - Have a centralized place for communication and updates.

Psychological First Aid

One resource is Psychological First Aid (PFA) (Ruzek et al., 2007), which is comprised of eight core actions designed at reducing post-traumatic distress and improving short- and long-term adaptive functioning for responders. The PFA Field Operations Guide is available online through the National Child Traumatic Stress Network (nctsnet.org). It is designed to be used by disaster mental health responders who may need to provide immediate support to survivors, and in any situation necessary for both individuals (Ruzek et al., 2007, Hobfoll et al., 2007) and in a small group format (Everly et al., 2006).

The eight core actions are based on the following principles (Hobfoll et al., 2007) of promoting:

- A sense of safety
- Calming
- A sense of self- and community efficacy
- Connectedness
- Hope

A sense of safety can be established through the provision of information. This allows for realistic appraisal of the current situation. Ongoing information that is constructive and directive is helpful. Structuring information so that people are clear as to what can happen, what has happened, and what they should do clarifies ambiguity. Another way might involve the removal of an individual to a safe environment and allowing them to regain a sense of biological and psychological "normalcy." When this is possible, the risk of anxiety and stress-related symptoms can be lessened. Further, connecting an individual with their network of support in some capacity can provide relief and a sense of having some control in the situation.

The promotion of calming is aimed at the biological and emotional impact of trauma-related stress and anxiety. This can ultimately generalize to many situations, potentially causing long-term effects on functioning. Teaching that these reactions are reasonable or normal reactions to abnormal situations can help reduce anxiety and some of the catastrophic thinking that can happen when one feels overwhelmed (i.e., "I can't manage this"). Activities that teach stress management, relaxation, cognitive restructuring, and deep breathing can be easily integrated into prevention and wellness activities.

To help develop self-efficacy, it is important that the individual has some belief in their ability to manage, cope, and have solid judgment and problem solving in the situation. It is important that individuals have a sense about their capacity to manage themselves and their situation. Helping set realistic and tangible goals can help this become a data-based effort, minimizing the "I feel" aspect of evaluation.

Connectedness and social support should be promoted, and research as discussed above has repeatedly shown that there is a better outcome for those with support than for those without support. This support should be at an individual, family, and community level. Further, it provides a forum for information exchange, problem-solving, norming of shared experiences, emotional understanding, and acceptance. Connectedness builds bonds that can rebuild communities and fosters a common "we" so necessary for mass recovery.

Instilling hope is the final component and the research for purpose, meaning, and hope is vast. In this sense, the authors defined this as a crucial component due to the increased likelihood of improved outcome because of their optimism, retention of hope for the future, feeling of confidence, expectance of positive outcomes, and other hopeful beliefs and definitions (e.g., God, religion, higher power). This allows for a sense of predictability in one's self and in life generally. Encouraging positive coping and meeting challenges helps minimize avoidance, withdrawal, and isolation.

The eight actions (Ruzek et al., 2007, Hobfoll et al., 2007) include

1. Contact and engagement. This involves rapid establishment of contact and rapport by initiating contact that is nonintrusive, compassionate, and supportive. It is important to ask if your presence or contact is wanted.
2. Safety and comfort. This involves addressing immediate and ongoing safety needs, and providing for physical (including medical) and emotional comfort. Providing fact-based information is important to help mitigate the impact of stress incurred because of false or inaccurate reporting.
3. Stabilization needs as warranted. This aspect involves calming and providing containment and orienting emotionally overwhelmed survivors.
4. Information gathering related to current needs and concerns, tailoring interventions, and responding to those needs and concerns.
5. Practical assistance involves addressing the immediate needs and concerns of the survivors and responders.

6. Connection with social supports by providing structured opportunities for brief or ongoing contacts with sources of support.
7. Information on coping, management, and support related to stress management, stress reactions, and coping strategies that prepares them to mitigate the effects of the incident and recovery process.
8. Linkage with collaborative services allows for the survivor to have resources available for present or future resource.

Critical Incident Stress Management

Critical Incident Stress Management (CISM) is defined by Everly and Mitchell (1997, 2000) as a system of crisis interventions that encompass the spectrum from acute crisis phase into the post-crisis phase of stress management and critical incident exposure. Although there has been a debate over its efficacy, when done according to the defined structure by trained facilitators, it has been found to be very helpful, educative, preventative, and supportive. There are many articles on this topic, and the reader is directed to the International Critical Incident Stress Foundation (ICISF) for these. It has specific interventions that can be applied to individuals, small to large groups, and to responders, families, organizations, and communities. It has as its main objectives:

1. Mitigation of the impact of the event through decreasing stress reactions.
2. Accelerate recovery process, by increasing normal recovery processes, in those experiencing stress reactions.
3. Restoring adaptive functioning.

The International Critical Incident Stress Foundation has many resources available for reproduction and use with agencies and responders. The literature has been conflicting on the benefits of CISM (Ruzek et al., 2007, van Emmerik et al., 2002, Mitchell, 2003, Everly and Mitchell, 2000). These criticisms were addressed very thoroughly and using a strict definition of CISM debriefing and techniques by Mitchell (2003), and by Everly and Mitchell (2000). In his research analysis, Mitchell clarified that a great number of the criticisms of the model came from situations where the techniques were in fact used by untrained individuals, misapplied to people for whom the techniques were not designed, and by individuals who used in a psychotherapy context.

From a CISM perspective, critical incidents are events that are sudden, often life threatening and time limited, and may overwhelm the capacity to respond adaptively (Flannery and Everly, 2000). Inherent in this definition, the authors point out that psychological homeostasis has been disrupted, usual coping mechanisms have failed to reestablish homeostasis and the resulting distress has caused some impairment in functioning. In the CISM frame, crisis intervention as defined by Flannery and Everly (2000) involves “provision of emergency psychological care to victims as to assist those victim's in returning to an adaptive level of functioning and to prevent or mitigate the potential negative impact of psychological trauma" (Flannery and Everly, 2000, p. 120). Mitchell (2003, p. 3) wrote, "the primary goals of the crisis intervention program entitled CISM are to mitigate the impact of a critical incident and to accelerate recovery processes of normal people who are having normal reactions to abnormal events."

Each component of the CISM Crisis Intervention model involves the following premise:

1. Intervene immediately to minimize the risk of maladaptive coping or responding.
2. Stabilize through mobilizing resources and supports to restore a semblance of order and routine.
3. Facilitate the understanding of what has happened. Gather facts, provide information, encourage expression, helping them understand the impact of the event and aftermath.
4. Focus on problem solving as part of the effort at regaining control and self-efficacy.
5. Encourage self-reliance to restore independent functioning, practical solutions to handling the situation and establishing a normal routine and balance.

There are seven core components of CISM designed to be used in a multiple pronged component approach, where each stage has a specific desired outcome intervention, timing, activation point, and format (Everly and Mitchell, 1997, 2000):

1. Pre-crisis preparation. Planning at this stage is primarily preventative. This stage involves stress management training and education, anticipating crisis/disaster response and mitigation.
2. Disaster or large-scale incidents, where it is necessary to include larger groups and types of groups (schools, churches, communities, etc.). These interventions generally include demobilizations, town hall types of meetings, informational briefings, and staff advisement.
3. Defusing intervention, which is a three-phased structured smaller homogenous groups of (primary affected) individuals with the exposure to the same evidence provided generally within hours (if not on-scene or immediately back in quarters), with the purposes of assessment, triage, and acute symptom mitigation.
4. Critical Incident Stress Debriefing (CISD), which is a seven-stage structured group discussion with individuals who are homogenous who had exposure to the same incident, designed at mitigating acute symptoms, providing education, support, clarification of thoughts, reactions, and physical responses that can occur as part of critical incident stress. This usually takes place 1–10 days post-incident.

5. One-to-one crisis intervention or psychological support can be offered at any point during the full range of the critical incident spectrum.
6. Family crisis intervention, education, and organizational consultation and support.
7. Follow-up and referral mechanisms for assessment and treatment or if necessary for identified personnel or organizations.

Briefing and Debriefing

- Be prepared in advance about CISM and education about the response. Many agencies are writing CISM services into their demobilization policies.
- Be prepared to provide education, such as in the form of Critical Incident Briefings, defusing (CISM) and educational briefings as part of demobilization. Demobilization is a process where the responders are provided with information about stress symptoms and potential problems that can arise from the deployment or service. It is also a brief and general education service. The International Critical Incident Stress Foundation has some very good handouts that can be used for this purpose.
- *Equally important, before debriefing, get your people back to their people.* Let them touch their people, rest in their home, be a person, before you bring them back as a responder or to participate in debriefing. To the best extent possible, make certain that your responders know when and where they are going home, and where their family are located if they have been relocated. When this has been attended to, then debrief them with the team they went out with initially. *Multijurisdictional planning may be necessary to coordinate this.*
- In the definition of CISM, the services are directed at groups who have had similar experiences/exposure and are peer lead.
- Before a disaster strikes, learn about any CISM teams in your region who may be available to provide services. There are federal, state, and regional teams who provide these services. In the true CISM model, the teams are made of volunteers who are also responders, so they are able to talk from a peer perspective. Some EAP services are also available, although their teams may not have peers, nor work from a traditional CISM model. The ICISF website provides a list of registered volunteer teams and their contact person.

Suggestions and Considerations

Have a policy about how long a shift can or will be, and whenever possible get that implemented and routine.

- Make defusing a part of shift change whenever possible or necessary.
- Other agencies are also recognizing the importance of having disaster mental health specialists available to use during the deployment as a specific support

staff whose job is to oversee the well-being of your responders, where their job is to interface with each person per shift (informally), support, oversee the nourishment, convey information in/to the field if needed, and so on. It is important to let your team know that making sure they get support at this juncture is to keep them able to do what they came to do. In some regard, these mental health support responders become like the Father Mulcahy character on MASH.
  - They can be identified as Support Services, because these services are not designed to provide therapy, but support and Psychological First Aid (Ruzek et al., 2007).
- Use the buddy system for support and safety.
- Rethink the length of and frequency of shifts/inservice.
- Set up debriefings within 2–7 days after leaving the scene, and debrief with the team they served with in their deployment.
- Allow people to go home and have a down-day with their people before coming back for a debriefing.
- Have a policy about how long a deployment can or will be, and whenever possible get that implemented and have a strategy for deployment into the situation and out of the situation.

There should be coordinated efforts for the CISM command staff that include briefing and information exchange per shift as well. Important information for them to discuss includes information about the changing situation, status reports, supplies, identification of potential problems, and other important personnel-related data.

Other supportive staff and personnel can be made available to help with provision of supportive services to the responders, including first aid, massage, food and beverages, clean/dry clothing, and rest areas.

Conclusion

Stress is a part of our daily lives. We can manage it while we live with it. When disaster strikes, it impacts the core of our existence and perception of safety and security. Different situations impact people's lives, physical and emotional health differently. It can have long-lasting impact on our communities and the people within them. This not only affects people, but also it impacts economics and resources within the community.

With consideration of the above factors in mind, it is important to develop plans that anticipate and plan for a wide range of possible community disasters. Keeping this in mind, you can anticipate the needs of both community and your responders. While critical incidents and disasters are unpredictable, they can be planned for with some degree of anticipation of the factors discussed above. While we plan based on the worst we have known, we need to plan for the worst we can image with impact greater than that we are afraid to contemplate. We also need to think outside the box, reaching for resources we have not considered, and plan to keep our personnel resources as healthy and supported as possible, not only because we need their professional skills to help us through the disaster but also because we care about them as valuable community members.

References

American Psychological Association (APA). 2004. Mind/body health: Did you know? *APAHelpCenter from the American Psychological Association*; http://www.apahelpcenter.org/articles/pdf.php?id=103.

Dale, K. and Fox, M. 2008. Leadership style and organizational commitment: Mediating effect of role stress. *Journal of Managerial Issues,* 20 (1), 109–134.

DeWolfe, D. J. 2000. Field manual for mental health and human services workers in major disasters. *National Mental Health Information Center, Center for Mental Health Services. United States Department of Health and Human Services, Substance Abuse and Mental Health Services Administration (SAMSHA).* DHHS Publication Number ADM 90-537. http://mentalhealth.samhsa.gov/publications/allpubs/ADM90-537/fmkey.asp

Everly, G. S. and Mitchell, J. T. 1997. *Critical Incident Stress Management (CISM): A New Era and Standard of Care in Crisis Intervention.* Ellicott City, MD: Chevron Publishing.

Everly, G. S. and Mitchell, J. T. 2000. The debriefing "controversy" and crisis intervention: A review of the lexical and substantive issues. *International Journal of Emergency Mental Health,* 2 (4), 211–225.

Everly, G. S., Phillips, S. B., Kane, D., and Feldman, D. 2006. Introduction to and overview of group psychological first aid. *Brief Treatment and Crisis Intervention,* 6, 130–136.

Fairbank, J. A. and Gerrity, E. T. 2007. Making trauma intervention principles public policy. *Psychiatry,* 70 (4), 316–320. *ProQuest Psychology Journals: Health Module.*

Flannery, R. B. and Everly, G. S. 2000. Crisis intervention: A review. *International Journal of Emergency Mental Health,* 2 (2), 119–125.

Galea, S. 2007. The long-term health consequences of disasters and mass traumas. *Canadian Medical Association,* 9, 176–178.

Goetzel, R. Z., Anderson, D. R., Whitmer, R. W., Ozminkowski, R. J., Dunn, R. L., and Wasserman, J. 1998. The relationship between modifiable health risks and health care expenditures: An analysis of the multi-employer HERO health risk and cost database. *Journal of Occupational and Environmental Medicine,* 40 (10), as cited in Colligan, M., Swanson, N., Hurrell, J., Scharf, F., Sinclair, R., Grubb, P., Goldenhar, L., Alterman, T., Johnston, J., Hamilton, and Tisdale, J. Stress at work. *National Institute for Occupational Safety and Health [NIOSH]–U.S. Department of Health and Human Services[USDHHS].* Public Domain Publications: DHHS-NIOSH Publication No. 99–101. Rockville, MD. http://www.cdc.gov/niosh/stresswk.html

Hobfoll, S. E., Watson, P., Bell, C. C., Bryant, R. A., Brymer, M. J., Friedman, M. J. Friedman, M. et al., 2007. Five essential elements of immediate and mid-term mass trauma intervention: Empirical evidence. *Psychiatry,* 70 (4), 283–316.

Jackson, B. A., Baker, J. C., Ridgely, M. S., Bartis, J T., and Linn, H. I. 2004. Protecting emergency responders: Safety management in disaster and terrorism response, Vol. 3. DHHS

(NIOSH) Publication Number 2004-114, RAND Publication Number MG-170. *Rand Science and Technology Policy Institute, Centers for Disease Control (CDC) and Prevention and National Institute for Occupational Safety and Health (NIOSH).* Cincinnati, OH: NIOSH Publications.

Mitchell, J. T. 2003. Crisis intervention & CISM: A research summary. *International Critical Incident Stress Foundation,* www.icisf.org.

Perrin, M. A., DiGrande, L., Wheeler, K., Thorpe, L., Farfel, M., and Brackbill, R. 2007. Differences in PTSD prevalence and associated risk factors among world trade center disaster rescue and recovery workers. *The American Journal of Psychiatry,* 169 (4), 1385–1395.

Ray, T. K. and Sauter, S. L. 2008. Work stress: Societal costs and organizational components. Poster #003. *The National Occupational Research Agenda (NORA), NORA Symposium 2008: Public market for ideas and partnerships,* http://www.cdc.gov/niosh/nora/symp08/posters/003.html.

Rogers, J. R. 2007. Disaster response and the mental health counselor. *Journal of Mental Health Counseling,* 29 (1), 1–3.

Ruzek, J. I, Brymer, M. J., Jacobs, A. K., Layne, C. M., Vernberg, E. M., and Watson, P. J. 2007. Psychological first aid. *Journal of Mental Health Counseling,* 29 (1), 17–50.

Sheehan, D. C., Everly, G. S., and Langlieb, A. 2004. Current best practices: Coping with major critical incidents. *FBI Law Enforcement Bulletin,* 73 (9), 1–13.

Van Emmerik, A. A. P., Hulsbosch, A. M., and Emmelkamp, P. M. G. 2002. Single session debriefing after psychological trauma: A meta-analysis. *Lancet,* 360, 766–771.

Chapter 4

Facility Vulnerability and Security

J. Lawrence (Larry) Cunningham

The Key to Effective Security Surveys: Accounting for Human Factors

This chapter supplements the compendium of emergency management perspectives in this book. It is intended to assist in the preparation of response personnel to a variety of emergencies, both human and natural. The goal is to help the responder understand the gap and in some instances, the gulf between the state of security preparedness and the reality of what is understood by people, and the resources identified to respond to and manage a given emergency. In brief, this chapter hopes to give you the insight to answer the question: do the security plan and resources we are relying on match the reality of what is realistically available for the emergency?

The security preparedness challenges facing our nation in this era are misunderstood, daunting, and multifactored. There are numerous proactive measures, if implemented, which could significantly mitigate, if not prevent many of the associated problems in emergency response. Critical among them is the integration of human factors and resources at every level. This needs to be recognized and implemented at the foundation level—the emergency planning phase.

The ultimate goal of this chapter is to offer insight into the security assessment and emergency planning process in an effort to develop effective plans that are collectively understood, rehearsed, and adapted by ALL concerned.

Identifying vulnerability and security planning continues to undergo analysis, scrutiny, and change since the 9/11 attacks, Hurricane Katrina, the Virginia Tech shootings, Hurricane Sandy, the Sandy Hook elementary school shootings, and numerous lesser known emergencies. Why do we continue to miss queues, misunderstand and/or underestimate the threat(s), fail to harness the right amount of resources, and be less effective than our experience would indicate? Where or what is the disconnect?

Before the 9/11 attacks, the nation's security posture and response philosophy was primarily reactive. That is to say many, if not most of the nations' security plans as well as those in the private sector, were structured to be implemented *after* an attack, *during or after* a storm or even after an alert.

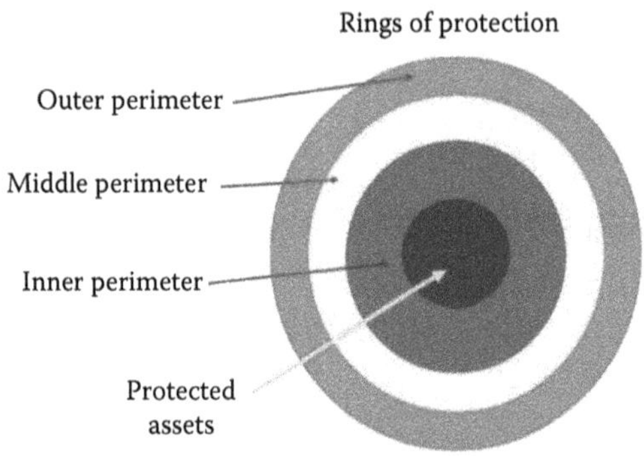

Far too little attention was paid (and sadly to this day) to *integrated, proactive* security response planning. Much more work to help integrate response planning and deployment efforts is critically needed. Historically, the security philosophy of *both* the public and private sectors subscribed to a parochial security perspective utilizing a perimeters approach to security planning and resource allocation. Indeed most of the security response training focused on the immediate effects of the emergency and resources required. Strategic planning to include planning involving contiguous jurisdictions, joint resource allocation, and assignment of roles—in advance—was not provided for in the most effective way. Basically we (planners, leaders, policy makers, responders) prepared and trained to respond to emergencies almost exclusively. The prevailing thinking was as long as we have our borders patrolled, our defenses alert and ready, and doors locked, we would be safe.

> By failing to prepare, you are preparing to fail....
>
> **Benjamin Franklin**

In March, 2001, The Carnegie Endowment of International Peace published a fascinating and portentous treatise *Strategic Warning* in which they identify numerous potential failings of our nation's preparedness philosophy. It cites the unfortunate lack of integration and structure to *proactively* prepare for most attacks, especially the ones festering beyond our borders. It predicted in an uncanny, almost ominous way that as a nation, we were ill prepared to accurately identify festering threats, assess their capabilities, and impact and effectively harness an efficient integrated response. This was *before* the Al Qaeda attacks!

In this post 9/11 era, the concept of sectors of security has been refined to include a more visionary approach. That is to say, security measures, if they are to be truly effective, must *anticipate* dangers and threats. One needs to look beyond the immediate security sectors and thousands of miles beyond our borders both literally and figuratively in order to identify and assess developing and/or imminent threats to be able to take effective protective action. Instead of looking at the potential business and resource value of the potentially volatile regions of the world, why not pay closer attention to festering threats? Why not try to understand the culture in other contexts? This visionary philosophy is being continually refined as the result of a greater understanding of the terrorists' asymmetric views of engagement and goals to inflict mass casualties on targets of interest (especially those that appear to be "soft targets") and their long-term surveillance methods. This visionary philosophy has extended to response preparedness to natural disasters, violence on campuses and the workplace, and in other sectors.

As most of you reading this are aware, the nation's state and local response community were tasked and mandated by the Government to implement the Incident Command System (ICS) to standardize the organizational structure of response with the implementation of The National Incident Management System (NIMS) following the in-depth studies of the emergency response to the 9/11 and other historical attacks and emergencies.

Along with this, the philosophy in the security assessment arena has changed to recognize how vital it is to recognize that each organization (public or private) is a unique entity with a unique culture with specific security concerns and needs that cannot be addressed with "a one size fits all" solution in the security planning process. In the context of the ICS structure, the mandated categories of Finance, Logistics, Operations, Planning, and the recent addition of Intelligence (sharing useful and timely information among responders), it is recognized that every jurisdiction needs the flexibility to adapt this structure to their people, culture, resources, and experiences. Other factors unique to a given geographical area, which impact the application of NIMS, are demographics, training capabilities, response history, size, and political structures.

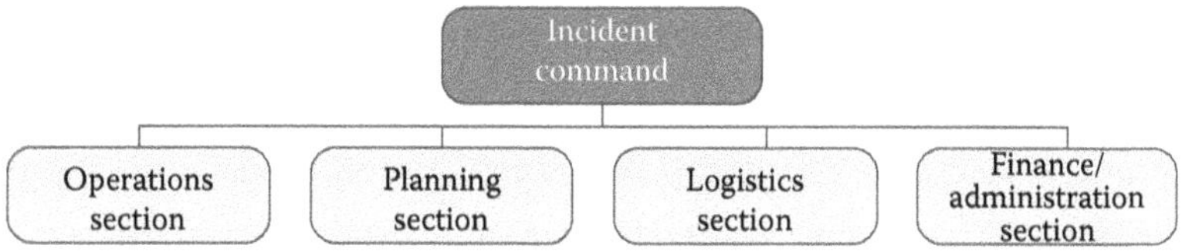

One must remember that *the most effective security is security implemented at a level that the organization needs, understands, and is willing to buy into.*

All members of any organization, especially management (policy makers and the check writers) need to be included in the developmental process, take ownership, and become part of the security network. In other words, for security to be effective, security awareness is shared by all personnel and integrated with local responders.

One underutilized way to lessen this "disconnect" between management and front-line members is to empower local jurisdictions beyond the general level identified in *The National Response Framework.* In it, the DHS has correctly assigned immediate emergency response responsibilities to local jurisdictions to prepare for and manage their emergency preparedness. The directive encourages local and state jurisdictions to identify potential hazards and prepare for them with their experiences, culture, response structure, and "playbooks" they successfully used in the past. The contention is that the locals know their houses, neighborhoods, cities, and regions better than outside response entities so they are ultimately best suited to respond to and manage manmade and natural disasters.

For this to work, each organization, facility, and responder jurisdiction needs to share and vet their respective response plan, with all who will be involved in the response to ensure its effectiveness.

The local authorities have institutional knowledge and experience with their own response organizations and structures and, importantly "know their own people." This element probably more important than anything else has and will continue to make the crucial difference with success and failure. Relationships nurtured by working together over the years; that special "esprit de corps" that can only be developed with working with one another for years is the magic ingredient. Much like any team worth its salt, regardless of its record, team members practice their plays endlessly and develop a deep level of communication enabling them to anticipate players' moves instinctively almost without thinking. Many of us are avid sports fans and can relate to the critical need to develop winning strategies and plays with everyone involved buys into and is willing to work on… incessantly. Do our responders train this way?

The Kennedy School of Government's Executive Session on Domestic Preparedness discussion paper *Winning Plays* aptly identifies the sports arena as a useful metaphor to explain how we as responders should prepare for emergencies—identify threats and TRAIN! What does prepare really mean? What does train really mean? Is it enough to train based on our institutional practices or do we train with as many departments that would be involved in the response? Do we have a system that affords planning *before* the disaster? Do we have the organizational structure and political will to include these entities? The answer most jurisdictions would give you is yes we do train and prepare. But how do they prepare?

Why then do we still often see delayed responses, inaccurate assessments of the disaster, less than efficient resource allocation, poor communication among agencies/responders, and inaccurate or incomplete information disseminated to the public? Many of you reading this will say that there are systems out there that prevent this or are supposed to minimize this. Namely, NIMS, ICS, UCS, etc. were instituted and refined since 9/11 that should take care of these missteps. *The problem or the solution is human factors.* How are these system protocols understood, structured, and applied? Is the biggest department or the loudest voice making these determinations?

I am reminded of the comments of the Postmistress of the Harvard University Post Office while conducting a security assessment for the university in 2002 after the 9/11 attacks. Part of the assessment included a review of the mail handling procedures for potential anthrax and ricin contamination. The Postmistress's first comment (after I introduced myself): "I hate consultants!" I responded: "fair enough." She went on to explain her painstaking efforts to compel the neighboring police departments to come together and collectively devise a mutually agreeable and doable plan for these and related threats. She shook her head and related how the command level officers were apparently too busy to attend themselves so less senior representatives were sent. She said they would happily drink her coffee and eat her donuts, devise reasonable plans, and leave. Her frustration piqued when no official commitment, follow-up, or plans were formalized after the fact.

Recognizing the vulnerabilities is the first step in this proactive process. Understanding the rationale, structure, and how to apply the elements of an effective security survey is next. Integrating the human factors with the tailor-made elements and structure will make the critical difference. Responders need to be aware of and to whatever extent possible, influence the security assessment and planning process to ensure the less obvious aspects are considered in security surveys. Their perspectives need to be considered. The assessment process is the key...the relationships that develop, integration mechanics are tested, and the "team" spirit that is fostered, prepares for the knowns and provides a foundation for adapting to the unknown.

Terrorists, criminals, and would-be attackers, especially those that conduct preoperation surveillance exercises, test, and assess their targets' weaknesses and make decisions based upon the target's perceived strengths. The presence of organized security personnel, technology, and security procedures is a deterrent. Arthur Bremer, in his book *An Assassin's Dairy** mentions stalking President Richard Nixon in Canada but abandoned the idea because of the threatening presence of the Royal Canadian Mounted Police (RCMP). He wanted to become famous but settled for "lesser prey." He chose Presidential candidate Alabama Governor George Wallace and shot him five times in Laurel, Maryland in 1972.

> It is senseless to go after tigers when there are so many sheep to be had.
>
> **Carlos the Jackal (Venezuelan terrorist, Figure 4.1)**

---

* *An Assassin's Diary* (ISBN 0-06-120470-6) is a book released in 1973, which was based on part of the diary of Arthur Bremer, the would-be assassin of Alabama Governor George Wallace. Bremer shot Wallace at the Laurel Shopping Center in Laurel, Maryland, while Wallace was making his third campaign for President on May 15, 1972.

Figure 4.1 Infamous Venezuelan terrorist Carlos the Jackal.

The Elements of an Effective Security Survey

We are all familiar with the frustration of going to the doctor seeking relief only to find out that the diagnosis is unclear or worse, the cure is uncertain. Typically the medical practitioner will listen to the patient, assess symptoms, and try to fit it into a known array of known conditions in the hopes to hit upon the right diagnosis and recommend a treatment. This process is often labeled a differential diagnosis,* essentially a process of elimination using a template, coupled with the practitioner's experience and ability to assess the *patient's unique physiology, environment, culture, and mental state* in an effort to arrive at an accurate diagnosis. Herein lies the critical piece—*the individual's uniqueness.* The template may give you 50% of the picture... but the part you really need is the integration piece...the piece that will take into account all of the unique factors that impact the individual and his/her ailment(s). So how can this be effectively done? I would offer—a carefully considered tailor-made approach identifying and accounting for important impacting factors.

The security assessment process, the development of appropriate recommendations, and the effective implementation of remedies are no different. There are countless versions of security survey templates out there with impressive detail and logical structure that purport to make you and/or your organization *feel* safe and protected. The reality is that many emergency plans are not accurate. The typical problems found in many emergency plans are as follows:

- Identified security goals are not prioritized
- The strategy and plan to get there are too narrowly focused

* A differential diagnosis (sometimes abbreviated DDx, ddx, DD, D/Dx, or ΔΔ) is a systematic diagnostic method used to identify the presence of an entity where multiple alternatives are possible (and the process may be termed differential diagnostic procedure), and may also refer to any of the included candidate alternatives (which may also be termed candidate condition) (Wikipedia).

- Vital pieces of information are missing
- Contains obsolete information (wrong names, wrong numbers, expired policies, etc.)
- Relies on capabilities and resources that are simply not available (or the same resources are committed to other organizations during the same emergency)
- Not shared or open for discussion with the rank and file
- Lack of common understanding among the entire staff
- Not integrated with other likely participants and responders in the community

A classic example of this was found in 2002 during the security assessment of the corporate offices of one of the largest sporting organizations in America (see Figure 4.2). The survey parameters included the elements identified in the template appended at the end of the chapter. Important security elements to include technology, lighting, emergency response considerations, and employee guidance were included in their existing plan.

Despite the good intentions of their security department, problems were found as follows:

- Security Plan locked in the Security Director's office
- Security Plan not vetted or validated with responders identified in the plan
- No corporate-wide awareness of the plan among the employees
- Police department identified in the plan did not have primary jurisdiction at the office address
- Police department was unfamiliar with the corporate office's location

Figure 4.2 NASCAR corporate offices—Daytona Beach, Florida. (Courtesy of Essential Security Strategies, LLC.)

- Police department response plan identified the sports track as the response location instead of the corporate office
- Access, lighting, and notification system were out of date and rendered ineffective due to lack of policy adherence
- Evening and night-shift personnel were not aware of the plan's security procedures

Many of the necessary security elements were in place however, the problems found were due to lack of effective configuration, communication, and understanding of how to effectively implement the plan. Any effective plan needs to be continually reevaluated; it is living and breathing. It needs to be supported and evaluated by management, the security staff, and vetted by the responders identified in the plan.

Management

Management plays a pivotal role in the security process with any organization. They set the tone, shape policy, write checks, and set the example. Several important areas where management can influence security policy and adherence include the following:

- Giving the security department the authority and backing required to enforce security policies
- Support employee hiring and security training of new and existing employees
- Actively formulating, endorsing, and supporting new security policies
- Setting the example by adhering to security policies (wearing ID, signing out equipment, enforcing security policies)
- Following technology and Intellectual Property best practices
- Avoid assigning "extraneous" nonsecurity duties to security staff, for example, vehicle maintenance, building maintenance, general answering service tasks, and so on
- Duties to security staff, for example, vehicle maintenance, building maintenance, general answering service task, and so on
- Supporting staff security meetings and policy changes

The following examples highlight the inherent risks organizations take when security planning and policies are not shared and integrated:

Organizational Structure Dysfunction

The organization chart reflects the structure of the organization and the relative interaction, by virtue of the respective division's or office's position (high or low) often dictates its power and financial resources and control. The position of the

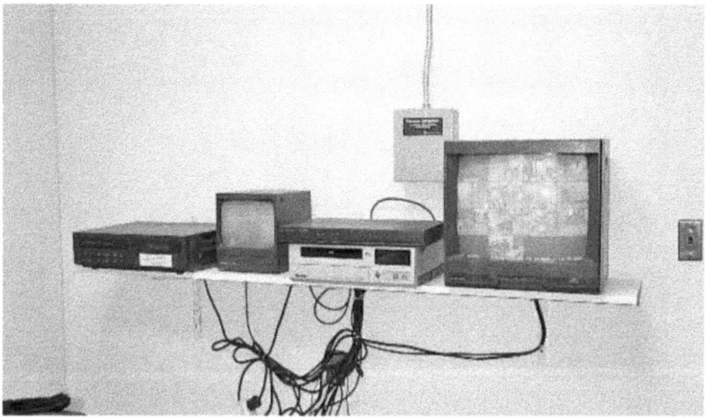

Figure 4.3 Unmanned multi-image CCTV monitor. (Courtesy of Essential Security Strategies, LLC.)

division and/or office (particularly in the private sector) that is responsible for security will have a bearing on its ability to establish and enforce security policies and protocols. Security resources will suffer if the security office is under the control and supervision of an administrative or planning arm that for a variety of reasons does not prioritize security.

An example of this dysfunction is seen here, due to security personnel budgeting, security positions suffer. Positions such as manning images from surveillance cameras, zone patrols, and so on are reduced or eliminated. Note the monitor with 12 images (see Figure 4.3). The capture of any suspicious images required a decision by a security guard. This room was locked.

Other examples of this have been documented at oil companies, major banks, and hospitals.

Figure 4.4 depicts a typical corporate Security Room Configuration.

Note the unmanned 24-image monitor being projected in the rear of the security command/dispatch center of this hospital in the photograph to the left.

Access Control

Access control is fraught with potential security lapses due to the multipurpose functions of the entry and exit points. Much of the problem lies in management's and the staff's lack of common understanding of the purpose and the role of security.

The defense contractor facility in this example prior to a security survey had *seven* active employee and two visitor entrances. Pictured to the left (Figure 4.5) is one of the employee entrances manned by a summer aide with no institutional authority to enforce the company sign-out and verification policy of proprietary documents and storage devices. The survey found that this "path of least resistance" was used most often by employees.

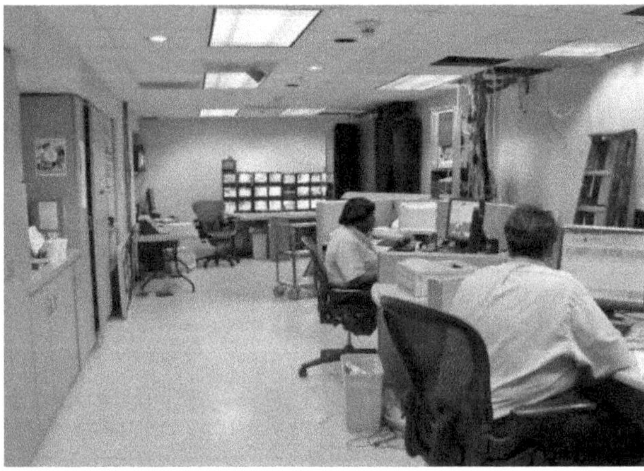

Figure 4.4 Typical Corporate Security Room Configuration. (Courtesy of Essential Security Strategies, LLC.)

In this example, the guard desk and camera console at the main entrance to this large facility obscures the view of the entrance placing the guards at risk and in a reactive position (see Figure 4.6). The survey also found these guards became the after-hours answering service taking general information calls during their evening shifts. It was determined that more than 5200 calls were answered by security guards yearly. In addition, they were tasked with the maintenance of the executive fleet of limousines.

Figure 4.5 Loosely Controlled Access Control Point. (Courtesy of Essential Security Strategies, LLC.)

Figure 4.6 Primary Security Access Control Point. (Courtesy of Essential Security Strategies, LLC.)

This chapter comprised and set forth the basis for rethinking typical security survey approaches being applied to both public and private facilities in the United States as of this writing. The questions raised here serve as the impetus to explore significant paradigm shifts with how security surveys should be conducted in the future.

Chapter 5

Immediate Response to Active Shooter Situations

Rick C. Mathews

Introduction

The U.S. Department of Homeland Security (DHS) defines an "active shooter" as "an individual actively engaged in killing or attempting to kill people in a confined and populated area; in most cases, active shooters use firearms(s) and there is no pattern or method to their selection of victims" (DHS 2008, p. 2). Situations involving active shooters can and have occurred in a variety of settings and circumstances.

Arguably one of the best analysis and reports dealing with active shooter incidents was compiled and reported by the New York City Police Department (NYPD). Their report "Active Shooter Recommendations and Analysis for Risk Mitigation" was first published in 2010 and was subsequently updated via its 2012 edition (NYPD 2012). The report is based on cases between 1966 and 2012, with 324 cases studies provided. Selected findings are illustrative of the typical active shooter incident (NYPD 2012):

- 98% of the attacks have been carried out by a single attacker; 97% of the attackers were male.
- The majority of the attacks occurred within organizations or communities familiar with the attacker, although in 265 of the cases, there appeared to be no relationship between the attacker and the target.
- NYPD found that the median in terms of the number of casualties were two fatalities and two injuries.
- 36% of the attacks involved more than one weapon.

Active shooter attacks typically end within a few minutes of the attacks' first shots (DHS 2008), yet all result in multiple casualties. Only 16% of the attacks ended without violence according to NYPD with 40% ending by attacker suicide and 43% ending through 3rd-party aggressive action (NYPD 2012). Although most attacks ended quickly, that which resulted in the highest number of casualties typically lasted longer than others. As one would expect, the vast majority of the casualties were injured or killed as the result of one or more gunshot wounds.

When discussing active shooter incidents, many people will quickly recall the Columbine or Virginia Tech attacks as they were so violent, involved educational institutions, received significant media coverage, and instigated changes in how agencies respond to such attacks. Others will look at the Aurora, CO attack in a movie theater as an example of an active shooter situation in a noneducational public gathering while others will remember it for the apparent delays in EMS being able to care for victims. The December 14, 2012 attack by Adam Lanza at Sandy Hook School in Newtown, CT will likely be remembered as one of the deadliest attacks on record and one that involves very young children as well as teachers and administrators (NYPD 2012). Anecdotal information from the response community indicates that it would appear everyone followed "the plan" yet there was still significant loss of life (26 dead, 2 wounded). As stories surface, it seems that the initial responding police officers formed a contact team and entered the school, saw the suspect who went into a room and apparently committed suicide. One could argue that the quick appearance of the law enforcement contact team led to the shooter stopping the attack through suicide. According to reports, Lanza came to the school with three weapons and plenty of ammunition (NYPD 2012). One lesson learned in this instance is that even when everything goes according to plan, casualties can and will occur. Another lesson learned is that the law enforcement response, based on the generally accepted best practices of the day, likely mitigated the number of casualties by causing the perpetrator to end his attack.

Reducing the Casualties from Active Shooter Attacks

The vast majority of injuries, fatal and not, resulting from active shooters are gunshot wounds. This stands to reason as the very definition of an active shooter situation is one in which an attacker is shooting other humans (DHS 2008). Strawder in an article appearing in *Joint Force Quarterly (2nd Qtr 2006)* describes the need or more aggressive medical care for combat casualties and states that approximately 67% of severe ballistic injuries (gunshot wounds) die within the first 30 min and approximately half of these are those bleeding to death (Strawder 2006). Arguably some of the best examples of physically fit and healthy humans are America's combat military personnel. Even so, 67% of those receiving gunshot wounds die, half bleed to death. Studies abound that support the concept that many deaths in combat are potentially salvageable if the wounded are provided appropriate care quickly enough. Out of this mind set, anecdotal evidence, and research, the Tactical Combat Casualty Care (TCCC) program evolved which primarily aims at reducing the delay between the time the wound occurs and effective emergency care is instituted, usually within minutes. In an effort to help transition the concepts of TCC into the civilian sector, the Committee on Tactical Emergency Casualty Care (TECC) was born and applicable guidelines produced. One of the key tenets of TECC is getting lifesaving care to the injured person as soon as possible—within the first minutes if at all possible (TECC 2013). The relationship between the TECC guidelines and the goal of reducing lives lost to active shooters seems fairly straightforward. The more quickly appropriate medical care can be provided to the casualties of active shooter attacks, the more lives can be saved. Accordingly it is important that not only must the attacker be neutralized as rapidly as possible but lifesaving emergency medical care must also be provided to the casualties as quickly as possible.

Immediate Response to Active Shooters

All one needs do is to conduct a short search of the web, looking for guidelines and protocols regarding the best way the law enforcement officers should respond to active shooter incidents. Virtually every article published says the same thing, so much so that what follows is generally accepted as being "common knowledge." *The police must rapidly respond to and then engage the attacker without waiting for superior firepower or specially trained teams.* Courses such as the Law Enforcement Active Shooter Emergency Response (LASER) developed by the National Center for Biomedical Research and Training at LSU are offered around the nation for law enforcement officers. This and courses like it are intended to help train law enforcement officers in the tactics, skills, and procedures necessary to rapidly engage a suspected active shooter/attacker, before special units can arrive on scene. The goal of this training is fairly straightforward, as soon as possible after arriving on the scene, law enforcement officers should neutralize the threatstop the attackers from shooting others. In courses such as LASER, the law enforcement officers are taught the basic skills needed to rapidly assess the situation and to organize into two or four officer (in most cases) contact/immediate reaction teams. These contact teams are expected to enter the building, move toward the sounds of gun fire, and engage the attacker, resulting in the neutralization of the threat. The more rapidly and skillfully this is accomplished, the less time will be available to the attacker to continue shooting others. Although specific techniques should not be discussed in public venues, the basic tenets are fairly clear. The officer must quickly determine whether the attacker is still engaged in shooting—the basis of the active shooter definition. As soon as additional officers arrive on the scene, a contact team is formed and entry can be made. As the contact team enters the building, they move toward the sound of gunfire. In so doing, they typically do not assess every room or hallway they encounter, rather they continue moving toward the sound. They make mental notes of other potential threats that may exist; they do not stop to care for the injured either. The single goal of the contact team is to stop the shooting followed by clearing the remainder of the building to ensure that all threats have been identified and eliminated. The older way of thinking was to fight the urge to enter a building and engaging the attacker. Rather, it was generally accepted that police should fight the urge and wait until they had sufficient resources to ensure they had superior firepower. Generally, if at all possible, the goal was to secure the perimeter, waiting for arrival of a specially trained and equipped tactical team to arrive and make entry. Although this older method was thought to be safer for law enforcement officers, it almost always required many minutes of time—maybe an hour or more. During this time, the attacker could continue shooting and casualties would not receive care.

Stopping the shooting is the first priority in any active shooter situation. This action will undoubtedly save lives. The second priority is to facilitate effective emergency care for the casualties. Any delay can minimize the effectiveness of emergency care in saving lives resulting in more lives being lost. Accordingly, after stopping the shooting, the next priority is to care for the casualties. The standard has been and in many localities continues to be for emergency medical services (EMS) personnel to stage at a location close to but a safe distance from the actual shooting. There they wait until law enforcement has determined that the building is safe which usually translates to waiting until the building has been searched and cleared. Although arguably safer for EMS, the stage-and-wait protocol almost certainly means a delay in medical care reaching the casualties and delays in getting them to definitive care. Applying the principles of Tactical Emergency Casualty Care (TECC) facilitates a better approach.

One of the key elements of TECC is dividing the response to an active shooter into three major action phases (TECC 2013):

- Direct threat
- Indirect threat
- Evacuation

The "direct threat" phase is the phase that encompasses the actual shooting actions of the attacker with respect to the area of the building where the shooting is occurring. Entering the area of the building by EMS where the shooting is still ongoing is generally not possible; it is just too dangerous. During this time however, law enforcement officers are also most vulnerable to being shot. If they carry appropriate supplies and are properly trained, they may be able to provide self-care or care to a partner officer even while being engaged by the attacker. Care provided during this phase is termed "direct threat care." Key aspects of direct threat care include:

- Hemorrhage control
- Airway management

- Sealing of sucking chest wounds
- Staying behind cover or moving to cover

As mentioned earlier, the care provided during the direct threat phase will be accomplished using whatever resources are readily available to the officer. This generally means the officer must carry the equipment/supplies needed to care him/herself or to another officer.

The second phase is termed "indirect threat phase" and immediately follows the "direct threat phase" and/or occurs in an area where the direct threat has apparently passed. During this phase, it is possible to bring EMS personnel into the "indirect threat zone" to provide care for casualties located there. During the "indirect threat phase," however, a threat could reemerge which means that law enforcement officers must maintain a security watch around the immediate area, being prepared to immediately neutralize any threat. Since officers will continue to pursue any direct threat (attacker still shooting), there must be additional officers available to secure the indirect threat zone. Bringing EMS into an indirect threat area to provide care for victims means emergency care is more quickly available, facilitating the potential saving of additional lives. Care provided during the indirect phase is typically limited to care immediately needed to save lives including hemorrhage control, sealing of sucking chest wounds, airway management, and proper movement of casualties to a safe evacuation area.

The final phase, in terms of casualty care, is the evacuation phase and involves the care and transportation of patients to medical facilities in manners consistent with "normal" practices. With that said, however, it must also be kept in mind that multiple casualties can and often do overload local medical facilities, so it is imperative that these institutions (emergency departments primarily) be looped into the communications process facilitating, when possible, direct communications with the EMS or law enforcement officers at the scene.

Putting the three TECC phase concepts together with the goal of stopping the threat creates the immediate response to active shooters approach that is needed to maximize the number of lives saved.

Training and Exercises

In order for emergency responders to develop procedures and tactics that can be applied in active shooters situations, it is essential that law enforcement, EMS, fire rescue, and other emergency response elements train and exercise together, employing the practices that will be needed should an active shooter situation occur. The training should include a blend of classroom principles, skill development, and scenario-based activities that facilitate the integration of theory and skills in a functionally integrated manner. Functionally integrated, scenario-based activities are designed with learning objectives guiding the activities. These activities resemble "exercises" in appearance but are actually different. Exercises, per se, are activities that help to assess and/or validate training and other preparedness activities that have occurred. They typically occur subsequent to "training" and are useful in designing future training efforts.

Scenario-based training that includes EMS and law enforcement can be a powerful tool to not only develop skill specific to active shooter situations but can also support better interagency relations, integrated responses to other types of emergencies, and related areas of interest. Role players can add the realism element to the training and can help dial in the training to the levels and specific areas of concern.

Exercises are essential to assessing the capabilities levels of emergency responder agencies when constructed appropriately, including comprehensive after-action reviews and reports. Drills are typically different from exercises in that they often are not as comprehensive or as detailed. Drills can be useful in a number of ways, depending on their purpose. First, they can be used to help acquaint emergency response agencies with specific jurisdictional facilities and organizations. For example, a drill held at a local school could help the participating agencies develop situational awareness for the school, its layout, doors, and so on. It can be useful to facilitate target-specific training with regard to staging, special needs, etc. Second, drills can be powerful tools for use in bridging the emergency responders with the jurisdictional agencies and organizations that they protect. A drill could be used to help inform school staff and teachers about how the law enforcement and EMS agencies will likely respond to an active shooter incident should one occur at their facility. It can help both sides better understand each other and what to expect should the worst happen. One caveat with regard to drills is that they can be seen as vehicles for good public affairs activities. While it is true that a drill can be illustrative of everything, a public official wants to demonstrate to the public affairs activities about the drill before it occurs can raise expectations or cause extreme artificiality to occur that can actually hinder preparedness activities. Consideration should be made to involving the press and media after the drill terminates or as it concludes. Good public affairs efforts can help the community to better appreciate its emergency responder agencies and their level of preparedness, provided the actual press and media inclusion does not get in the way of the drills intended purpose.

DHS recommends that training for facilities and organizations should, to the extent possible, include "training exercises" or drills (DHS 2008).

Within that context, DHS recommends that the "training exercises" incorporate the following:

- Recognizing the sound of gunshots
- Reacting quickly when gunshots are heard and/or when a shooting is witnessed
- Evacuating the area
- Hiding out
- Acting against the shooter as a last resort

- Calling 911
- Reacting when law enforcement arrives
- Adopting the survival mind set during times of crisis

From "Active Shooter: How to Respond" from the U.S. Department of Homeland Security (DHS 2008).

Conclusion

Preparing communities for possible active shooter attacks involves the development of effective procedures, protocols, and tactics that can be immediately employed by the initial responding units to the scene. These must include not only the law enforcement threat mitigation efforts needed to stop the shooting but also the blend of TECC principles that enable emergency care to be provided during both the direct and indirect threat phases, reducing delays to that care resulting in more lives saved. Finally, the efforts must include appropriate training, exercises, and drills to help responders and the community they protect to better prepare.

References

Committee on Tactical Emergency Casualty Care: http://c-tecc.org/guidelines

NYPD. *Active Shooter: Recommendations and Analysis for Risk Mitigation,* 2012 Edition. http://www.nypdshield.org/public/SiteFiles/documents/Activeshooter.pdf.

Strawder, G.S. The Golden Hour Standard: Transforming combat health support. *Joint Forces Quarterly*, 41(2nd Quarter), 2006, 60–67.

TECC, 2013. http://c-tecc.org/tactical-emergency-casualty-care-guidelines.

U.S. Department of Homeland Security. Active Shooter: How to Respond. October 2008.

PUBLIC HEALTH CONSIDERATIONS

III

Chapter 6

Coordinated Terrorist Attacks and the Public Health System

Raymond McPartland and Michael J. Fagel

Introduction

Most areas of the health care system have been preparing for what terrorism experts call "low probability, high consequence" events. These events, such as the detonation of a radiological dispersion device or the dissemination of the plague virus involve the public health system simply by their nature and intent. Public health's preparation has been somewhat centered on events of a scientific nature where the release of some form of contamination would directly involve the response of the medical system and all its functions. A public health outbreak was considered the most dangerous and, at the post-September 11 timeframe, the most likely. But as history revealed, high consequence terrorism has not evolved as predicted. Other than a few isolated incidents that resulted in minimal casualties, the direct involvement of the public health system through an intentional release of biological or chemical contaminate has been hardly evident. With the public health world's lack of involvement followed the difficulty of obtaining a clear and updated operational picture and current threat perspective.

Terrorist groups today still express interest in using weapons of mass destruction whether it is biological, nuclear, radiological, or chemical in nature. However, interests aside, their mainstay to inflict something of a near mass causality incident

(MCI) has involved the use of less scientific weapons like explosives and small arms or submachine guns. A central goal of most terrorist groups is the infliction of mass destruction in the form of deaths and injuries but they must weigh the successful deliverance of an attack against their capability and availability of resources. It is much easier to acquire, train, and execute a plan of attack that does not involve any scientific application or complex delivery system. The difficulty in creating and delivering something so dangerous as a biological or chemical weapon, while at the same time not having the ability to accurately target and control its delivery has caused problems for terrorist groups. That has pushed them into a "simpler" yet, equally effective method—**Complex, Coordinated Attacks**.

They have learned the effectiveness of making their attacks more complex by combining various simplistic delivery systems such as the use of firearms, improvised explosives, and multiple assailants; *coordinated attacks carried out using small unit tactics and firearms is the delivery of choice in the twenty-first century*. This was never more evident, nor more effective, than the Mumbai, India attack of 2008. By using 10 trained men armed with small arms and explosives, the Lashkar-e-Taiba terrorist organization out of Pakistan was able to hold the city of Mumbai at bay for nearly 3 days.

Case Study: Mumbai, India November 26–29, 2008

The attacks in Mumbai, India beginning November 26, 2008 and ending the afternoon of November 29 can only be described as India's 9/11. After suffering from more than 10 coordinated attacks throughout the city and peninsula, Mumbai suffered more than 172 causalities at the hands of 10 gunmen armed with small arms, makeshift explosives, and the will to kill.

Directed through the leadership of Lashkar-e-Taiba (LeT), one of Pakistan's most active and militant terrorist organizations, armed gunmen were able to hold a city of more than 13 million at bay and send a message to the world that such a low-tech, inexpensive attack is possible with little way to defend against it.

The City of Mumbai

Formerly named Bombay, Mumbai is the most populated city in India and the second largest city in the world. With a growing populous of over 13 million, projected to rise to 28 million by 2020, Mumbai acts as India's Mecca for entertainment and commercial exploration. Similar to New York City, it acts as a symbol of economic indulgence and cultural power drawing tourists and businessmen alike. Financially, Mumbai generates 5% of India's GDP, accounts for 25% of their industrial output, 70% of maritime trade in India, and 70% of capital transactions to India's economy. Essentially, Mumbai acts as India's economic hub. Their seaport handles half of India's cargo traffic.

Along with its financial footprint, Mumbai houses such attractions and soft targets as the City's movie and television industry, famous hotels and tourist attractions, multinational corporations, the stock exchange, and the Reserve Bank of India. Mumbai does not have a single face and all things combined, becomes a melting pot of occupations, attractions, communities, and varying wealth that attract the meagre sightseer to the corporate mogul.

Besides Mumbai's economic output and their cultural footprint, another crucial facet of their existence that needs mention is its reliance on an extremely populated urban transportation system. Statistics show that roughly 88% of Mumbai's residents rely on the train and bus system to move about the City on a daily basis. During their average rush hour a new train arrives about every 3 min and within seconds it is filled to three times capacity leaving little room to ride let alone make a mistake; Mumbai loses an average of 10 passengers per day to train travel and transportation-involved accidents. This fact of overcrowding is pivotal considering the target selection by the 10 men tasked with assaulting the city.

Pre-Assault Preparations

Some of the most notable aspects of the attack on Mumbai were its complexity and preplanning. Plan building for the attack began mid-2007 leaving substantial time to gather intelligence through digital and personal means. Operatives knew careful pre-planning was crucial to the overall success of a mission involving multiple teams led remotely through what strategists call a "mothership" operation. Assaulters were provided information from other cell members designated solely as surveillance operatives on their various targets. Because they had to maneuver themselves through an unfamiliar city at night while engaging targets, they needed to be as familiar as possible with their environment. Information and digital images were provided via CD by the mission handlers to the attackers in order to accomplish this.

Operators of LeT formed a team of assaulters trained in small arms tactics, demolitions, and waterborne assaults. Narrowed from an initial group of 100 men, 10 men were selected once the training was completed.

Water Incursion and Landing

Because Mumbai is a peninsula surrounded by water on three sides, the choice of a water incursion was considered optimal. The terrorist assault team departed Karachi, Pakistan on November 22, 2008 at approximately 8:00 am. They departed aboard a Pakistani vessel named *Al Huseenni*, a vessel owned and operated by one of LeT's chief operatives, Zaki-Ur_Rehman Lakhvi. They traveled roughly

30 hours, and then transferred to another boat where their weapons and equipment were waiting. Fearing discovery in Indian waters piloting a Pakistani ship, they ordered the hijacking of a third fishing vessel off the coast and the killing of its crew at around 3:00 pm, November 23. Once aboard the hijacked Kuber shipping vessel, they navigated Indian waters using GPS devices with preprogrammed waypoints. The planners understood a water landing afforded them the ability to bypass security checkpoints on land and air and approach the southern and poorest part of the city under the cover of darkness. Once within striking distance of the shoreline, handlers instructed the team to divide into five teams of two and board two inflatable rafts destined for two separate landing locations. They were to arrive undetected by sea on November 26 as dusk fell and begin their assault at the height of the commuter rush hour.

Armament

Each individual member of the assaulting team was given an AK-47 assault rifle, a 9-mm pistol, hand grenades, a substantial amount of ammunition, and dry fruits or rations. Each team was handed an improvised explosive device (IED) built of RDX explosive and a satellite phone.

Deployment

Once on shore, the assaulters separated into groups based on their preplanned attack schedule. The terrorists divided themselves into four attack teams, one with four men (Team One) and three with two members each (Teams Two, Three, and Four).

- **Team One** boarded taxis headed for the Leopold Café and Taj Mahal Hotel.
- **Team Two** moved on foot to the Nariman House, a commercial-residential complex run by the Jewish Chabad Lubavich movement.
- **Team Three** headed to the Trident Oberoi Hotel, and
- **Team Four** began their assault on a major transportation hub, the Chhatrapati Shivaji Terminus (CST).

Those members who took a taxi to their location, before leaving the vehicle, armed and secreted a small improvised explosive device consisting of RDX high explosive under the front driver's seat. They were set to detonate nearly an hour later.

The Leopold Café and Bar—21:15 Hours

The Leopold Café and Bar has been in operation since 1871. It is considered a popular tourist site frequented by foreigners and westerners that visited Mumbai.

Team One arrived on target via taxi, quickly assessed if the target was viable and open for attack, and began firing at patrons and workers. One grenade was detonated. The assault on the café was over quickly, lasting only 5 min. Team One then moved quickly on foot towards the Taj Hotel.

The CST Attack—21:20 Hours

The CST Railway Station is the headquarters to the Central Railways of India. More than 3.5 million passengers pass through the station every day. Security was minimal at best, littered with mostly unarmed police, ill prepared for what was to come. Team Four, consisting of two men, Mohammed Ajmal Amir Kasab and Ismail Khan, entered the CST through a newly built entrance and began firing indiscriminately at travelers waiting to depart. They were allowed to roam for nearly 90 min and engage targets at will before being forced to move on due to an increasing police presence. The resulting death toll was 58 dead, 104 injured.

Once the attack on the CST was concluded, Team Four navigated back alleys and entered the Cama and Albless Hospital where they began firing again. It is unclear if this stage of the attack was preplanned or simply a target of opportunity, but nevertheless, the assaulters capitalized on the find of a soft and unprepared target.

As they departed the hospital, they encountered four members of the Indian counter-terrorism force arriving via vehicle. They ambushed the officers killing three, and stole their marked police vehicle. Blending easily with a marked police vehicle, they drove slowly firing at bystanders and reporters and even a movie house.

After switching vehicles once theirs became inoperable, Team Four hijacked yet another vehicle and began heading towards the Trident Oberoi Hotel. However, police radioed ahead and were able to set up a roadblock capturing one of the terrorists and killing the other. This team of two men was responsible for roughly a third of all deaths during the attack.

The Taj Mahal Hotel—21:40 and 22:10 Hours

The Taj Mahal Hotel was constructed in 1903 and is considered an historical icon and piece of local heritage. It consists of two wings with nearly 300 individually constructed rooms in each.

Team One entered the Taj through the north court entrance, avoiding security in the front of the building. Within the first few minutes, at least 20 patrons were killed. They linked up with the other two members of the team arriving from the Leopold Café attack and together armed their IEDs by the front entrance of the hotel. They traveled to the 6th floor and began engaging civilians and setting fires to slow rescue attempts and create chaos. Team One members were able to maintain control of the upper floors of the hotel for nearly a full 40 h before Indian Special Forces made entry and neutralized them.

The Trident-Oberoi Hotel—21:50 Hours

The Trident-Oberoi Hotel was a more modern structure consisting of two towers connected by a walkway and formerly operated and owned by the Sheraton and Hilton. It was built with an atrium-like feel to its interior and had the most modern conveniences known in everyday hotels. Total number of rooms amassed 877.

Team Three landed via a raft at an alternate site relatively close to the Trident-Oberoi Hotel and entered the Trident through the front engaging guests and staff immediately. They moved their way through the hotel and to the upper floors, specifically the 16th and 17th floors, where they were able to hold the location for nearly 42 hours before being killed by responding Indian Special Forces operative. They were able to kill at least 33 people.

Taxi Explosion—22:00 Hours

The IED consisting of RDX explodes in the first taxi in the Vile Parle area of the city, north of the attacks areas. The blast kills the driver and passenger and injuries a number of bystanders.

The Nariman House—22:25 Hours

The Nariman House is a contemporary structure of five floors purchased 2 years earlier by the orthodox Jewish organization called Chabad Liberation Movement of Hasidic Jews. Renamed the Chabad House, a Rabbi and family lived there with the task of accommodating Jewish visitors to the city.

Team Two approached the Nariman house on foot and before making their assault lobbed grenades into a nearby gas station to cause panic and distraction. They assaulted the front of the house and were able to gain control of the building after taking 13 hostages, five of whom they killed. Once inside they fortified their location and commandeered the high ground awaiting the police response. However effective this group was in grabbing media attention and making demands, they only accounted for eight fatalities and held rescuers at bay for more than 30 hours.

The Chabad House was the only location where demands were made and negotiations attempted. Entry coverage into the building was made available to the attackers through live feed television news crews delivered via satellite phone to the attackers from their remote handlers.

Taxi Explosion—22:45 Hours

The second IED explodes killing the driver, passenger, and two bystanders. At this point, the Indian police officials felt they were under attack by more than 100 men and that the entire peninsula was targeted due to the explosions now happening in the North.

Conclusion

At the conclusion of the attack there were more than 175 people killed and more than 300 wounded from explosives and small arms fire. This style of attack is known as a "swarm attack" using "Fedayeen," or self-sacrificing, tactics. This is not to be confused with a suicide attack. A Fedayeen raid requires the fighter to fight as long as possible, killing as many as possible, only dying at the hands of the enemy when the mission is complete and/or to avoid capture.

Their goal was the killing of as many people as possible while attracting as much attention from the rest of the world. The entire attack was watched on national television allowing the handlers thousands of miles away the ability to direct and redirect their attackers. Not only were the assaulters using satellite phones to communicate but also any means available taken from tourists and victims. There was no escape plan and very little if any negotiations. Police and emergency responders, including the public health system, were overwhelmed by the time the second attack began. Ill prepared logistically from the start, the police force of nearly 30,000 was initially stunned, which allowed the movement of the ten member attacking team free access.

Outside of direct contact with small arms and explosives, the emergency responders were forced to handle multiple tasks of varying disciplines simultaneously whether it was IED disarmament, fire fighting, medical rescue, and counter-assault tactics. emergency medical personnel (EMS), whether staged or deployed, were involved in all matters. In order to handle the situation, responders needed to handle each incident from the operational level individually while the command executives managed the larger picture of how all of the events were unfolding. A task more difficult than imagined.

Swarm Attack Characteristics

The following characteristics of Mumbai-style attacks was compiled by analysts of the New York City Police Department's Intelligence Division and circulated throughout the response community in hopes of aiding with recognition and disruption.[1]

- Use of firearms, mainly military grade, and portable explosive devices;
- Extensive intelligence collection and analysis prior to deployment in order to aid in the group's ability to select targets and prepare attacks;

- Transportation to and within the target area while minimizing contact with the public and security forces;
- Multiple teams attacking several targets simultaneously;
- Diversions drawing security forces away from the prime target(s);
- Deployment of IEDs at entrances and exits or in or about target areas eliminating the need for an extra security team;
- Tactics such as arson, IEDs, covering fire, and grenades allowing terrorists to reposition themselves;
- Attackers reaching targets, seizing hostages, and prolonging the situation as much as possible;
- Innovative use of communications technology hindering intelligence collection; and
- Handlers using open-source media to plan responses against security forces.

Terror Medicine

It is important for first responders to assess their calls for service differently than ever before. What was once considered routine must now be scrutinized as atypical. This scrutiny comes from the understanding that the new threat spectrum is considered multitiered and littered with potential hazards not traditionally seen by first responders. Those in the medical profession must now contend with the lone gunman or multiple school shooter, a possible improvised explosive device and possible secondary devices, and even multiple armed men acting in unison using military-style weapons.

Emergency medical responders and public health staff must maintain a heightened sense of awareness while maintaining appropriate and effective medical care. The ideal scenario of treating everyone equally may have to transition to only doing what must be done in order to save those deemed savable. This transition, or application of an *altered standard of care*, will happen after responders evaluate all the necessary avenues using a risk–benefit analysis. The combination of patient criteria and medical response must also weigh in the current threat and operational picture. *Terror Medicine*, a term coined by leading emergency management practitioner and public heath responder Shmuel C. Shapira, may now become the norm more than the rarity.

Terror medicine consists of four main areas; *preparedness, incident management, mechanisms of injuries and responses, and psychological consequences.*[2] All four areas must constantly keep in mind the duality of the threat presented as well as the overall threat or operational picture.

Preparedness can fall into the operational or procedural lane. Training in treatment, establishing protocol, and gathering supplies are all examples that fall under the preparedness umbrella.[3]

Incident Management consists of the system or organizations' ability to manage a crisis. Establishment of an incident command system, even one modified to fit the needs of the public health system, is critical to quicker scene stabilization and recovery.[4]

The practitioner's third area of concern is their ability to effectively understand the nature and **mechanism of injuries** and how best to **respond**. Traditional responders need to familiarize themselves with injuries from blast devices and large-caliber weapons.[5]

Finally, the EMS responder's ability to understand and manage the **psychological consequences** of a terrorist attack becomes paramount. Early psychological intervention is essential. If not appropriately treated during the first 6 months after an incident, patients may suffer irreversible stress disorders.[6]

Question...

How would a coordinated assault using tactics similar to the 2008 Mumbai attacks affect a major metropolitan area's public health and response system? Public health as a whole is often overlooked due to the immediate need to prepare the law enforcement response instead. Experts agree, however, that without a functioning and effective public health response, a mass casualty incident of this magnitude will surely compound and cost additional lives even after the shooting stops.

To better assess this problem, an incident such as this along with its ramifications needs to be evaluated from both the *micro and macro levels of review.*

- The **macro viewpoint** is that of the public health system as a whole and all the interconnecting nodes that allow it to function effectively.
- The **micro viewpoint** focuses primarily on individual responders, a single hospital, or isolated function within the public health system.

Macro Level: Public Health System Issues When Facing a Coordinated Attack

According to the United States Department of Human Health and Services or HHS, the Public Health System in the United States is defined as

> ...the complex network of organizations that work towards fulfilling the public health mission of assuring conditions for a healthy population (HHS, 2011).[7]

Most laypersons see the public health systems as just a single hospital, clinic, or even physician they visit. Their recognition is normally based on their personal experience and interaction but few recognize the magnitude and complexity of how the system operates each and every day. It is a network consisting of various functions including response and transport, treatment, and recovery that rely on the operational capacity of each other. If one node, or connection point, in the system were to deteriorate, the effectiveness of the entire system will be affected and eventually its overall ability to respond to and care for patients in an effective manner will collapse.

Even robust systems like those in major cities like New York, Chicago, and Los Angeles, no matter how prepared, will initially be stressed to the point of failure during such an attack. It may be hours before a proper response and recovery can be put in place. For the sake of discussion, it is those systems we will be referring to. The assumption is that the same issues in response, recovery, and mitigation will be faced even more so and quicker in smaller jurisdictions.

The following is only a partial listing of problems a public health system may encounter:

Lack of immediate medical response due to the severity of the attack and the report of multiple attackers. Emergency medical personnel are normally not equipped with any form of ballistic protection nor are they trained in a tactical response to an event of this magnitude. Most agency protocols require an area to be safe before a nonlaw enforcement agent can make entry. With the reports of multiple attackers and the possibility the attackers could still be at large, EMS personnel will hold at the outer-perimeter and begin to stage, waiting to receive the victims. This delayed response, although understandable due to available protective equipment and training, will undoubtedly cost additional lives.

Access to the scene needs to be established and secured as quickly as possible in order to begin the triage process. With every second passing, those injured that were initially viable can become unsavable.

Immediate recognition that the incident they are facing is part of a multitiered coordinated attack. When members of the emergency medical services respond to an incident, their concern is for the care of the patients immediately in front of them. Like many first responders, EMS responders tend to operate in an unintentional vacuum. Understanding that the MCI they are facing may be linked to other incidents around the city may not come immediately or even in the immediate future. What may become recognizable is the severity of the wounds and similarity to other more serious events in the jurisdiction.

No jurisdiction, even the busiest, is confronted with multiple incidents involving multiple gunshot wounds from high-caliber weapons all within minutes to hours of each other. Couple this with multiple attackers that appear organized and possibly using explosive devices, and the picture becomes a little clearer as long as communication is occurring between responders from different areas within the jurisdictions. Seeing these incidents in a repeated fashion should indicate that the attacks are not random and are more organized allowing the responder to increase their situational awareness. Recognition also lends for better future preparation in transport and receiving as similar incidents begin to unfold.

Triage and Casualty Collection Location. Once the first isolated incident concludes and injury assessments and prioritization of care begins, a *triage location* or *casualty collection point (CCP)* will need to be established by first responders. It is very likely, due to the initial inability to predict another coming incident, that the triage location will begin on site or very near to where the victims were injured. Without proper force protection of the medical responders, treatment will be minimized. As additional events begin to unfold and the bigger picture is revealed, triage locations will become more designated, tighter run, and may be further away in order to have established area security. This will hamper response and slow the treatment process. Security and site evaluation must be done when creating an initial and future CCPs.

Transport and Tracking of Victims. As victims are triaged, the need for immediate transportation will be necessary. Some immediate life-saving treatment may be done on scene but will be temporary and the victim will need to seek hospital care. Taking into account the style of current coordinated attacks, there is a strong possibility of another mobile attacker still in the immediate area or even that the transport vehicle, if unescorted, could enter into another attack area unknowingly. As events progress there are areas of the city that may become unnavigable. Security becomes an issue and requires the immediate establishment of safe routes to designated receiving points. But no matter how important, these safe routes may not be established prior to the areas being declared clear by law enforcement.

Transport of the victims corresponds easily with the tracking of who is being moved and to what location. Tracking becomes critical for three reasons; first, the injured are crime victims who require contact with law enforcement; second, they may have information or actionable intelligence that could further aid the responders and need to be debriefed as soon as possible; third, tracking of who injured are being sent to what hospital will determine whether that area hospital will be at the brink of their surge capacity.

Proper transport and tracking is contingent upon solid communication between the field teams and those receiving. Information such as the number of patients moved and their injury classification, locations being used for casualty collection, and areas in which entry is prohibited is key to keeping the incident manageable.

Lack of Experienced Medical Personnel and Adequate Trauma Hospitals. Many metropolitan areas have various levels of trauma receiving areas or trauma hospitals. These hospitals are medical receiving points designated for the most serious of wounds. They may have their own designated operating rooms and emergency staff trained to deal with high-level emergencies. Given the severity of the attacks, the variety and severity of the injuries, and the overwhelming number of victims, the local trauma hospitals will be overwhelmed in the first hour if not the first few minutes. Once those hospitals are filled to capacity and can no longer receive emergency room patients, incoming victims will be diverted to another nearby location. The staff at this location, although qualified medically, may not be equipped or experienced enough to manage the injuries they are receiving. This is even truer if the incoming victims are arriving en masse.

Public vs. Private Response and Individual Deployment. A majority of the public health emergency response community consists of volunteers and/or private sector-based staff. The level of experience and training for each responder possessing even a common medical baseline, will be different. Many volunteer responders are there for injury management and transport, not critical care under fire. Whether or not these resources will be deployed will be dependent on the incident specifics and command control decisions. Self-deployment of responders without a full understanding of the operational picture is also a dangerous issue. Private resources responding with the best of intentions may find themselves entering the kill-zone complicating matters even further.

Overall Diminished or Altered Standard of Care. Many plans established to respond to and mitigate a mass casualty event normally assume that the standard of care administered will be comparable to what is done in most other emergencies. But an event that unravels this quickly, involves these types of serious and immediate injuries, as well as unimaginable number of victims will compromise even the strongest of systems. Local hospitals are not mirror images of army field hospitals. The standard of care will be different and the ability to operate as current protocols dictate will become evident.

In August 2004, a panel of medical experts in the fields of bioethics, emergency medicine, emergency management, health administration, health law and policy, and public health was convened to discuss matters concerning altered standards of care to an MCI. Their key findings are summarized as follows:

1. The goal of an organized and coordinated response to a mass casualty event should be to maximize the number of lives saved.
2. Changes in the usual standards of health and medical care in the affected locality or region will be required to achieve the goal of saving the most lives in a mass casualty event. Rather than doing everything possible to save every life, it will be necessary to allocate scarce resources in a different manner to save as many lives as possible.
3. Many health system preparedness efforts do not provide sufficient planning and guidance concerning the altered standards of care that would be required to respond to a mass casualty event.
4. The basis for allocating health and medical resources in a mass casualty event must be fair and clinically sound. The process for making these decisions should be transparent and judged by the public to be fair.
5. Protocols for triage (i.e., the sorting of victims into groups according to their need and resources available) need to be flexible enough to change as the size of a mass casualty event grows and will depend on both the nature of the event and the speed with which it occurs.
6. An effective plan for delivering health and medical care in a mass casualty event should take into account factors common to all hazards (e.g., the need to have an adequate supply of qualified providers available), as well as factors that are hazard-specific (e.g., guidelines for making isolation and quarantine decisions to contain an infectious disease).
7. Plans should ensure an adequate supply of qualified providers who are trained specifically for a mass casualty event. This includes providing protection to

providers and their families (e.g., personal protective equipment, prophylaxis, staff rotation to prevent burnout, and stress management programs).

8. A number of important nonmedical issues that affect the delivery of health and medical care need to be addressed to ensure an effective response to a mass casualty event. They include:
   a. The authority to activate or sanction the use of altered standards of care under certain conditions.
   b. Legal issues related to liability, licensing, and intergovernmental or regional mutual aid agreements.
   c. Issues related to effective communication with the public, special needs population and patient transport.

— Altered Standards of Care in Mass Casualty Events. Prepared by Health Systems Research Inc. under Contract No. 290-04-0010. AHRQ Publication No. 05-0043. Rockville, MD: Agency for Healthcare Research and Quality. April 2005.

Micro Level: Untraditional Response Protocols

Because the results of a coordinated attack will resemble more a battlefield than a city street scene, it forces the responder to behave more like a soldier than a traditional first responder. In order to do so, lessons need to be taken from past real world battlefield engagements and transplanted into the civilian environment.

The result—the creation of a new set of guidelines that balances the threat, civilian limitations of medical practice, a holistic look at the civilian population, and what they represent as victims and equipment and resource limitations. Based on the military version of combat medical response known as *Tactical Combat Casualty Care or TCCC*, comes the civilianized version, *Tactical Emergency Casualty Care or TECC.*

TECC tries to maintain a realistic look at the medical response to an event reminiscent of a military attack. It provides guidelines on care management of preventable deaths located close to, if not in, the area of attack. This is done while constantly reassessing the scene and allowing the responder to maximize their response potential while minimizing their risk.

Established by responders for responders, TECC is an avantgarde look at medical response by civilian responders. Traditionally, medical response and immediate treatment would come to a halt awaiting an all-clear notice by law enforcement. TECC guidelines provide responders with the understanding that in an event of this nature, waiting may take longer than average due to the attack's scope. Treatment needs to be administered once the area is no longer declared a *direct threat area*.

TECC discusses three phases of care defined by the relationship between the responder and the impending threat:

1. Direct Threat Care or Care Under Fire
2. Indirect Threat Care or Tactical Field Care
3. Evacuation Care[8]

Direct threat care involves minimal medical actions taken because the threat is still active and immediate. There is an emphasis on responder safety, evacuation of the wounded, and mitigation of any heavy bleeding if feasible.[9]

Indirect threat care involves more involved medical actions and more thorough patient assessment. The threat may be terminated with the possibility of other unknowns or may simply be no longer on scene. The area is not completely secure. Assessment and treatment priorities include major hemorrhage control, airway, breathing/respirations, circulation, head and hypothermia, and anything else that is problematic.[10]

Evacuation Care involves moving the aided to a treatment facility. This phase more closely resembles normal EMS operations.[11]

It is important to note that these phases are fluid. An area that was deemed an indirect threat area allowing EMS personnel to begin their work can suddenly become a direct threat area. This unstable and potentially life-threatening environment combined with the altered standard of care is something current first responders may be unprepared for.

Conclusion

Medical responders need to constantly reassess their current threat perspective. Intelligence analysis is no longer just a law enforcement responsibility and must be done by all responders, particularly those at the epicenter of the incident. By assessing the current operational picture and allowing the ability to engage in altered standards of care when the need arises, public health personnel will be in a better position to handle a complex, coordinated attack like that of Mumbai, India 2008.

References

1. NYPD Intelligence Division, *Terrorism Awareness Bulletin*, New York City Police Department, 2008.
2. S. C. Shapira, Terror medicine: birth of a discipline, *Journal of Homeland Security and Emergency Management* 3(2), 2006, Article 9. PDF.
3. Panel on Terror Medicine and Domestic Security, May 2009. Hadassah Hospital, Jerusalem.
4. Environmental Public Health Systems and Services Research, June 2011. Capt. John Sarisky, Senior Environmental Health Officer, Environmental Health Services Branch, Division of Emergency and Environmental Health. http://www.cdc.gov/nceh/ehs/Docs/JEH/2011/June_Sarisky_Gerding.pdf

5. D. W. Callaway, E. R. Smith, M. J. F. Shapiro, and G. Shapiro. The committee for tactical emergency medical care (C-TECC): Evolution and application of TCC guidelines to civilian high threat medicine. *Journal of Special Operations Medicine* Vol. II, Edition 2 (Spring/Summer II). PDF, 2011.
6. M. J. Fagel, (Ed). *Principles of Emergency Management: Hazard Specific Issues and Mitigation Strategies.* CRC Press, Boca Raton, FL, December 2011.
7. World Health Organization (WHO). The elusive definition of pandemic influenza. March 31, 2011. http://www.who.int/bulletin/volumes/89/7/11-086173/en/
8. Scientific American. Deadly Pandemic Bird Flu Details Finally Are Made Public. *Scientific American* June 21, 2012. http://www.scientificamerican.com/article.cfm?id=pandemic-bird-flu-studies-public&WT.mc_id=SA_CAT_BS_20120622
9. Brown University. A SMART(er) way to track influenza. June 11, 2012. http://news.brown.edu/pressreleases/2012/06/smart
10. The White House. The National Security Strategy, May 2010. http://www.whitehouse.gov/sites/default/files/rss_viewer/national_security_strategy.pdf
11. CIDRAP. Studies: Antiviral prescribing reflects CDC guidance, flu activity. June 11, 2012. http://www.cidrap.umn.edu/cidrap/content/influenza/swineflu/news/jun1112antivirals.html

Chapter 7

Emergency Management, Public Health, and Private Sector Healthcare

New Opportunities for Collaboration

Ruth Cover

> Political leaders, health officials, hospital CEOs and emergency management officials have a vested interest in maintaining public confidence in their respective institutions before, during, and after an extreme event. Regional partnerships can help this to be realized....
>
> *U.S. Department of Health and Human Services, Assistant Secretary for Preparedness and Response, 2007*

Background

During the last century, practitioners of emergency management, public health, and private sector healthcare entities frequently maintained a narrow track within their disciplines, seldom interacting with or planning with members of other disciplines. Certainly, there were few incentives and even fewer mandates to do so. Events in the early 21st century, however, changed this. Legislation, programs, and funding designed to better prepare the nation for catastrophic events, whether naturally occurring, accidental, or humanly engineered, impacted all three disciplines. These programs have evolved, outlining more specific goals and tasks even while federal funding declined. The March 2011 issuance of Presidential Policy Directive on National Preparedness (also known as PPD-8) marked a turning point for tighter alignment of emergency management, public health, and private sector health-care preparedness activities. Three grant programs that were immediately affected by PPD-8 were the various subgrants of the Homeland Security Grant Program (HSGP) overseen by the U.S. Department of Homeland Security (DHS) and two programs overseen by components of the U.S. Department of Health and Human Services (HHS). Those were the Public Health Emergency Preparedness (PHEP) and Hospital Preparedness Program (HPP) cooperative agreements. As funding continues to decrease for all three programs, increased collaboration between the three disciplines will be critical as each strives to attain the mandated goals of the separate grant and of the National Preparedness Goal.

In 2009, the DHS Grants Program Directorate (GPD) published a report confirming that expenditures during the first 4 years of Homeland Security FEMA Preparedness Grants were concentrated on very few capability areas. Although the DHS grants were to be used on capabilities similar to those defined in the HPP and PHEP focus areas, none of those capabilities were among the top five DHS spending categories, as illustrated in Figure 7.1.

Histories of the Programs, Including Funding

HSGP

In 2003, several projects that had been funded by the Office of State and Local Government Coordination and Preparedness were consolidated and became overseen and administrated by the DHS. Originally, five programs were included in the HSGP: State Homeland Security Program (SHSP), Urban Areas Security Initiative (UASI), Operation Stonegarden, Metropolitan Medical Response System (MMRS) program, and Citizen Corps Program. As will be described later, the UASI and MMRS program goals were most closely related to the objectives of the HHS PHEP and HPP goals. Over the years, the eligibility for the UASI status changed, first with the addition of some second-level urban areas and later eliminating some of those. The FY-12 federal budget cut direct funding for MMRS program, although each awardee's State Administrative Authority was allowed to fund existing MMRS programs using part of the state's SHSP award. The MMRS activities corresponded very closely with requirements for the HHS HPP and PHEP cooperative agreements, namely, strengthening capabilities for medical surge, mass prophylaxis, Chemical, Biological, Radiological, Nuclear, and Explosive (CBRNE) activities, interoperable communication, and information sharing; expanding regional collaboration; and activities relating to mass care, public information and warning,

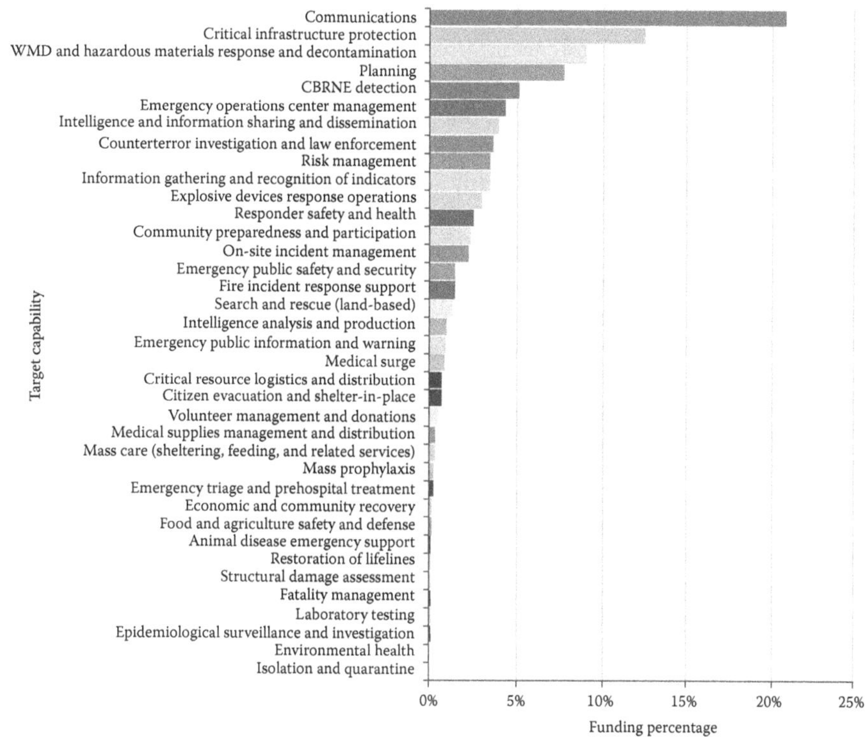

Figure 7.1 DHS grant funding by target capabilities 2003–2007. (Adapted from U.S. Department of Homeland Security, Federal Emergency Management Agency. 2009. *FEMA GPD Grant Program Accomplishments Report: Summary of Initial Findings (FY03-07)* [Monograph]. Retrieved from http://www.fema.gov/pdf/government/grant/GPD_grant_accomplish_report.pdf)

prehospital care and triage, and management of medical supplies, fatalities, and volunteers. Between 2003 and 2007, recipients of HSPG funds concentrated on equipment purchases and planning (DHS, 2009).

PHEP

The Centers for Disease Control and Prevention (CDC), part of HHS, first offered the PHEP cooperative agreements to 62 public health entities in 2002. These recipients were, and remain, all 50 states, 4 separately funded major metropolitan areas (Chicago, District of Columbia, Los Angeles County, New York City), and 8 U.S. territories and freely associated states (American Samoa, Guam, U.S. Virgin Islands, Northern Mariana Islands, Puerto Rico, Federated States of Micronesia, Republic of the Marshall Islands, and Republic of Palau). The purpose was to develop capabilities and capacities of public health departments to effectively respond to public health consequences in an all-hazards environment of natural disasters, infectious disease outbreaks, terrorist threats, and chemical/biological/radiological/nuclear (CBRN) incidents. All awardees received base plus population funds, Real-Time Disease Detection funds, and later Pandemic Influenza Supplement Phase III funds. Select awardees were eligible for Cities Readiness Initiative (CRI) funds, Level 1 Surge Capacity Chemical Laboratory funds, and Early Warning Infectious Disease Surveillance (EWIDS) funds. Original PHEP activity areas (focus areas) were Preparedness Planning and Readiness Assessment; Surveillance and Epidemiology Capacity; Laboratory Capacity—Biologic Agents; Laboratory Capacity—Chemical Agents; Health Alert Network/Communications and Information; Risk Communication and Health Information Dissemination; and Education and Training.

As might be expected, for most of the first 11 budget periods, PHEP recipients spent a good portion of the awards on equipment and systems and added enough staff to support those activities. During the first years, there were also several cross-cutting critical benchmarks (CCCBs) shared with the HPP. These were CCCB #1—Incident Management; CCCB #2—Joint Advisory Committee for PHEP and HPP Cooperative Agreements; CCCB #3—Laboratory Connectivity; CCCB #4—Laboratory Data Standards; and CCCB #5—Jointly Funded Health Department/Hospital Activities.

HPP

Also in 2002, the Agency for Healthcare Research and Quality (AHRQ), another part of HHS, offered the HPP cooperative agreement to the same 62 entities that were eligible for PHEP funding. Although administered by the government-level awardees, the focus of this funding was to prepare private sector healthcare systems and other partners to provide effective, coordinated care during public health emergencies and to victims of bioterrorism or other terrorism. Awardees decided how to distribute the funds to the private sector, either through individual subawards to facilities, through a centralized regional committee for the good of all healthcare entities in the region, or through some other method. Original priority-area activities for HPP were Administration; Regional Surge Capacity; Emergency Medical Services; Linkages to Public Health Departments; Education and Preparedness Training; and Terrorism Preparedness Exercises.

As with PHEP, HPP recipients were required to attain certain benchmarks related to surge capacity: beds, isolation capacity, healthcare personnel, advance registration system for healthcare volunteers, pharmaceutical caches, personal protective equipment, decontamination, behavioral health, trauma and burn care, and communications and information technology. And again, as with the DHS and PHEP awards, much of this funding was spent on equipment and supplies to build capacity and capability, along with adding staff to support the programs. Beginning in 2006, the original priority areas changed. The new categories were overarching capabilities that had to be demonstrated in all Level 1, Level 2, and additional consideration activities. Overarching capabilities included National Incident Management System (NIMS) principles, Education and Preparedness Training, Exercises, Evaluations and Corrective Actions, and Needs of At-Risk Populations. Five Level 1 capabilities were required to be funded and addressed in the work plans of all awardees. These capabilities were Interoperable Communications System, Bed Tracking System, Emergency System for the Advance Registration of Volunteer Health Professionals (ESAR-VHP), Facility Level Fatality Management Plans, and Hospital Evacuation Plans activities. Level 2 capabilities could be funded and addressed only when the awardee could ensure that all Level 1 capabilities could be met. The five optional Level 2 projects included Alternate Care Sites, Mobile Medical Assets, Pharmaceutical Caches, Personal Protective Equipment, and Decontamination.

Additional considerations of the awardee's choice such as partnership development, critical infrastructure protection, and other projects identified by hazard vulnerability assessment or gap analysis could be addressed and funded only when the recipient could ensure completion of all Level 1 requirements.

Pandemic and All Hazards Preparedness Act of 2006, Public Law 109–417

The passage of this legislation affected both the PHEP and the HPP programs. The activity focus changed from infectious disease outbreaks, terrorism, and bioterrorism to preparedness in an all-hazards environment. In 2005, Congress appropriated $350,000,000 for pandemic influenza purposes (HHS/CDC, 2011). This money was allocated to the 62 PHEP awardees in phases. Phase One ($1,000,000) was distributed in March 2006 followed by $225,000,000 Phase Two funding in July 2006. The remaining $25,000,000 was offered in 2008 via competitive grants for public health demonstration projects related to pandemic influenza. The Pandemic and All Hazards Preparedness Act of 2006 (PAHPA) caused the oversight of the HPP to be moved from AHRQ to another HHS office, that of the Assistant Secretary for Preparedness and Response (ASPR). As mentioned above, it prompted the change from HPP priority-area activities to Overarching, Level 1, Level 2, and other considerations capabilities. In 2009, $90,000,000 in pandemic preparedness cooperative agreements was distributed to recipients of HPP grants. The alignment of both programs' activities began in 2008 with the integration of public/private capabilities; increasing preparedness, response and surge capabilities with expansion to all types of healthcare facilities; consideration of at-risk populations, with specification of what *at-risk* meant; coordination with other activities and alignment with the National Response Framework, NIMS, the National Preparedness Goal, and the *Medical Surge Capacity and Capability: A Management System for Integrating Medical and Health Resources* (MSCC) handbook; and more emphasis on continuity of operations (COOP).

At this point, collaboration with emergency management was indirectly implied, but not clearly delineated, in the funding requirements for both grants.

Table 7.1 Funding History of Selected Preparedness Grants and Cooperative Agreements

Funding Year (FY)	*Public Health Emergency Preparedness (PHEP) Cooperative Agreements (Base)*[a]	*Hospital Preparedness Program (HPP) Cooperative Agreements*[b]	*Homeland Security Grant Programs—All Components*[c]
2002	$918,000,000	$125,000,000	Not available
2003	$970,000,000	$498,000,000	Not available
2004	$849,600,000	$498,000,000	$2,220,000,000
2005	$962,800,000	$470,755,000	$25,187,763,121
2006	$991,400,000	$450,000,000	Approximately $1.7B
2007	$896,700,000	$415,032,000	Not available
2008	$704,800,000	$398,059,000	$1,698,959,000
2009	$618,830,835	$362,017,984	$1,714,172,154
2010	$616,741,225	$390,500,000	$1,786,359,956
2011	$537,215,590	$352,605,175	$1,289,296,132
2012	$554,803,057	$351,644,731	$339,500,000

[a] http://www.cdc.gov/phpr/coopagreement.htm
[b] http://www.phe.gov/Preparedness/planning/hpp/Pages/funding.aspx
[c] http://www.fema.gov (various annual program guidance documents)

As can be seen in Table 7.1 and in Figure 7.2, funding began to decline in 2007, although requirements were expanding and becoming more specific.

In addition to these three main emergency preparedness grants and cooperative agreements, both HHS and DHS have offered other funding to help jurisdictions and the private sector prepare for emergencies and disasters. In 2007, a competitive grant distributed approximately $15,000,000 via the Healthcare Facilities Partnership program, in parcels ranging from $500,000 to $2,500,000. As previously mentioned, additional funding for pandemic influenza preparedness was distributed to HPP and PHEP awardees in 2006–2009.

On the emergency management side, the HSGP contained several separate grants related to emergency management and health-related initiatives, notably the UASI and the MMRS. These two grants, as the names imply, were available only to larger urban areas and not to the majority of U.S. municipalities. At peak, only 124 cities were eligible for MMRS funds and only 64 urban areas held UASI designation. Designed to sustain and further enhance integration of local emergency management, health, and medical systems into a coordinated, sustained local capability

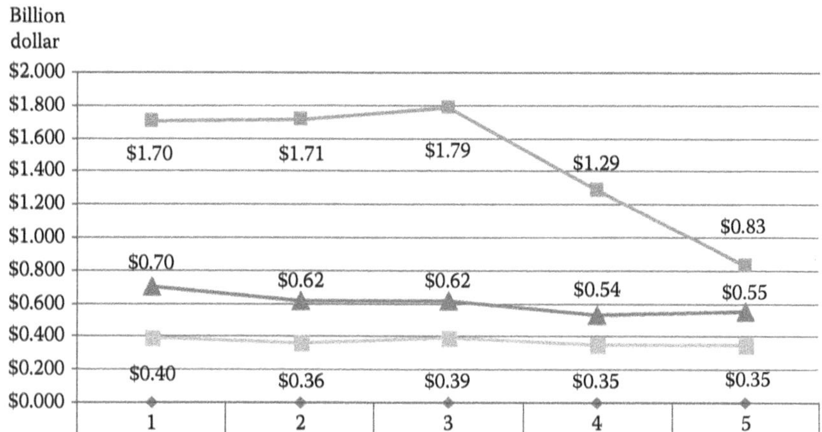

FY	2008	2009	2010	2011	2012
HSGP	$1,698,959,000	$1,714,172,154	$1,786,359,956	$1,289,296,132	$830,790,000
PHEP	$704,800,000	$618,830,835	$616,741,225	$537,215,590	$554,803,057
HPP	$398,059,000	$362,017,984	$390,500,000	$352,605,175	$351,644,731

Figure 7.2 Preparedness grants funding 2008–2012.

for all-hazards mass casualty preparedness and response, the MMRS was the program more likely to bring those entities to the same collaborative table.

Since FY 2010, MMRS recipients were also required to effectively demonstrate coordination of mutually supportive program requirements from HHS and other relevant federal agencies. UASI, on the other hand, was designed to focus on multiagency planning for building and sustaining capabilities to prevent, protect against, respond to, and recover from threats or acts of terrorism. The President's FY 2012 HSGP budget eliminated MMRS funding as a separate grant but did allow MMRS activities to be funded at the discretion of the recipient State Administrative Agency. UASI funding remained a separate grant under the President's FY 2012 budget proposal. Tribal governments have not been neglected. The Tribal Homeland Security Grant Program (THSGP) made up to $26,000,000 available in 2010 through 2012.

Healthcare Coalitions, Medical Surge Capacity and Capability

The idea of healthcare capabilities and coalitions is not new. In 2002, Barbera and Macintyre conceptually addressed the complex health and medical issues associated with large-scale medical incidents in their work, *Medical and Health Incident Management (MaHIM) System*. In 2004, the CNA Corporation, Institute for Public

Research, expanded on MaHIM and prepared the first edition of MSCC for the HHS. This handbook, commonly known as the *MSCC Handbook* and revised in 2007, outlines a management methodology using the Incident Command System principles and other valid emergency management tenets. It describes six tiers of healthcare participation and coordination during a major health-affecting event as shown in Figure 7.3.

As shown in Figure 7.3, the intersection of emergency management and healthcare coalitions occurs at tier 3. Also during 2007, the funding opportunity announcement for the Healthcare Facilities Partnership Program contained a significant and prophetic statement:

> Political leaders, health officials, hospital CEOs and emergency management officials have a vested interest in maintaining public confidence in their respective institutions before, during, and after an extreme event. Regional partnerships can help this to be realized through brokering of relationship building, engendering trust among otherwise traditionally competitive entities and maximizing coordination and cooperative planning among healthcare entities and other partners (HHS/ASPR, 2007, p. 2).

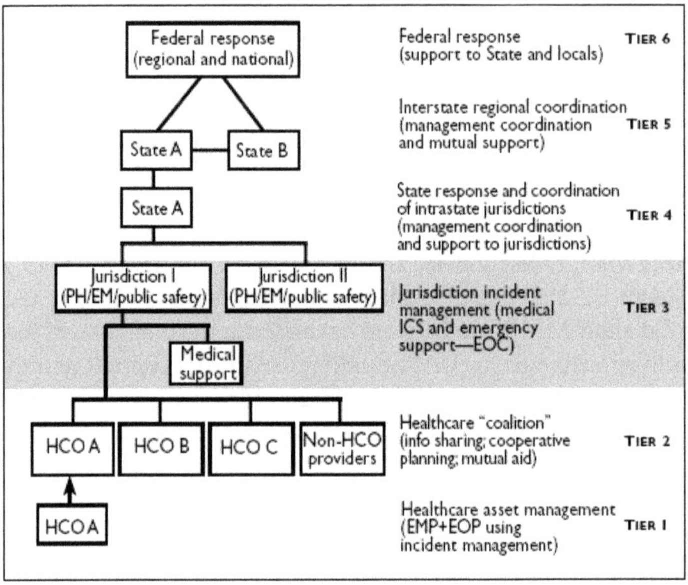

Figure 7.3 Tiers of medical response: facility to federal levels. (Adapted from U.S. Department of Health and Human Services. 2007. *Medical Surge Capacity and Capability: A Management System for Integrating Medical and Health Resources During Large-scale Emergencies* (2nd ed.). Retrieved from http://www.phe.gov/Preparedness/planning/mscc/handbook/chapter4/Pages/default.aspx)

Moving toward Health Preparedness Grant Alignment

In the first few years following the passage of PAHPA, the directors of public health preparedness, most of whom had ultimate responsibility for administrating both the PHEP and HPP funds within their jurisdictions, began to realize that the requirements for the cooperative agreements were similar yet disparate enough to cause duplication of efforts in activities as well as in data collection and performance reporting. Some were also concerned that the requirements of the HHS grants had little correlation to the DHS Target Capabilities and Universal Task Lists. They began to voice these concerns to the federal-level HHS program leaders but for several years saw little movement toward resolution of their concerns. Some of this changed in March 2011 with the funding opportunity announcement for the FY12 PHEP cooperative agreement. At the same time, HHS issued a new publication, *Public Health Preparedness Capabilities: National Standards for State and Local Planning* (HHS/CDC, 2011). These two documents implemented a sea change in the PHEP program planning and activities, bringing in 15 Public Health Preparedness Capabilities and associated functions and resources: (1) Community Preparedness, (2) Community Recovery, (3) Emergency Operations Coordination, (4) Public Information and Warning, (5) Fatality Management, (6) Information Sharing, (7) Mass Care, (8) Medical Countermeasure Dispensing, (9) Medical Material Management and Dispensing, (10) Medical Surge, (11) Non-pharmaceutical Interventions, (12) Public Health Lab Testing, (13) Public Health Surveillance and Epidemiology Investigation, (14) Responder Safety and Health, and (15) Volunteer Management.

In January 2012, HHS issued a similar document related to the HPP award, *Healthcare Preparedness Capabilities: National Guidance for Healthcare System Preparedness* (HHS/ASPR, 2012). This document also contained eight similar capabilities for healthcare organizations: (1) Healthcare System Preparedness, (2) Healthcare System Recovery, (3) Emergency Operations Coordination, (5) Fatality Management, (6) Information Sharing, (10) Medical Surge, (14) Responder Safety and Health, and (15) Volunteer Management. Although the titles of the capabilities were similar, the functions with each capability were not (see Table 7.2).

At this point, the set of capabilities did not match well with the old DHS Target Capabilities List or with the Preliminary Targets found in the September 2011 DHS publication, *National Preparedness Goal*. The *Crosswalk of Target Capabilities to Core Capabilities* (DHS/FEMA, 2011) did not clarify the nature of the relationship between emergency management and health-related capabilities. During 2011, HHS reorganized and moved the oversight of funding for both programs to the CDC in Atlanta, Georgia. In March 2012, a joint funding opportunity announcement was issued that contained new requirements for both programs with a project period of July 1, 2012 through June 30, 2017.

Current PHEP and HPP Program Collaboration Requirements

The FY 2012–2013 program requirements clearly mandate collaboration between the awardees within a jurisdiction with inclusion of emergency management. There must be a joint senior advisory committee that includes jurisdictional emergency management (HHS, 2012, pp. 28–29). Each public health jurisdiction must conduct a joint risk assessment that should incorporate findings from "their state or regional threat hazard identification risk assessment (THIRA)," an assessment done by the emergency management agency (HHS, 2012, pp. 22–23).

Additionally, awardees were required to describe within their applications planned activities for resource elements and how they would work with the HPP to inform healthcare organizations and coalitions of their public health preparedness and response roles (HHS, 2012, pp. 42–49). Very complex and detailed budget period 1 performance expectations were also outlined (HHS, 2012, pp. 217–222).

For the HPP portion of the application, awardees had to address all planning and resource elements for each capability, outline how they would develop healthcare coalitions, and identify existing coalitions (HHS, 2012, pp. 31–42). Although Partnership Development had been a part of the Level 2 and Other Considerations allowable activities since FY 2009, the FY 2012 funding announcement was the first year that formalized activities related to healthcare coalitions. As part of the application, awardees were required to perform a self-assessment on where the state, territory, or freely associated state stood on the various capabilities. Following the actual granting of the award, they were directed to work with their regions or existing health-related partnerships to assess the characteristics and functions of each region, partnership, or existing healthcare coalition. The results of that assessment became the content of much of the mid-year progress report and established the baseline against which progress during the remaining project period years would be measured. Provisional long-term performance metrics to be met by June 30, 2017 included 100% compliance for coalitions to have formalized agreements and demonstrated ability to function and execute the HPP capabilities (Preparedness); have a process for short-term recovery and health system continuity of operations plans (Recovery); use the Incident Command System to coordinate operations and resource sharing (Emergency Operations); have systems to manage mass fatalities per defined roles within the jurisdiction (Fatality Management); define, monitor, and share electronically elements of essential information to support a common operating picture (Information Sharing); have mechanisms to provide appropriate care, including providing 20% bed availability above the daily census within 4 h of a disaster (Medical Surge); have systems to preserve healthcare systems operations and protect coalition member employees (Responder Safety and Health); and have plans to manage volunteers supporting a public health or medical incident (Volunteer Management) (HHS, 2012, pp. 214–215).

Collaborative Opportunities

Historically, there has been, and continues to be, strong interest and rationale for emergency management, public health, and private sector healthcare to strengthen their partnerships. One recent example was the 1995 formation of Metropolitan Medical Strike Teams in the National Capital Region, a joint effort between the Washington Metropolitan Council of Governments and U.S. Public Health Service. These strike teams became the model for the MMRS in 2001 (Medicom, n.d.). In 2002, two physicians from George Washington University researched mass casualty incident response and suggested a systematic, functional management approach, stating: "Medical and Health emergency management is identical to all other studied components of emergency management in that effective development of response capacity requires a 'bottoms-up approach' and promoting interdisciplinary coordination and integration at the local level as 'most important'" (Barbera and MacIntyre, 2002, p. 94). This was followed in 2004 with the HHS releasing the *MSCC Handbook* that further outlined a tiered, multidisciplinary response model. In 2005, the passage of the Pandemic and All Hazards Preparedness Act Public Law No. 109–417 (PAHPA) paved the way for more interest in collaborative emergency management, public health, and private sector healthcare relationships. PAHPA effected changes in the FY 2006 and following HHS, PHEP, and HPP cooperative agreements. It also supported funding allocation for the Healthcare Facilities Partnership grant, a special, short-term competitive grant for research and development of collaborative demonstration projects.

During the same time period, the local- through federal-level responses to Hurricane Katrina provided evidence that the health-related emergency planning and medical surge response contained large gaps. Researchers, policy advisors, and private enterprises did not miss the importance of leveraging HSGP, HPP, and PHEP funding and assets to mitigate these gaps. In 2007, the Federal Emergency Management Agency commissioned a work group to define the basic principles of emergency management. The result was the confirmation of eight principles. The three that were most supportive of collaborative partnerships were integration, collaboration, and coordination. In other words, emergency managers ensure unity of effort within levels of government and within communities, create and sustain relationships that promote team building, encourage trust, facilitate communication, and achieve a common purpose by synchronizing stakeholder activities (International Association of Emergency Managers, 2007). Consulting firms published articles about the role of jurisdictional emergency management in community health preparedness. One stated that emergency managers possessed the leadership and communications skills, subject matter expertise, and access to resources to allow them to become the facilitators of integrated partnerships (JVR Health Readiness, 2008).

In 2009, HHS/ASPR issued the *National Security Strategy of the United States of America,* reinforcing the idea that strong partnerships must exist between health systems and the emergency services system composed of emergency management, law enforcement, fire services, and emergency medical services (HHS/ASPR, 2009, p. 5). In 2012, the Homeland Security Policy Institute at George Washington University reexamined the issue and concluded that although there had been many advancements and accomplishments during the previous 10 years, "The legacy missions of public health and emergency management must be synchronized ... to be effective. To achieve this ... stakeholders must develop a comprehensive and integrated approach," and recommended institutionalizing that approach across governmental levels for both routine and catastrophic events (Barishansky et al., 2012, p. 2). Stronger collaboration between public health and emergency management communities should be a "top priority" (Barishansky et al., 2012, p. 11). Another recommendation, the ability to share information and resources before and during an event, is in concert with the Emergency Operations Coordination and Information Sharing Capabilities (capabilities 3 and 6) of the public health and healthcare capabilities (HHS/CDC, 2011 and HHS/ASPR, 2012), as well as with the National Preparedness Goal Core Capabilities of Operational Coordination, Information and Intelligence Sharing, and Situational Assessment (DHS, 2011), and the third strategic object of the National Health Security Strategy, Ensure Situational Awareness (HHS/ASPR, 2009, p. 10).

Conclusion

Maintaining safe and secure infrastructures, resources, homeland, and communities is a critical joint responsibility of the government, the private sector, and individuals. Directives in the form of Presidential Policy Directives, legislation, national security strategies, and grants and cooperative agreements from the U.S. Departments Homeland Security and Health and Human Services validate the importance of collaboration. Much has been accomplished since 2001 but much remains to be done, particularly pertaining to building and sustaining strong, integrated systems that include emergency management, public health departments, and private sector healthcare. In many regions and jurisdictions, these partnerships have been long established, functional, and tested. In areas where the integration is not as mature, the FY-2012 HPP and PHEP cooperative agreements provide a ready opportunity for the disciplines to reach out to each other. It is likely that health departments of states, major metropolitan areas, U.S. territories, and the freely associated states that are the recipients of HPP-PHEP funds will initiate contact with private sector healthcare and emergency management agencies, as awardees are required to establish those relationships to accomplish the program goals by June 30, 2017. However, emergency management professionals should not hesitate to make the first move, whether to advance existing partnerships or to implement new ones. By leveraging relationships, financial and material resources, and expertise, whole-of-community preparedness, response, and recovery from incidents that impact health will benefit, and all partners should realize benefits that cannot be attained in unilateral endeavors.

Table 7.2 Appendix: Comparison of Healthcare and Public Health Capabilities

Healthcare/Healthcare Organizations (HC/HCO) Capabilities and Functions	*Similar Public Health Capabilities and Functions*	*Other Public Health Capabilities and Functions*
1 HC System Preparedness	**1 Community Preparedness**	**4 Emergency Public Information and Warning**
F 1—Develop, refine, sustain coalitions	F 1—Determine risks to health of jurisdiction	F 1—Activate emergency public information system
F 2—Coordinate planning to prepare for disaster	F 2—Build community partnerships to support health preparedness	F 2—Determine need for JIC
F 3—Identify and prioritize essential HC assets and services	F 3—Engage with community organizations to foster public health, medical, and mental or behavioral health social networks	F 3—Establish and participate in information system operations
F 4—ID gaps in HC preparedness and ID resources to mitigate	F 4—Coordinate training or guidance to ensure community engagement in preparedness efforts	F 4—Establish avenues for public interaction and information exchange
F 5—Coordinate training to develop necessary response skills	**2 Community Recovery**	F 5—Issue public information alerts, warnings and notifications
F 6—Improve response through coordinated exercises/evaluation	F 1—Identify and monitor public health, medical, and mental/ behavioral health system recovery needs	**7 Mass Care**

(Continued)

Table 7.2 (continued) Appendix: Comparison of Healthcare and Public Health Capabilities

Healthcare/Healthcare Organizations (HC/HCO) Capabilities and Functions	*Similar Public Health Capabilities and Functions*	*Other Public Health Capabilities and Functions*
F 7—Coordinate planning for at-risk and special medical needs	F 2—Coordinate community public health, medical and mental/behavioral health system recovery operations	F 1—Determine public health role in mass care operations
2 HC System Recovery	F 3—Implement corrective actions to mitigate damages from future incidents	F 2—Determine mass care needs of impacted population
F 1—Develop recovery process for the HC delivery system	**3 Emergency Operations Coordination**	F 3—Coordinate public health, medical and mental/behavioral health services
F 2—Assist HC organizations to implement COOP	F 1—Conduct preliminary assessment to determine need for public activation	F 4—Monitor mass care population health
3 Emergency Operations Coordination	F 2—Activate public health emergency operations	**8 Medical Countermeasure Dispensing**
F 1—HC multiagency representation and coordination with emergency operations	F 3—Develop incident response strategy	F 1—Identify and initiate medical countermeasure dispensing strategies
F 2—Assist and notify stakeholders of HC delivery status	F 4—Manage and sustain public health response	F 2—Receive medical countermeasures
F 3—Support HC response efforts through coordination of resources	F 5—Demobilize and evaluate public health emergency operations	F 3—Activate dispensing modalities
F 4—Demobilize and evaluate HC operations	**5 Fatality Management**	F 4—Dispense medical countermeasures to identified population

Table 7.2 (continued) Appendix: Comparison of Healthcare and Public Health Capabilities

Healthcare/Healthcare Organizations (HC/HCO) Capabilities and Functions	*Similar Public Health Capabilities and Functions*	*Other Public Health Capabilities and Functions*
5 Fatality Management	F 1—Determine role for public health in fatality management	F 5—Report adverse events
F 1—Coordinate surge of deaths and remains at HCOs with community fatality management	F 2—Activate public health fatality management operations	**9 Medical Material Management and Dispensing**
F 2—Coordinate surge of concerned citizens with family assist agencies	F 3—Assist in collection and dissemination of antemortem data	F 1—Direct and activate medical management and distribution
F 3—Mental/behavioral health at HCO level	F 4—Participate in survivor mental/ behavioral health services	F 2—Acquire medical material
6 Information Sharing	F 5—Participate in fatality processing and storage operations	F 3—Maintain updated inventory management and reporting system
F 1—Provide HC situational awareness that contributes to incident command	**6 Information Sharing**	F 4—Establish and maintain security
F 2—Develop, refine, and sustain interoperational communication system	F 1—Identify stakeholders to be incorporated into information flow	F 5—Distribute medical material
10 Medical Surge	F 2—Identify and develop rules and data elements for sharing	F 6—Recover medical material and demobilize distribution operations
F 1—Coalitions assist coordination of HCO response during surge incidents	F 3—Exchange information to determine a common operating picture	**11 Nonpharmaceutical Interventions**

(*Continued*)

Table 7.2 (continued) Appendix: Comparison of Healthcare and Public Health Capabilities

Healthcare/Healthcare Organizations (HC/HCO) Capabilities and Functions	*Similar Public Health Capabilities and Functions*	*Other Public Health Capabilities and Functions*
F 2—Coordinate surge operations with EMS operations	**10 Medical Surge**	F 1—Engage partners and identify factors that impact nonpharmaceutical interventions
F 3—Assist HCOs with surge capacity and capability	F 1—Assess nature and scope of incident	F 2—Determine nonpharmaceutical interventions
F 4—Develop crisis standards of care	F 2—Support activation of medical surge	F 3—Implement nonpharmaceutical interventions
F 5—Assist HCOs regarding evacuation and shelter in place operations	F 3—Support jurisdictional medical surge operations	F 4—Monitor nonpharmaceutical interventions
14 Responder Safety and Health	F 4—Support demobilization of medical surge operations	**12 Public Health Lab Testing**
F 1—Assist HCOs with more pharmaceuticals to protect HC workers	**14 Responder Safety and Health**	F 1—Manage lab activities
F 2—Assist HCOs with access to more PPE for workers during response	F 1—Identify responder safety and health risks	F 2—Perform sample management
15 Volunteer Management	F 2—Identify safety and personal protective needs	F 3—Conduct testing and analysis for routine and surge capacity
F 1—Participate in planning to determine need for volunteers in HCOs	F 3—Coordinate with partners to facilitate risk-specific safety and health training	F 4—Support public health investigations
F 2—Volunteer notification for HC response needs	F 4—Monitor responder safety and health actions	F 5—Report results

Table 7.2 (continued) Appendix: Comparison of Healthcare and Public Health Capabilities

Healthcare/Healthcare Organizations (HC/HCO) Capabilities and Functions	*Similar Public Health Capabilities and Functions*	*Other Public Health Capabilities and Functions*
F 3—Organization and assignment of volunteers	**15 Volunteer Management**	**13 Public Health Surveillance and Epidemiology Investigation**
F 4—Coordinate demobilization of volunteers	F 1—Coordinate volunteers	F 1—Conduct public health surveillance and detection
	F 2—Notify volunteers	F 2—Conduct public health and epidemiology investigations
	F 3—Organize, assemble, and dispatch volunteers	F 3—Recommend, monitor, and analyze mitigation actions
	F 4—Demobilize volunteers	F 4—Improve public health surveillance and epidemiology investigation systems

References

Barbera, J. and Macintyre, A. 2002. *Medical and Health Incident Management (MaHIM) System: A Comprehensive Functional System Description for Mass Casualty Medical and Health Incident Management*. Washington, DC: Institute for Crisis, Disaster, and Risk Management, George Washington University. Supported by a grant from the Alfred P. Sloan Foundation. Retrieved from http://www.gwu.edu/ ~ icdrm/publications/MaHIM%20V2%20final%20report%20sec%202.pdf

Barishansky, R., Bourne, M., Darnell, D., Kadlec, R., Kaniewski, D., Paczkowski, J., Roman, D., and Thiel, A. June 7, 2012. *Public Health and Emergency Management: Challenges and Opportunities* [Monograph]. Washington, DC: Homeland Security Policy Institute, George Washington University. Retrieved from http://www.gwumc.edu/hspi/policy/report_phem.pdf

International Association of Emergency Managers. September 11, 2007. *Principles of Emergency Management* [Monograph]. Retrieved from http://www.iaem.com/publications/documents/EMPrinciples091107.pdf

JVR Health Readiness. 2008. *Resources: The Emergency Manager's Role in Healthcare and Public Health Readiness.* Retrieved from http://www.jvrhr.com/The-Emergency-Manager.php

Medicom. (n.d.). *History of MMST/MMRS.* Retrieved from ftp://mediccom.org/ndmsinfo/ndmsconfmmrsoverview.pdf

U.S. Department of Health and Human Services. 2007. *Medical Surge Capacity and Capability: A Management System for Integrating Medical and Health Resources During Large-scale Emergencies* (2nd ed.). Retrieved from http://www.phe.gov/Preparedness/planning/mscc/handbook/chapter4/Pages/default.aspx

U.S. Department of Health and Human Services. April 2012. *Hospital Preparedness Program (HPP) and Public Health Emergency Preparedness (PHEP) Cooperative Agreements Announcement.* Retrieved from http://www.grants.gov/search/announce.do;jsessionid=6cpTQhhLy275jBLQGjxVHQvtzQKB6qkJ3PFGKjNx47GNGjyFdsdl!-1502138492

U.S. Department of Health and Human Services, Assistant Secretary for Preparedness and Response (ASPR). *Funding for Hospital Preparedness Program.* Retrieved from http://www.phe.gov/Preparedness/planning/hpp/Pages/funding.aspx

U.S. Department of Health and Human Services, Assistant Secretary for Preparedness and Response (ASPR), Office of Preparedness and Emergency Operations (OPEO), Division of National Healthcare Preparedness Programs (DNHPP). 2007. *Announcement of Availability of Funds for Healthcare Facilities Partnership Program.* Retrieved from http://www.phe.gov/Preparedness/planning/hpp/Documents/2007partnershipguidance.pdf

U.S. Department of Health and Human Services, Assistant Secretary for Preparedness and Response. 2009. *National Health Security Strategy of the United States of America.* Retrieved from http://www.phe.gov/Preparedness/planning/authority/nhss/strategy/Documents/nhss-final.pdf

U.S. Department of Health and Human Services, Assistant Secretary for Preparedness and Response. 2012. *Healthcare Preparedness Capabilities: National Guidance for Healthcare System Preparedness.* Retrieved from http://www.phe.gov/preparedness/planning/hpp/reports/documents/capabilities.pdf

U.S. Department of Health and Human Services, Centers for Disease Control and Prevention. 2011. *Public Health Preparedness Capabilities: National Standards for State and Local Planning.* Retrieved from http://www.cdc.gov/phpr/capabilities/DSLR_capabilities_July.pdf

U.S. Department of Health and Human Services, Centers for Medicare and Medicaid Services [CMS]. 2012. *Funding for Public Health Emergency Preparedness.* Retrieved from http://www.cdc.gov/phpr/coopagreement.htm

U.S. Department of Homeland Security, Federal Emergency Management Agency. 2011. *Crosswalk of Target Capabilities to Core Capabilities.* Retrieved from http://www.fema.gov/pdf/prepared/crosswalk.pdf

U.S. Department of Homeland Security, Federal Emergency Management Agency. May 2009. *FEMA GPD Grant Program Accomplishments Report: Summary of Initial Findings (FY03-07)* [Monograph]. Retrieved from http://www.fema.gov/pdf/government/grant/GPD_grant_accomplish_report.pdf

Chapter 8

Hospital Management and Disaster Planning

Robert Muller

Introduction

It is critical that hospitals be prepared for any and all disaster situations. While this might sound like an oxymoron, this is truly not the case. While hospitals are generally prepared for the everyday community emergencies of dealing with trauma and medical events, the majority are really not well prepared for any significant major disaster of any nature.[1] Many hospitals have chosen to channel their limited funding in many directions other than emergency preparedness.

We have seen in Hurricanes Katrina, Rita, Gustav, Isaac, and now Sandy that hospitals are just not prepared to handle the internal situational problems that develop and can be anticipated, just from lack of planning and mitigation.[2] The most common problems seem to arise from infrastructure failures and will be addressed in a later section.

The reason for this is the fact that disaster preparation requires a significant amount of planning, preparation, and budgeting along with a very significant commitment from the administration in terms of mitigation. Hospitals desire total productiveness from every available square foot of floor space. Hospitals no longer work on high profit margins and therefore must maximize every aspect to enhance their productivity.

Disaster preparedness mitigation does not necessarily fall within this realm, and only until recently in the past several years has this become a requirement of The Joint Commission (TJC) to present a plan of emergency management and disaster planning for periodic review at the time of inspections. Since the world-changing events of September 11, 2001, and the advent of mitigation grants from the CDC for hospital preparedness (HRSA grants), along with the TJC requirements, hospitals have been able to adequately prepare, with proper leadership and expertise, for disaster preparedness, from weather and natural events, to the reality of bioterrorism.

Hospitals should and must have both a short- and long-term plan for emergency management and the plans should be adequately budgeted on an annual basis with long-term future goals for a completely self-sufficient operational system. Hospitals *must* have a dynamic leader, educated and trained in emergency management, who is capable of organizing a program, conducting education and training for the facility, and assuming the role of Incident Commander (IC) in times of crisis. This individual may be the hospital president, or chief executive officer (CEO), but is probably better served by an individual specifically trained in these aspects and having the administrator as the general overseer rather than the IC. While this concept may not fit the traditional administrators' ego process, it is definitely what is best for the hospital, unless specific criteria are met. Hospitals must conform to the National Incident Management System (NIMS) concept and train all their key personnel in the structure and relationship of the Hospital Incident Command System (HEICS).[3,4]

Hospital *must* plan, prepare, train, and exercise in emergency preparedness and disaster management to be able to present themselves to their communities as "disaster ready" to handle any potential threats to the population.[5]

Hospital Inspection and Analysis

One should begin the evaluation process by a complete review of all current related emergency and disaster plans and protocols. Carefully read over and make notes of areas of possible deficiencies and other aspects of the plan that may involve further investigation. For example, the plan may state the use of an internal radio system based on a repeater within the building. Investigate how the system operates, the condition of the batteries, if the *repeater is on the emergency power system*, should there be a power failure, and check if the system has a current valid license with the Federal Communications Commission. Evaluate the entire process and play the devils' advocate.

Closely evaluate the communication system, both internal and external, in every detail and evaluate for and include all possible redundancies should there be a system failure. Look at how staff will be notified in case of a disaster—telephone lines may be destroyed, and beepers will not work. Are there *alternate plans* for notification or protocols in place for the staff to automatically report for duty?

The most common infrastructure failure during the past hurricanes and weather-related disasters, including the most recent hurricane Sandy, is the failure to plan and mitigate for power failures. One of the most important and most expensive of plans is for emergency generators, cross-over switching, and refueling.[6] This same process should be carried out in every aspect of the plan, protocols, hospital departments, administration, and external factors. At no time should there be a "that won't happen attitude." Plan that it will happen and mitigate.

Upon completion of the analysis, rewrite the plans and protocols incorporating all the new found information and mitigating factors. Compose a checklist for future use for all items that need periodic follow-up. Reanalyze the process at least annually, preferably every 6 month. Make all the necessary changes as situations change and or new situations develop, such as alterations in the physical plant layout, administrative or departmental personnel changes, or contractual changes, just to name a few.

Joint Commission

"Environment of care (EC) standard EC.1.4 requires hospital, ambulatory care, behavioral health, home care, and the environment of care. Standard EC.2.4 requires these organizations to *implement* the emergency management plan. Standard EC.2.9.1 requires them to *execute* the plan by conducting emergency management drills.

Although not required by Joint Commission of Accreditation Hospital Organization (JCAHO) standards, it would be prudent for other types of healthcare organizations to plan for disasters given today's environment." The requirements of standards EC.1.4 and EC.2.4 vary among the accreditation programs. Refer to the TJC accreditation manual to identify what specifically is required of your organization.

As part of the four phases of emergency management activities, organizations may be required to

- Identify and implement processes to conduct a hazard vulnerability analysis.
- Establish, in coordination with the community emergency planning, priorities among the potential emergencies identified in the analysis.
- Identify procedures to mitigate, prepare for, respond to, and recover from the priority emergencies.
- Define and integrate the organization's role with that of community emergency response agencies, including identifying a community command structure.
- Define a common command structure (for all hazards) within the organization that links to the community structure.
- Cooperate with healthcare organizations within a contiguous geographic area to establish a process to share information about the essential elements of a command structure and emergency control centers, such as
  - Names, roles, and phone numbers of individuals in the command structure.
  - Resources and assets to share or pool in a community emergency response.
  - Timely identification and location of names of patients and deceased individuals following an emergency.
- Describe how, when, and by whom the plan is activated.
- Identify which personnel are responsible for which activities during emergencies.
- Initiate response and recovery phases.

- Notify external authorities of emergencies, including possible community emergencies such as evidence of a bioterrorist attack.
- Notify care providers and other personnel when emergency procedures are initiated.
- Identify personnel during emergencies.
- Assign available personnel to cover all necessary positions under emergency conditions.
- Manage patient/resident care activities, staff and family support activities, logistics of critical supplies, security, and communication with news media during emergencies.
- Evacuate the facility if necessary.
- Establish an alternate care site(s) that can meet patients' clinical needs.
- Transfer patients/residents or transport and track patients/residents, staff, and equipment to an alternate care site as needed.
- Communicate with the alternate care site.
- Reestablish and/or continue operations following the disaster; provide alternate means of meeting essential building utility needs to provide continuous service.
- Establish backup internal and external communication systems.
- Identify radioactive, biological, or chemical isolation and decontamination sites (ambulatory care and hospitals only).
- Clarify alternate responsibilities of personnel, including to whom they report during a disaster, in a command structure consistent with that used by agencies in the local community.
- Establish an orientation and education program for staff, including licensed independent practitioners, who participate in implementing the plan.
- Monitor ongoing performance in drills and real emergencies, and determine how an annual evaluation of the plan's objective, scope, performance, and effectiveness will occur.

The emergency management standards also address the needs of your staff. In an actual emergency, staff will naturally be concerned for the safety and well-being of their colleagues and loved ones. Accordingly, standard EC.1.4 calls for the management of staff activities—including housing, transportation, and incident stress debriefing—and staff and family support activities.[7]

These points will be addressed in the following section.

Mitigation, Preparation, and Planning

The hospital administration Emergency Management Team faces numerous situations and events both internal and external in coping with disasters. The preparation and scenarios differ from the daily events in hospital operations and will be dealt with herein on an individual basis. This is not meant to be all inclusive but to serve as a primer to help develop some thought for emergency planning.

Types of Disaster for Planning Purposes

External:

- The most common disaster events are those that can take place from natural community situations and include the following: building collapse, explosions, floods, food-borne exposure, accidents involving mass casualties, that is, bus, plane, ships, barges or trains; radiological accidents, chemical leaks, drug lab accidents, school shooting, and environmental exposure accidents to cover the more common ones. The natural disasters include earthquakes, thunderstorms, and tornadoes, hurricanes, and snow storms.
- The preparedness factor is time dependent and some can be instantaneously occurring while others will provide hours to days of warning, but each must be adequately prepared for in terms of spatial event occurrence.
- It is essential to perform a complete vulnerability analysis in conjunction with the county or state office of emergency preparedness for not only the community but also the surrounding counties or even neighboring state if so warranted.

Internal:

- Internal disasters can take place within the framework of the hospital walls as well as its campus; some can be life threatening, others can be of minor as well as major threats to the facility, staff, and patients.
- Some good examples of internal hospital disaster planning would include things such as bomb scares, infant abductions, hostage situations, infectious and radiological exposures, as well as mechanical disaster as sewerage backup and contamination, electrical failure, fires, explosions, and HazMat scenarios.

Hospital Preparation

Hospital preparation for disasters must start with internal planning of the facility, its capabilities, as well as its limitations. A SWOT (strengths, weaknesses, opportunities, and threats) analysis is an excellent way to proceed to properly determine the types and degree of planning for the facility. Not all facilities will need to have the same level of planning and preparation based upon this analysis, their size, component staff, location, and ability to respond to community or regional needs. The degree of planning should be coordinated with hospital administration, hospital association, emergency management, department of health and hospitals (local and state), local as well as regional businesses, fire, and police.

Staffing

The key to the continuity of operations (COOP) of a hospital is its entire staff: administrative, medical, nursing, paramedical, technical, and support; without a complement of staff, a facility cannot function, especially in an emergency situation with the possibility of dramatically increased volume and the probability of failure of various systems as well as multiple system failures. There must be planning for multiple shifts as well as staff embedding if the ingress and egress of staff are inhibited due to various reasons. There should be advanced planning as to how staff will be paid in a crisis situation for overtime, embedding time, hazardous exposure, and so on.

Staff should be trained and kept with some mode of communications for emergency contact, be it the simplicity of telephonic to the use of notification trees and any and all in between. The staff should be advised to always keep a "Go Bag," which should be kept in the trunk of their car or within the facility in employee lockers. (Some facilities supply such an item to their employees or give them as presents during special events (hospital week) or holidays).

A Go Bag is always a *must*. This should be a medium-size soft bag or suitcase, and should be part of any emergency responder's armamentarium. It should contain all personal items such as underwear (buy cheap or disposables), medications (2–4 week supply), hygiene products (plastic bag for item disposal), extra clothes, uniforms, shoes, and coat (in case of evacuation or displacement), cash ($100 minimum—in small as well as mixed bills—$1, $5, $10, and $20—as many business may not be able to make change following a disaster and credit cards may not work if there is no electricity or satellite antennas have been damaged); cell phone and charger, extra batteries as needed for any devices and any other items that may be appropriate to the individuals personal situation.

The staff should be assured that every effort will be made for them to maintain contact with the families and will be discussed further in the communications section. The staff should have a family plan as well as plans for their pets should a mandated evacuation become necessary.

The hospital should maintain within the facility a cache of "cash" for staff payroll advances should an event become extended and the community facilities be handicapped in the use of credit due to lack of power, computerization, satellite reception, and so on. The hospital likewise may need to have cash on hand, unless preagreements have been made within the community for credit should *recovery* materials and supplies need be purchased.

Cash on hand to supplement or provide payday loans to employees can be a very nice but very dangerous as well as a laborious process. If large amounts of cash are held on premise, strict confidence must be kept until the last moment when this service is announced to the staff, and significant security must be provided due to the nature of this situation and the vulnerability it presents.

The human resources department should be well versed as to their role in preparedness staffing, and should maintain rosters of ancillary personnel as well as alternate personnel contacts. The possibility for the use of volunteers to staff some positions vacated by absenteeism should also be a consideration in preparedness. The ability of personnel to act in multiple roles should be entertained and personnel skill levels should be maintained as well as cross-matched as alternates for preidentified positions.

Personnel Pool

Create a plan and a location for a personnel pool. During and after a disaster, there are many volunteers that show up at the hospital facility, that is, staff families, friends, relatives, and outsiders from the community as well as those that come to your aid from out of state, including movie stars and celebrities (John Travolta, Nick Lashea, Brad Pitt, Matthew McConaughey—to name a few). See volunteer agreement.

Make advanced assignments where anticipated staffing needs may require additional personnel.

Good examples may be *security* (internal and external) and may utilize those with previous law enforcement or military experience, or federal agents who may volunteer; *traffic and parking* direction, helicopter landing zone security (use those with prior military experience preferably in aviation)—a *heliops manager*; outside *animal control*; outside *logistics* (to procure barricades, tents, etc.); staff *babysitting* (teenage family members are excellent for this function); if there is only one access road into the facility staff, a *checkpoint* to help alleviate unnecessary traffic and movement around the facility and secure the area that may be utilized a landing zones (LZ) for helicopters; others may include *PBX operators, food preparation/disbursing assistants; receiving and supplies; and so on.*

Postdisaster donations in the form of food and clothing can rapidly become problematic; think about assigning at least one person to be in charge of *donation management.* (There is also an excellent Federal Emergency Management Agency (FEMA) Emergency Management Institute (EMI) course on the subject.)

A pet manager is essential. This person needs to be responsible to register and keep track of all pets within the facility area. See policy.

A *hotel manager* is necessary to assign rooms to those that will be embedded within the facility. This position is best served by a hospital employee rather than a volunteer. This position requires someone that is not only totally familiar with all the areas of the facility that can be utilized for embedding, but also the personnel that require embedding; designated areas may also be defined as to staff, volunteers, and so on.

Identify and list potential needs in the facility's disaster plan where they can be easily referenced and assignments made rapidly as qualified volunteers become available.

All volunteer personnel should have some form of ID. Generic IDs can be made and kept for pool use or the use of color wrist banding will likewise serve the same function.

Staff Education and Training

"When emergencies strike, the casualties are not the only victims. Staff members in your organization will be on the frontline responding to any emergency your community faces. The best way a healthcare organization can prepare staff members to meet the challenges of an emergency is to educate them in all aspects of working and protecting themselves in such a situation."[10]

Training is essential. The absence of training is the absence of preparation. Poorly prepared and educated staff are of little value to the organization as well as a potential liability and could even pose a threat and danger to themselves as well as their coworkers. Training may take place in many forms, including self-education through numerous websites, inservice lectures, community lectures, and guest lectures. Many subject matter experts (SME) are available for training and resources may include the local as well as state agencies, communications, information technology, hazard materials, and posttraumatic stress experts to name only a few.

The FEMA EMI Independent Study (IS) program (http://training.fema.gov/IS/crslist.asp) has an excellent array of online courses available. Upon successful completion, a certificate may be printed. (A permanent record of training is also kept at EMI under the individual social security number for future reference.) All hospital personnel should understand the basics of the Incident Command System (ICS) and thus course number IS 100.HCb should be mandatory. Additional ICS courses, 200, 300, and 400, are significantly more advanced and should be required of those with incident command assignments on both the command staff and general staff levels. The IC should have at least the 300 level of training and someone capable of in-service training should obtain the 400 level to be qualified as an instructor. Most IS courses carry the availability of college credits, and in addition, may be used by many certifying boards and organizations for continuing education credits.[5] In addition, the staff should be trained on IS 700 and IS 900. All staff member training records must be kept either as hard copy or on computer or preferably both, for inspection by TJC should there be an inquiry.

Training cost should be a part of all hospitals' budget. However, emergency management, incident command, technology, hazardous materials, chemical, radiological, explosive, and bioterrorism training cost may be reimbursable through various grants, including the present Health Resources Services Administration (HRSA) grant from the Center for Disease Control (CDC) through state Departments of Health and Hospitals. This was created after the events of 9/11 and has been available up until the present time (2012); its perpetuity is unknown for future planning.

Remember that *there is no such thing as overtraining* and various members of the staff should be trained for multiple positions should they be needed; a good example would be to have someone trained in helicopter landing safety procedures besides their regular duties, and the possibility exists that this scenario may become a reality. Remember, just because a hospital does not have an official Federal Aviation Administration (FAA), heliport does not preclude a nearby landing in a field, vacant property, or the street should a dire emergency situation exist. (This is for an example only, and many other similar dual situations may likewise have more value.)

Communications

The most vital link in any hospital emergency scenario is the communications network that has been previously established. It is essential to perform an analysis of not only the hospital needs, but also likewise those of the community. A communications worksheet should be completed and shared with affiliated hospitals and agencies. Worksheets ("Communications Sheet") are provided for your benefit at the end of this chapter.[8]

When performing the analysis, several areas should be evaluated:

Internal communications:

- The hospital's internal needs are dependent upon the hospital size as well as the physical plant layout as to hospital design.
- Multiple types of communications should be utilized to create a redundancy of systems in case of one or more system failures.
- Internal communications should likewise be linked to the emergency power system at all locations of base radios as well as radio chargers systems. Internal repeater systems should likewise be on the emergency power system.
- Internal communications can be linked to internal phone systems such as a Spectra link type system, internal as well as external pagers, cellular, direct connect systems such as Nextel and Verizon, and direct connect system such as portable handheld radios on VHF, UHF, CB, or business band frequencies; *or* preferably a multiple combination of two or three of these systems.
- Before deciding which type of system is best for the hospital, make sure the system is adequately tested throughout the hospital in all possible locations, as some facilities have shielded blocking effects that do not allow certain frequencies to penetrate in various areas due to types of construction, and so on.
- As a general rule, the least reliable system for internal use is the CB (citizen band) radio and generally only effective with limited range externally on flat topography.

External:

- External communications links are dependent on how sophisticated the hospital desires its system to be relevant to the community. It is imperative that an external link be established utilizing a HAM (amateur) radio system, configured to the design of the hospital. This system requires a federally licensed operator and this person can either be trained through a training course or be preferably someone within the community that does this as a full-time hobby.
- These people use their equipment on a regular basis, stay up on all the recent sites and developments and know and have multiple contacts within their

brotherhood. Check with a local HAM club to see if anyone is interested in performing this service during an emergency for your hospital. Once this person is established, he or she should then become a part of the hospital emergency preparedness planning team and should be included in all preparedness meetings.

- This person should be allowed access to the facility and to evaluate the facility for the best location to place the equipment for optimum operation. Antennae placement and distance are critical to the performance of the system. Arrangements may be made with the individual to bring his/her own equipment and utilize preestablished antennae plug-type connectors, or for the hospital to purchase a system for a permanent instillation. The advantage of the latter is that the system is ready to go upon arrival of an licensed operator without any set up delay, which could take up to several hours.
- Vital community links may be established with law enforcement (state, county, and local), fire, EMS, county and city officials, and surrounding entities as backup plans may dictate. Their frequencies and specific type of equipment may be purchased with their permission and may be able to be purchased through the agency and their municipal discount structure with their vendors. This equipment must be kept secured at all times and not be available for any form of casual use or for monitoring purposes and this assurance must be conveyed to the agency for their assurance that this equipment will not and cannot fall into misuse.
- There is *no* such thing as having too much communications redundancy. Communications is the backbone to disaster survival and contingency planning.

Antenna Systems

Communications systems are totally dependent upon their antennae systems. Communication range and clarity of communications are dependent upon antennae connections, types of cable, cable lengths, connectors, and antennae height and location placement. All forms of communications are dependent upon antennas from television to HAM, with the exception of the Internet, which is dependent upon telephone or satellite links.

Owing to the extreme significance of the antenna to communications, it is imperative that the antenna locations be carefully planned. Antenna should be placed with some form of protection such as wind-resistant webbing, and so on or even better, placed in such a way that the system can be folded or cranked down and then be brought back up after the event. This of course only holds for those weather-related events than can be anticipated. In addition, in case of damage to the system in unanticipated events, backup systems should be available with rapid connectors that can be easily replaced.

Remember, the most sophisticated communications system is totally worthless without the proper working antenna system.

Command Structure

Most hospitals are organized with a CEO or chief administrative officer (CAO) as the lead executive in charge of the hospital. He/she may have various assistants depending on the size of the hospital and may include a chief operating officer (COO), chief nursing officer (CNO), human resource officer, and chief financial officer (CFO). Many hospitals have a president and vice presidents in charge of various operational aspects. In smaller hospitals, these functions may be combined.

Hospitals also have department heads or chiefs that are responsible for the operational aspects of sections of the hospital, and, again depending upon size, may likewise be combined. Examples include dietary, engineering, laboratory, x-ray, cardiology, and so on.

It is imperative that no matter what the particular hospital structure is called or what the composition, it functions as an entire team, and be structurally organized to be carried out in the most cost-efficient manner. It is imperative that the team be trained in the HEICS and that each member of the team knows and understands their role and function within this structure.[8]

Many hospitals choose to have the CEO or president assume the role of the IC within this structure and this may be a perfect fit for many hospitals. One suggestion would be if the hospital has a designated emergency management person, that this person assume the role of the IC and the CEO be placed as a general overseer of the entire hospital function. In this way, he/she is able to roam the hospital for potential problems and not be held within the constraints of the command center.

Whoever is the IC, he/she must have the training and experience to carry out this very demanding and highly sensitive function and should possess at least ICS 200 level education.[4,8] The designated emergency manager is usually better trained in this function and thus better prepared to handle the scenarios, and has the know-how for community liaison. This is not a time for egos to dominate and this should all be predetermined by a command structure established in the planning phase long before any incident should arise. Remember the old saying, "the time to learn to dance is not fifteen minutes before the party."

Emergency Operations Center

The Emergency Operations Center (EOC) is a vital part of the success of any emergency operation. It should first be organized properly, well in advance of any potential event; second, it should be activated at a time appropriate to the event, either immediately with an event such as a tornado or several hours to days in advance of a hurricane or pandemic like event, and so on. The effectiveness of a HEICS is enhanced when personnel have access to an EOC where they can coordinate information and resources to navigate the hospital and medical staff through the crisis.

The EOC should be designed with several factors in mind. First, it should be readily accessible and located so that it is physically protected to sustain its function. If possible, it should be located in the central building core and possibly on the second floor, depending on the local flood plain. Also, an alternate site should be identified, should for some reason the primary site sustained damage or requires being moved to another location. It should be located where there is limited and controllable access into the area so that security is easily provided with a limited number of persons being allowed access. Its location should be marked with signage and designated to the internal staff. The controlled access into the EOC can be easily accomplished utilizing brightly color-coded wrist bands, similar but different from the patient registration bands.

The design and content of the EOC are very important to its operational capability. The EOC should be designed so that an area can be easily converted to an EOC. Having a full-time designated EOC space is truly a luxury that most facilities cannot afford due to premiums of space limitations. Identify an area that is used daily for some other function, that is, board room, lecture room, auditorium, physical therapy, and so on. The chosen space should be large enough to allow ease of movement without being overcrowded and thus limiting its functional capability. If possible, a good rule of thumb would be to allow a minimum of 10 square feet per person while 25 would be even better and 50 is the ideal. Each facility will have its limitations, but the important thing to remember is to maintain *functionality*.

Smaller adjoining rooms can also be utilized for other functions but all incident command functions should be coordinated through the main EOC. For example, the finance section could utilize an adjoining room whereas operations and logistics need to definitely be all coordinated within one common area to function properly within the ICS.

The EOC should be set up to contain a large erasable board on one wall and if possible a local as well as a state map on another wall. A large diagram of the hospital and surrounding property is also desirable. The room should be prewired to allow for the plug-in of multiple telephone lines and or have the capability of having the lines drop down out of the ceiling if removable tiles are available. A large board room-like table or several folding tables should be placed so that they are in the center of the room and can be easily walked around and allow for the use of all four walls, so that all four walls are easily viewed by the command staff. The room should have access to high-speed computer connections and/or Wi-Fi if available and a wireless card (as an alternate system), to allow additional mobility and eliminate additional cords running in and around the room. There are also additional satellite Internet services available as alternate backup systems that can be hardwired to a roof top antenna. Communications is the heart of an EOC. The room should be outfitted with one or more flat screen televisions to keep track of local stations as well as national feeds such as The Weather Channel, MSNBC, CSPAN, and so on and therefore cable or satellite access is highly recommended.

Internal communications within the hospital should be by internal sources, that is, handheld VHF or UHF portable radios, direct connect cell phone systems (Nextel and Verizon), Spectra link like systems, and so on.

External communications is a vital link with the hospital to the community, law enforcement, fire, county EOC, state EOC, vendors, and other area or regional hospital facilities.

This communications link should be VHF (HEAR), UHF, 700/800 MHz, and amateur (HAM) radio that is utilized within the facility's operating area and region as well as state hospital linkages. A local communications specialist used as a consultant will be able to assess your communications needs. It is also important to seek approval from the area law enforcement, fire, EMS, and various EOCs to access their frequencies as well as being assigned a call number or title, that is, unit # 2300 or just this is XYZ Hospital calling, and so on. Also, they may choose to limit only one channel for hospital communications use; if this is the case, make sure that in time of emergency someone will be monitoring this channel in their EOC should the hospital be calling them. Many agencies require monthly radio checks to assure operability and monitoring of their systems.

The stress of working in an active EOC can be equated to the stress encountered by air traffic controllers. It is ever present and constantly demanding. There is little or no options for failure and in the EOC situation, there will always be some form of minor or major conflict with other coworkers. When assigning roles for positions in the incident command structure, the most important factor is the person's qualifications and expertise; the second most important is personality—can the person handle the job as well as the stress that goes along with its responsibility?

To determine these two factors, it is best to create exercises that test both the competency and the human factor reactions to situations. It is imperative that the personnel be given adequate breaks, kept well hydrated, and provided proper foods, and that sleep time be determined with alternating changes of shifts during the height of an event, unless a general and command staff briefing is planned to be done at one time. If a facility has the luxury of having a staff psychiatrist or psychologist available during an event, it is worthwhile to have him/her attend staff meetings at least once a day to pick up on any potential staff escalating stress levels.

Generally, the longer and more complex the event, the more likely will there be a command or general staff conflict. This is also more common when these staff members are not well trained and do not feel competent in their job positions, as well as those who feel they are better qualified to assume an another position level, that is, a general staff person desiring a command staff position.

It is of great value to bring this topic up at the very first staff meeting so that the subject is breached and that all members are well aware of a potential situation and what to expect in case a situation does arise. Upon completion of the event during the demobilization process, a qualified Critical Incident Stress Management (CISM) professional should meet with the team and discuss the signs and symptoms of posttraumatic stress syndrome and provide information as to how, when, and where they may obtain future help and counseling should it be needed.

PIO

A working relationship with the media will help during an incident. Establish a media contact list with after-business hours contact information. Keep media aware of all preparedness/awareness campaigns. Invite local media to the EOC, or other areas prior to any incident or planned event to show them the location and to answer questions about how information will be disseminated during an incident or planned event. Positive media relationships built during normal day-to-day activities will be valuable during emergency situations. Do not wait until an incident to make first introductions to the media.

A PIO should be involved in all phases of exercises:

- Planning
- Development
- Participation
- Evaluation

It is also recommended to involve the local media in drills and exercises, and encourage them to role play during those drills and exercises in addition to covering the incident.

PIOs should be able to gather, verify, prepare, coordinate, and disseminate information to all audiences, including those with disabilities, special needs, or language requirements. It is important to have materials translated into common non-English-area languages and to utilize other formats such as Braille, large print, audio, and so on. Contacts should be established to translate emergency information.

Know the local media; there may be specialized newspapers or radio stations in the community that reach specific audiences. These audiences may need to be targeted during awareness/preparedness campaigns. Review and update all contact lists (e.g., media, PIO, and other agencies) every 6 months and include basic information such as telephone numbers (e.g., office, home, and cell), fax numbers, e-mail addresses, and websites.[8]

Go Kits for the PIO

It is important for the PIO to have the tools and resources available for utilization during an incident. Although this is not a complete list, a Go Kit might include

- Office supplies such as pens, paper, stapler, tape, and so on
- Laptop computer and portable printer with an alternate power source(s), including accessories (e.g., memory stick, CDs, and mouse)
- Maps
- Television, radio, and/or broadcast recording equipment
- Cell phones/personal data assistants (PDAs)

- Fax machine
- Agency letterhead
- PIO and other emergency operations plans
- Camera
- Contact lists
- Battery-powered radio
- Prescripted messages and template releases

Prior to an incident or planned event, establish agreements with businesses or agencies that can assist with the operations; examples would be contracts with translation services or individuals; printing companies (publish brochures, fact sheets, or other emergency documents); and telephone companies to install hardline telephones.

Decontamination Team

Each hospital should have some type of decontamination plan in association with their community they serve and the risks and hazards of their particular service area. Each hospital does not need to have a full decontamination setup and team but should have some sort of Decon plan no matter how small their risks and vulnerability.

In many small communities with volunteer fire departments, the hospital may be the primary Decon facility, whereas in other communities with larger departments and full professional training, the primary role for Decon will be at the scene rather than at the hospital. In this case, the hospital must have total confidence on the primary Decon process or should make plans to have a secondary Decon protocol. In any event, mitigation and planning for this scenario should generate some type of protocol appropriate for the situations that may present themselves.

Unless there are otherwise well-trained personnel, level 3 suits should be sufficient for hospital decontamination. Personnel should be trained as well as drilled on a regular basis in the use of suits as well as the techniques of decontamination. If this is not feasible due to any reason, then this job should be turned over to another group within the public service or military community. (An excellent training program exists for hospital personnel at the Department of Homeland Security (DHS) facility in Anniston, Alabama and information may be obtained by contacting your state Office of Emergency Preparedness.)

Improper training and lack of exercising can lead to breaches, resulting in compromising the safety of hospital personnel.

The hospital should also have an adequate budget to provide for the necessary equipment as well as maintenance of the equipment. This can be expensive on an annual basis, but it is necessary to maintain competence and safety.

Members chosen from hospital personnel for a Decon team should not include any professionals (MD, RN, respiratory, etc.) as these people will be needed for triage and treatment; the use of ancillary personnel best serves this function. A good section to recruit from would be engineering, business office personnel, housekeeping, and food services. These people may likewise enjoy being part of a somewhat "clinical" experience that is beyond their normal routines.

Facility Management

Agreements

The hospital may consider agreements for embedding of *key holding* personnel from community stores that may be of benefit to the hospital, that is, grocery stores, Wal Mart, Home Depot, and so on; another excellent PR as well as community service is a verbal offer to the Weather Channel or other news media for their visiting personnel.

Local or state law enforcement may appreciate an offer to house their personnel during the height of a disaster or to give relief in shifts to personnel that have been brought in for the disaster from outside the area, as these people live a distance and usually many times lodging facilities are unavailable. This of course should be preplanned and on as a "as space available" basis. Ancillary campus facilities such as ambulatory surgery suites may be a good place for some additional housing space.

Community agreements with local outpatient facilities (than may not be operable for services, but may be an excellent space for embedding) should be considered.

Nursing home/rehabilitation facilities—Agreements with these facilities to house their patients should be discouraged due to the need for acute care space following a disaster, as well as the demand for increased staffing due to the acuity of this category of patients. If agreements are made with these facilities, then agreements should be made with them to also send their own staffing to accompany their patients.

Parking—Agreements with local governmental agencies to house their vehicles.

Transfer agreement—Agreements should be made with local, regional, and distant hospital facilities to send or accept patients in transfer in the event of an impending disaster or an internal event such as a fire, explosion, contamination, and so on. Agreements should be made appropriately regarding sending or not to send staffing with the patients, or likewise to provide staffing services to distant shelters or off-site facilities or campuses (clinics), as it will become necessary to keep sufficient staffing within the facility to deal with the possible influx of acute care patients postdisaster in the affected area.

Food Preparation

Hospital facilities should have alternate plans for the preparation of food as well as alternate food supply sources.

In the event of loss of power or gas, alternate preparation sites should be preidentified. The utilization of local school cafeterias, restaurants, or local catering companies that have adequate facilities may be alternatives. On a limited basis, the use of such things as propane tank grills and or barbecue pits can be utilized. However, this is not feasible for the large-scale preparation of foods.

Additional food sources should be identified within the community, especially areas that may be evacuated and or abandoned, including large grocery stores, Wal Mart, Sam's Club, and restaurants than may store large supplies of meat and food stores, that is, Outback, Applebee's, and so on.

Preplanning within the community should involve these potential suppliers and their general as well as their store managers. Complete listings of these persons contact numbers and alternate means of communication should be established, that is, text messaging, contact via a headquarters phone number or email address, and so on. A good community activist program is essential for all the local sources that may be beneficial to the hospital facility.

Hospital Identification System

All staff hospital personnel should have some form of picture identification card, including administrative staff, physicians, and board members; frequent hospital contractors may likewise be given a different color ID with or without a picture but redeemable upon demand as well as a predetermined expiration date. After the expiration date on ID cards, a different color card should be used. If it is not hospital policy for ID cards to be returned upon completion or termination of employment, then a 6-month revalidation process must take place. This can easily be done by placing a validation sticker on the front of the ID card with a different color (that can be easily recognized from a distance of 6–10 feet) with an expiration date printed on it. This keeps all ID cards current without the expense of remaking new ID cards for all employees.

Generic ID cards (without a picture) can be printed in advance and kept for use by volunteers, family members, and law enforcement or security personnel. They should be of a particular color or design with large distinct lettering on them, such as "LEO" (law enforcement officer), "V" (volunteer), "G" (guest), or "FM" (family member) for some examples. See the example in the addendum. Colored wrist bands may be used as an alternate, particularly for guests and visitors, with specific color designated for different floors or areas of the hospital, that is, blue—first floor, orange—emergency room, and so on.

Everyone who is in the hospital during a disaster and particular any type of national security event should have some form of ID visible to internal hospital security. Anyone who does not have some form of authorized ID should be reported, and stopped and questioned by *security.*

All plain-clothed LEOs (local, state, or federal) should have their badge displayed on their shirt, hanging on a lanyard around their neck, or displayed on their belt facing forward so they are readily visible. All experienced law enforcement will have some form of badge or ID in their possession at all times. If someone should present claiming they are an LEO and do not possess any form of valid ID and or badge should be thoroughly questioned before assuming their authority.

Par Values

Par values should be established for the operation and internal sustainment of the facility for a minimum of 96 h longer, if the facility is located in an environmentally vulnerable zone with high probabilities of occurring events (i.e., a hospital located along coastal areas, earthquake areas, etc.). The values should likewise take into account the patient *volume* of the hospital as well as its *service area* and *support* from other surrounding facilities in the community and or region (i.e., the only hospital in a coastal community versus a hospital 30 miles inland with two other hospitals in the service area; a hospital in a community that is totally evacuated versus one in a community of a sustainable population of 50,000, etc.).

All critical supplies should be maintained by computerized inventory as well as manual inventory and should be assessed preceding seasonal environmental events as well as National Security Alert elevations to a higher level. Each hospital's par value supply will be dependent on its locale, accessibility, supply chain, and physical facility limitations on storage. Par values should include all critical areas of the hospital from dietary to pharmacy.

Always plan for the worse and do not assume that your supply chain will always be easily accessible. Multiple contracts should be in place to establish a redundancy of the supply chain, in case of failure for whatever reason, of one or more of the primary suppliers.

Parking

Garages—those facilities that have parking in garage facilities should have some form of security in terms of either manned security or automated gated security to both enter and exit the facility. In a disaster, parking garages may become inundated with locals who desire to secure their vehicles from flying debris and or rising water, and so on, especially if it is a free parking garage. Also, garage space may be reserved for police and fire agencies as well as public works to safeguard their vehicles for immediate utilization postdisaster. Prearranged agreements with the various agencies are highly recommended. Also, hospital or clinic entry points from a garage facility should have some form of planned security or closed to prevent unauthorized or undetected entry.

Parking lots should likewise be reserved and secured for facility personnel by utilizing locking barricades or some other form of permanent delineation as to the designated site. Metal pipes should be secured in concrete well in advance of any type of event and utilized to chain the barricades to prevent them from becoming projectiles. These areas should be inspected for any loose objects or objects that may become projectiles. In situations of wind-related disasters, it may be suggested in advance for employees to purchase some form of locking car covers to help preserve their vehicles from flying debris. (A small pebble at 100 miles per hour can do severe damage to windows and painted surfaces.)

Pharmacy

The pharmacy should be well stocked for potential anticipated disasters. In most states, following 9/11, federal grants called HRSA (Health Resources and Service Administration) have been allocated to hospitals via the state departments of health and hospitals, for the purpose of stockpiling for disasters. A percentage of each of the grants over the past years has been for stockpiling in the pharmacy.

In anticipation of seasonal events such as hurricane season (June 1–November 30) or other weather-related events, increased inventories of such drugs as tetanus and snake bite antivenom should be evaluated, depending on the hazard vulnerability of the facility as well as the surrounding locale. Agreements by memorandums of understanding (MOU) should be executed with other hospital pharmacies in the community as well as the regional area for mutual aide exchange of supplies.

While in most disaster scenarios, pharmaceuticals may be readily available through pharmaceutical warehousing and delivery services; this may be precluded in environmental disasters such as earthquakes, tornadoes, and hurricanes due to the inability to communicate or access the facility. The Strategic National Stockpile is a cache of drugs that is housed in secret locations strategically located around the United States for delivery to requesting areas within 12 h. This should be requested through the state regional coordinators of the department of health and hospitals or directly through the county/parish Emergency Operations Command Center; (should additional drugs, vaccines, or supplies become critically necessary and unavailable from the contracted supply sources or previous MOU facilities).

If not already established by the healthcare facility, the name and contact information of the regional coordinator or state contact should be made and recorded as a vital part of the facility emergency management plan. Some states have special disaster laws that take effect upon orders of the Governor during declared disasters. Some of these laws allow hospital pharmacies to dispense quantities of medications to patients either via a clinic attached to the hospital or to hospital patients being discharged. (Many hospitals do not have a license to dispense medications on an outpatient basis.) Some of the laws also allow for community pharmacies to dispense/refill patient medications without physician authorization for a period of 1 month based on the situation and patient need. Check with your local state pharmacy board regarding dispensing authority associated with disasters so that you may be well informed as well as properly inventoried.

Surge Capacity

Hospitals must be prepared for surge as the result of any type of mass casualty event, including those related to pandemics or bioterrorism. The following is recommended from the CDC to help prepare a hospital for surge.[9]

- *Surge capacity:* Beds—calculated as 1 per 2000 population or 500 per 1 million populations of both adult and pediatric cases.

- *Surge capacity:* Isolation—regional hospitals should be able to support at least 10 adult and pediatric patients at a time in negative pressure isolation.
- *Surge capacity:* Personnel—1:4 to 1:6 (for 50 beds, there should be a minimum of 12 healthcare personnel).
- *Surge capacity:* Pharmaceutical caches—a "sufficient supply of antibiotics to provide prophylaxis for 3 days to all hospital personnel, emergency first responders, and their families, as well as the general community."[9] (The Strategic National Stockpile (SNS) contains enough antibiotics for 13 million people for a 60-day course.)
- *Surge capacity:* Decontamination—portable or fixed decontamination units sufficient to be able to decontaminate all patients and providers within 3 h from the onset of the event. Personal protective suits (PPE) sets should be calculated at eight per provider.
- *Surge capacity:* Behavioral—plans should be developed to deal with surge addressing various issues from quarantine to death of victims.
- *Surge capacity:* Trauma and burns—should be prepared to handle at least 50 severe trauma or burn victims per 1 million population with up to 30% of the trauma victims being burn patients. (For every 200 trauma victims, expect 50–60 acute burn injuries.)
- *Surge capacity:* Communications—secure and redundant communications system that ensures connectivity during a public health emergency between healthcare facilities and state and local departments, emergency medical services, emergency management agencies, public safety agencies, neighboring jurisdictions, and federal public health officials.
- *Hospital laboratories:* Hospitals must have hospital lab protocols for the rapid referral of clinical samples and associated information to appropriate labs operating with guidelines in CDC Focus Area C.

Forensics

From time to time, there are incidents that are of a forensic nature and require proper training and handling to be compliant in the transfer of potential evidence. The most common incidents in hospitals are victims with gunshot or knife/ice pick wounds with the cartridge or object readily visible, loose, or impaled. These objects should be treated as evidence and handled appropriately with the proper use of gloves and proper tagging, initialing, and bagging to be turned over to the requesting law enforcement agency.

Sexual assault victim's clothing should be collected and properly bagged according to predesigned protocols, and labeled and sealed with evidence tape in *paper bags* only (never use plastic bags). All evidence gathered should be *initialed* by the person who collected the evidence. *Initial* the waistband of clothing with a permanent marker, scrape your initials on the blade or handle of a knife, gun, bullet, and so on (An ice pick is excellent for this purpose). This will always answer the question asked in cases that go to court of "how do you know this is the same piece of evidence that you collected?" Answer? "By my initials marked on it."

The chain of evidence must be maintained and potential evidence should not be allowed to be laid on carts or tables, and so on. Once the evidence is collect, each person to whom the evidence is transferred must be required to sign an evidence transfer sheet to maintain the chain. *At no time and to no one should evidence be handed unless the proper transfer sheet is signed and maintained with the evidence, and a copy retained in the patient's hospital records (chart).* The original evidence transfer sheet should follow the evidence and is to be signed and given to law enforcement. The last signature that should appear on the evidence transfer sheet in the patient record should be that of the LEO accepting custody of the evidence; beyond that point, it becomes the LEO's responsibility to continue the chain. If for some reason the evidence is not able to be immediately transferred to a LEO, then it should be secured in a locked safe, closet, or narcotics locker with limited key access by staff members. At no time should it be "secured" in an employee's car or trunk.

Bioterrorist and hazardous materials events mandate special handling of victim's clothing not only from an evidence standpoint *but* also from a decontamination standpoint. Great care must be taken as well as proper training and drilling of staff required to deal with such incidents prior to their occurrence. The necessary supplies and PPE must be available to deal with such incidents. A standard operating procedure (SOP) must be written in conjunction with local law enforcement and fire officials as to who becomes responsible for victim's clothing for evidence purposes as well as proper disposal. This is a very specialized field and the necessary training must be obtained to not only protect the victim's privacy but also to protect the hospital staff. There are numerous other areas that cannot all be covered within a single chapter, but this will give you an idea of the width and breath of the material that needs to be planned for and mitigated.

Below is a sample of some of the type of checklists and charts that should be created on an individual basis for your facility, well in advance of any type of event. They should be created both in a computer-adaptable file and on paper and kept within a ring binder with plans, contracts, agreements, and so on. Also see the references for websites for additional information.

Hospital Disaster Checklist

Task	*Responsible*	*Time/Date Assigned*
Fuel Supply	John Smith	0800 6/7/12
Security	Frank Johnson	0900 6/7/12
Pet Manager	Sue Labadore	0915 6/7/12

Hospital Identification System

	Department/Manual: Hospital Wide, Administrative		
	Original Date: 1/31/01	Revised: 6-21-2007	Page Number 1
	Subject:		Prepared by Rob Muller, CEM
Number	**Hurricane Plan** **PET POLICY**		Approved: Jack Jones, CEO

Employees are encouraged to make advanced arrangements for the long term care of their pets outside of the hospital; with relatives, friends or their local veterinary clinic for boarding should disaster preparation become imminent.

In the event that this is totally impossible and it becomes a necessity to bring your pet to the hospital, NorthShore will provide a place for sheltering only. Each employee MUST provide care to his or her pet as outlined below.

1. You must check-in and register your pet upon entering the hospital.
2. You must have a 3" X 5" color picture of you and your pet, taken and printed prior to arrival, to be turned in at the time of your pet registration.
3. You must complete and sign the consent form agreement (to which your picture will be attached).
4. You must bring your pet in a secure plastic airline type approved cage so your pet can stand, sit and turn around.
5. You must bring proof of inoculations or current veterinarian issued tags.
6. You must provide all food, water (a supply may not be available depending on the type of emergency condition), blankets/toys and clean up supplies.
7. You must bring a collar and leash.
8. You must bring a supply of 30 zip ties with a box cutter, side cutting pliers or large scissors to cut them free from securing of the cage door when removing your pet for personal attention. The door must be re secured with zip ties each time your pet is removed.
9. You must be responsible for pet visitation, feeding, walking, relief and cleaning of your pet, cage and area, if applicable.
10. There will be NO PETS allowed into the building for any reason.
11. Employees are encouraged to prepare now and to secure all necessary papers, supplies, cage, etc.

Communications Sheet

Name of Hospital/Facility ___________ Parish/County _________ Facility
Tier Level I II (circle one)
Form Completed By ________________ Licensed Bed Capacity ____________
Please Complete the Communications Available at your Hospital/Facility: (Circle Choices)
HEAR Radio: Base Handheld Power (watts output) ____________ Roof Antennae: Y N (circle)

VHF Frequency: Transmit ____________ Receive ________________ PL Code ____________

UHF Frequency: Transmit ____________ Receive ________________ PL Code ____________

Repeater: Internal External Radio Range ___________ mile(s)

800 mhz Analog Digital TYPE: Motorola Erickson Johnson Icom Other _______________
HAM Frequency: Transmit __________ Receive __________ Antennae Type: External (roof) Portable
Nextel/Verizon Direct Connect Code # ____________________________
Emergency Telephone Numbers: (include proper area codes)
Command Center Telephone Numbers
Administration ____
Emergency Room ______________________________________
Hospital Emergency Lines _______________________________
Emergency Cellular Lines _______________________________
Satellite Phone Number ________________________________

FAX Telephone Numbers
Command Center _______________________________________
Administration ____
Emergency Room ______________________________________
Does your Facility have Telephone Priority System (TPS)? Y N
Does your Facility participate in the Satellite Emergency Communications System? Y N

DISASTER SERVICES VOLUNTEER RELEASE AND WAIVER OF LIABILITY

I desire to volunteer my time and efforts in a nonclinical capacity during the current disaster. I do so freely and voluntarily with the understanding that I am not an employee of XYZ Hospital. As such, I am performing these services without the promise, expectation, or receipt of compensation or other benefits and will conduct myself in a reasonable manner and remain personally responsible for my actions.

I acknowledge there are inherent and other risks associated with the volunteer activities that I perform. These risks may include, but are not limited to, the possibility of personal injury or property damage or loss arising from my volunteer activities. I understand and acknowledge the potentially dangerous environment I may be exposed to and freely assume all of the risks associated with my volunteer activities.

I, for myself (or for my child if under 18) and anyone entitled to act on my or my child's behalf forever discharge, waive and release XYZ Hospital Health System, its members, affiliates, subsidiaries, officers, directors, employees, agents, successors, and assigns from all claims of damage, loss, or liability of any kind or nature arising out of my participation as a volunteer for XYZ.

I further grant and convey to XYX all right, title, and interest in any and all photographic images, video, or audio records made by XYZ Hospital or its agents during my volunteer work at XYZ Hospital, including, but not limited to, any royalties, proceeds, or other benefits derived from such photographs, images, or recordings.

Volunteer: ______________________________________ Date: _________________
Printed Name: ___________________________________
Parent/Guardian: _________________________________ Date: _________________
(if volunteer is under 18)
Printed Name: ___________________________________

Bibliography

1. Hospitals Ill Prepared for Mass Casualty Events, Amednews.com (last accessed October 9, 2013).
2. Toner, E., Waldhorn, R., Franco, C. *Descriptive Framework for Healthcare Preparedness for Mass Casualty Events.* Prepared by the Center for Biosecurity of UPMC for the U.S. Department of Health and Human Services under Contract No. HHSO100200700038C. 2008.
3. Federal Emergency Management Agency Emergency Management Institute Independent Study Courses: www.fema.gov/emi (last accessed September 15, 2012)
    - IS 29 PIO Awareness
    - IS 100HCb Introduction to ICS for Hospitals
    - IS 200HCa Advanced ICS for Hospitals
    - IS 700 National Incident Management System (NIMS)
    - IS 704 NIMS Communications & Information Management
4. NIMS National Response Plan (NRP): www.dhs.gov (last accessed September 15, 2012).
5. Emergency Management Principles and Practices for Healthcare Systems. The Institute for Crisis, Disaster, and Risk Management (ICDRM) at the George Washington University (GWU); for the Veterans Health Administration (VHA)/US Department of Veterans Affairs (VA). Washington, D.C., June 2006. Available at http://www1.va.gov/emshg/.
6. United States Guard Incident Management Handbook; U.S. Coast Guard COMDTPUB P3120.17, April 11, 2001; pp. 8–12, available at: http://www.uscg.mil/hq/nsfweb/download/IMH/IMH-2001.pdf.
7. Joint Commission Prospective, Vol. 21, Number 12, December 2001.
8. Healthcare Resources and Service Administration (HRSA/DHHS). *Emergency System for Advance Registration of Volunteer Health Professionals (ESAR-VHP),* available at: http://www.hrsa.gov/bioterrorism/esarvhp/guidelines/(last accessed September 15, 2012), NIMS National Response Plan (NRP): www.dhs.gov (last accessed September 15, 2012).
9. CDC Mass Casualties |Preparedness and Response to a Mass Casualty Event Resulting from Terrorist Use of Explosives. http://www.bt.cdc.gov/masscasualties/terrorist_explosives.asp (last accessed September 15, 2012).
10. Federal Emergency Management Agency, FEMA, *State and Local Guide (SLG) 101: Guide for All-Hazard Emergency Operations Planning*: http://www.fema.gov/plan/gaheop.shtm (last accessed September 15, 2012).
11. National Bioterrorism Hospital Preparedness Program, FY 2004, CFDA # 93.003. United States Department of Health and Human Services Special Programs Bureau.
12. NFPA 1600, Standard on Disaster/Emergency Management and Business Continuity Programs 2004 Edition: http://www.nfpa.org/assets/files/pdf/nfpa1600.pdf (last accessed September 15, 2012).
13. Shaw, G. and Harrald J. The Identification of the Core Competencies Required of Executive Level Business Crisis and Continuity Managers. *The Electronic Journal of Homeland Security and Emergency Management.* Berkeley Electronic Press, January 2004, p. 6.

Chapter 9

Hospital Business Continuity

Linda Reissman and Jacob Neufeld

Historical Prospective

Early Hospital Preparedness

Prior to the events of 9/11 and Hurricane Katrina, traditional hospital preparedness planning was focused on mass casualties and emergency department readiness. Very little planning focused on scenarios in which the hospital was directly impacted by the disaster, impeding the ability to continue care. Internal and community-level plannings were limited in scope, and concepts such as the hospital incident command system (HICS) were not employed.

Since 2002, federal support for hospital emergency planning has been provided through the Department of Health and Human Services Hospital Preparedness Program (DHHS-HPP) grants. Early funding was primarily directed at bolstering hospital capabilities to manage the threat of terrorism and pandemic. Then, in 2005, the devastating consequences of Hurricane Katrina explicitly bore out the need for improved hospital and public health preparedness. The traditional hospital disaster plan was not effective for catastrophic events that impacted the hospital, the community, and the critical infrastructure supporting both.

After Hurricane Katrina, a number of governmental reports issued assessments and recommendations for methods for ensuring the continuity of operations, the coordination of communications, and standards for strengthening mitigation measures. A report by the ANSI Homeland Security Standards Panel concluded that organizations of all sizes, in both the public and private sectors, would be well served by complying with NFPA 1600 and using it as a guideline for their disaster/ emergency management and business continuity programs.

Subsequently, there were dramatic changes made to the scope of hospital preparedness. As an example, in 2008, the Joint Commission revised its standard on a hospital's ability to sustain in place. Previously, the standard had been that a hospital emergency operations plan (EOP) must address an assessment of its ability to sustain in place for 72 h. After the revision, this was increased to 96 h. This standard does not require the hospital to sustain itself for 96 h, but rather have a process to assess capabilities during escalating incidents to determine its ability to do so, and then plan and respond accordingly. In addition, the hospital EOP must address the following critical areas:

- Communications
- Resources and assets
- Safety and security
- Staff responsibilities
- Utilities
- Patient clinical and support activities

Joint Commission Accreditation Requirements: Emergency Management (EM)

EM.02.02.01 As part of its Emergency Operations Plan, the hospital prepares for how it will communicate during emergencies.

EM.02.02.03 As part of its Emergency Operations Plan, the hospital prepares for how it will manage resources and assets during emergencies.

EM.02.02.05 As part of its Emergency Operations Plan, the hospital prepares for how it will manage security and safety during an emergency.

EM.02.02.07 As part of its Emergency Operations Plan, the hospital prepares for how it will manage staff during an emergency.

EM.02.02.09 As part of its Emergency Operations Plan, the hospital prepares for how it will manage utilities during an emergency.

EM.02.02.11 As part of its Emergency Operations Plan, the hospital prepares for how it will manage patients during emergencies.

Although the term "business continuity" is not specifically termed in the Joint Commission Emergency Management standards, the elements of performance do address it in terms of continuity of operations, patient care, and other critical functions. The Joint Commission's standard do address disaster recovery in IM.2.30, and requires that the hospital develop and maintain a disaster recovery plan identifying the most critical information needs for patient care, treatment, and services and impacts if information systems are disrupted. The plan should also outline alternative means for processing and providing data recovery.

Joint Commission Accreditation Requirements: Information Management (IM)

IM.01.01.03 The hospital plans for continuity of its information management processes.

INTRODUCTION TO STANDARD IM.01.01.03

The primary goal of the information continuity process is to return the hospital to normal operations as soon as possible with minimal downtime and no data loss. The hospital needs to be prepared for events that could impact the availability of data and information regardless of whether interruptions are scheduled or unscheduled (due to a local or regional disaster or an emergency). Interruptions to an organization's information system can potentially have a devastating impact on its ability to deliver quality care and continue its business operations. Planning for emergency situations helps the organization mitigate the impact that interruptions, emergencies, and disasters have on its ability to manage information. The hospital plans for interruptions by training staff on alternative procedures, testing the hospital's Emergency Operations Plan, conducting regularly scheduled data backups, and testing data restoration procedures.

The healthcare industry has become increasingly dependent on technology to computerize almost all aspects of patient care, ranging from automated provider order entry to sophisticated image-guided surgery systems, billing and accounts management, and center-wide communications. The unavailability of clinical and communication applications is untenable to clinicians left without the critical electronic resources they rely on to provide care. New generations of technology-oriented clinicians will have little, if any, experience with handwritten documentation and downtime processes. Therefore, a hospital's information technology (IT) infrastructure must be available to provide an uninterrupted flow of data and is a critical component in the delivery of care, the well-being of its patients, and its reputation.

Therefore, it is critical for healthcare organizations to conduct ongoing assessments, disaster recovery, and clinical and business continuity planning.

Why Business Continuity?

In addition to the continuity of information management, hospitals must ask themselves, "What if facilities or their contents became inaccessible" or "What if 30% of your workforce is unavailable (e.g., pandemic)?" or "What if key suppliers and partners can't fulfill obligations?" or "What alternate process do we use if IT systems are unavailable?" These are the hard-line questions a healthcare institution needs to ask itself. Some of these events may not even be precipitated by a disaster per se. Today's healthcare environment relies on IT to support clinical and business operations. In addition to protecting human capital, institutions must consider the protection of research assets and other intellectual property that are irreplaceable. The critical infrastructure provided by external or municipal sources can fail despite all efforts of an organization to insulate itself from such events through mitigation practices.

A business continuity program can considerably improve clinical and business recovery (CBR) capabilities for healthcare organizations. Knowing the upstream and downstream impacts of critical systems provides leadership with key information required to make decisions quickly that will minimize disruptions and impacts to life safety and lessen financial losses.

Business continuity, emergency planning, disaster recovery, and IT security management are mutually exclusive planning processes, yet interdependent. A comprehensive business continuity program incorporates all three planning platforms and is directed at maintaining an organization's ability to sustain an acceptable level of care despite the emergency. Although there are similarities in terminologies, processes, and response steps, it is important to articulate the differences between emergency management, disaster recovery, and business continuity when attempting to implement a program and presenting a proposal to leadership. Statements such as "but we have an emergency plan or we have a disaster recovery plan" are commonplace when introducing the concept of business continuity to an organization. The organization's leadership must believe there is a value rather than another project burden that has no "end game." That being said, cultural acceptance of a business continuity program is integral to its success, so hospital leadership and the organization as a whole must understand the value of such a program. You will need to provide clear and concise definitions and descriptions to differentiate the three as well as include the value of such a program in day-to-day operations. The following definitions clarify these differences:

- *Emergency planning*: The procedures and steps taken immediately after an event. It is a component of a comprehensive business continuity program.
- *Disaster recovery*: The steps to restore some functions to resume some level of services (IT), including security management.

- *Continuity*: Restoration planning to get the organization back to where it was before an interruption.

Hospital EOP is usually designed under one of two different models: (1) all-hazards planning or (2) event-specific planning. Business continuity planning is "class specific" and applies mitigation and response strategies that can be applied for any vulnerability, regardless of the event that caused it. These can be classified as the 5 S's of clinical and business continuity planning. They include

1. Space
2. Stuff
3. Staff
4. Systems
5. Services

A clinical and business continuity program focuses on the ability of a healthcare institution to fully identify and sustain core operations, and functions, and services through a formal recovery process. Unlike short-term crisis planning, continuity planning addresses *extended* disruptions. This will require specific and detailed planning and recovery strategies not within the purview of traditional emergency management. They include

- Maintaining/rapidly restoring critical systems/services
- Developing the capability to self-sustain when resources are scarce
- Providing a safe level of care, treatment, and services, and/or developing strategies to prioritize or safely discontinue care, treatment, and services
- Identifying, documenting, and exercising practical work around procedures

A business continuity program and its focus on the 5 S class vulnerabilities dovetail with the Joint Commission's six critical areas for hospital emergency management. Essentially, a business continuity plan is a road map for recovery. However, during the implementation of a continuity program and plan development, it will become apparent that equally important to establishing documented and codified processes is the planning process itself. The process provides a platform for information sharing among individuals who may have never had the opportunity to interact and discuss interdependencies and decisions that could impact their respective department and/or functions. Participants will soon see the everyday operational value, and the process flow of day-to-day business usually improves as a result. At the departmental planning level, specific informational components must be collected.

These can be done via a survey process manually or with the business continuity tool:

- Identify critical services and functions
- Gauge disruption impacts

- Identify the length of time until an impact becomes unacceptable (recovery time objective [RTO])
- Identify dependencies/interdependencies
- Codify/validate recovery strategies and tasks
- Contact information (staff, vendors, etc.)
- Identify needed recovery resources

Below is a high-level "business continuity program design checklist" that can be used to initially scope a clinical and business continuity program.

1. Scope business continuity/disaster recovery project.
2. Collect business continuity/disaster recovery data and information.
3. Complete business impact analysis (BIA).
4. Formulate business continuity/disaster recovery strategies.
5. Design and activate recovery and crisis management organization.
6. Document agreements and action plan to institutionalize business continuity and finalize pilot preparations.
7. Prepare a business recovery template using a business continuity planning tool or process.
8. Prepare pilot launch support material.
9. Execute a pilot and report on results.

What Is the Business Impact Analysis?

Business Impact Analysis

Through the enterprise BIA, you will communicate with specific senior leadership representatives (stakeholders) about scalable options of how CBR planning can be accomplished across the institution:

- What granularity of planning detail?
- Which functions and services must participate in the program?
- Which resources will be doing the work?
- In what time frame are initial CBR plans completed?
- At what cost (time and money) to the organization?

Here, you set your hospital's goals and lay out what the hospital will need to reach those goals. The work carried out here allows leaders to see the effect that scaling up goals will have on resources.

Through the CBR program design workshop, you get management to commit to a course of action that gets a sustainable CBR program underway by developing a practical, achievable implementation plan composed of

- Program scope—who and what are included in the CBR program
- Program operations—how the CBR program will be conducted
- Program rollout—when will the CBR program be deployed?

Stakeholder conversations during the enterprise BIA provide necessary insights into what is appropriate within your institution. You can now roll out a program in which it is inherent that reassessment and adjustments are made much like the cycle of continuous improvement.

Below is a detailed outline of specific tasks and processes associated with the full implementation of a business continuity program.

1. Engage leadership teams and employees with business continuity responsibility to conduct a self-assessment to establish or validate organizational business continuity objectives, develop a business continuity planning structure, and provide appropriate resource allocation.
2. Prepare for pilot deployment of an integrated business continuity program.
   a. Conduct an enterprise BIA.
      i. Engage executives to define critical impacts and "RTOs" guideline.
      ii. Review BIA results and engage leadership in developing an implementation plan utilizing enterprise standard methods and tools.
   b. Conduct a business continuity program design workshop.
      i. Assemble a business continuity program design team composed of a selected subset of participants from the business continuity self-assessment workshop.
      ii. Define who, what, when, and where of business continuity program deployment. Develop the draft charter for business continuity.
3. Develop a formal business continuity awareness program and delivery vehicle for new employees (as part of orientation) and for existing employees. Delivery vehicles can include websites (with online training modules), periodically scheduled training classes, and access to business continuity knowledge sources.
4. Use the results of the BIA and the agreed-upon business continuity program scope to confirm and authorize resource requirements for implementing and sustaining the business continuity program (budget, staff, and tools).
5. Develop, document, and distribute enterprise-level business continuity policies, standards, and procedures. Standards will cover such things as business continuity plan content, testing and maintenance requirements, and vendor and supplier expectations.
6. Establish a business continuity program governance office responsible for enforcing business continuity policies and standards, and assuring that business continuity goals will be achieved.
7. Develop a system of metrics (reviews, reports, and enforcement) to assure compliance in meeting organizational and enterprise-wide business continuity requirements, objectives, and goals. This includes business continuity-related regulatory requirements.
8. Consider including manager/employee performance relative to business continuity-related goals and objectives on their annual performance reviews.

9. Require that existing change control processes include checks for any changes that may have impacts on clinical or business, incident management, technology recovery, or security management-related recovery plans or strategies.
10. Utilize the emergency notification system to develop departmental- and enterprise-level business continuity calling trees for notifying employees, internal and external business partners, and service providers following a disruptive event.

Physical Risk Assessment Process

The scope of this assessment is to identify internal and external vulnerabilities, the availability and use of resources and controls to eliminate or mitigate risks, and plans and procedures to address emergency events. The following threat parameters provide the baseline for risk identification:

- *Acts of nature*: Earthquakes, rain, wind, ice, and so on that threaten facilities, systems, personnel, utilities, and physical operations, as well as hamper or deny access to the sites.
- *Hazardous conditions*: Fire, chemical and nuclear spills, biological events, structural instability, and so on that threaten facilities, systems, personnel, and operations. These may be the result of natural events, environmental control failures, human errors, and/or violent acts, as well as hamper or deny access to the sites.
- *Dependency failures*: Failure of a system or service outside the direct control of your organization that harms the system and/or affects its ability to perform. Examples include public utility failures, or the failure of a service or system controlled by an external company or public agency.
- *Environmental failures*: Failure or lack of a protective control that disrupts, harms, or exposes the system to harm or loss. Examples include lack of heating/air conditioning/cooling (HVAC) for cooling servers, which can be affected by loss of power and generator.
- *Public safety actions*: Actions taken by law enforcement, fire, regulatory, administrative, and/or other parties to safe-guard the community that may inadvertently result in harm to the organization or the system. Examples include mandating the complete shut-down of IT systems and generator to safeguard occupants during external or internal hazardous incidents.
- *Prior events*: An analysis of prior events that threaten facilities, systems, personnel, and operations, as well as activities to mitigate the effects of prior events.

Advantages of Using a Business Continuity Planning Tool

One of the most challenging steps in the development of a new clinical and business continuity plan or improving upon one that is already implemented is assessing the maturity of a business continuity management (BCM) program and is achieved by assessing the organization's current level of BCM capability by identifying gaps and risks and establishing a process improvement process. There are a number of tools and processes on the market that can do this. Two examples are the Gartner IT Score and the Virtual Corporation's Business Continuity Maturity Model® (BCMM®). The evaluations usually consist of rating scales with descriptive and detailed elements of assessment. In the below example, upon the completion of the assessment, the organization will achieve a level to determine their organizational resiliency and program maturity.

Level 1—Self-Governed

- *Business continuity planning attribute*: No formal planning; the business reacts to disruptive events.
- Business continuity has not been formally implemented anywhere within the organization. Few, if any, documented business continuity plans exist for any of the three business continuity disciplines. The organization reacts to disruptive events when they occur. The state of preparedness is low across the enterprise.

Level 2—Supported Self-Governed

- *Business continuity planning attribute*: Planning is limited to a few areas that prepare alone.
- A few functions or services, on their own, have developed and maintained business continuity plans within one or more of the three business continuity disciplines. There is little or no cooperation or coordination of planning activity between these groups. The state of preparedness may be moderate for participants, but remains low across the majority of the enterprise.

Level 3—Centrally Governed

- *Business continuity planning attribute*: Participating functions share methods and tools.
- Participating functions or services have instituted common business continuity practices, tools, and support resources in one or more of the business continuity disciplines. The interest in leveraging the work of these groups is being promoted as a business driver for launching an enterprise business continuity program. Some functions or services have achieved a high state of preparedness. However, as a whole, the enterprise is still largely ad hoc.

Level 4—Enterprise Awakening

- *Business continuity planning attribute*: All functions implement initial, self-interested plans.

- An enterprise business continuity program is deployed using standard methods and tools integrating all three business continuity disciplines, typically supported by a centralized support business continuity program office. These initial business continuity plans encompass most critical functions and services across the enterprise, although with little attention on protecting critical dependencies. Most business continuity plans are tested and updated routinely, as dictated by enterprise business continuity policy.

Level 5—Planned Growth

- *Business continuity planning attribute*: Most critical dependencies are incorporated into plans. All critical enterprise functions and services have developed and tested business continuity plans, including their critical internal and external dependencies. A communications and staff training program continuously measures business continuity planning competency. Audit reports no longer highlight business continuity shortcomings. Competitive advantages achieved from business continuity planning are highlighted in internal and external communications.

Level 6—Synergistic

- *Business continuity planning attribute*: Comprehensive plans are tied to changing business needs. Continuous process improvement keeps this organization at an appropriately high state of preparedness to stay current with the dynamic business environment. Over time, innovative policy, practices, processes, and technologies are piloted and incorporated into the business continuity program. Cross-functional business continuity capabilities are measured and codified.
- How will employees maintain open lines of communication?
- Where will employees go if the disruption involves an inability to use the existing office space?
- Fundamentally, how will employees be able to keep doing their jobs?
- How will the company's employees continue to be paid on a regular basis, with no interruptions?

Key Continuity Definitions

Business continuity: A formal planning process that minimizes or eliminates the impact of disruptive events on critical business operations, functions, and services.

Clinical and business recovery (CBR): The discipline of business continuity planning that provides advanced planning and preparation to help ensure the continuity of critical clinical and business functions in the event of a disaster. It includes identifying impacts to the business environment, implementing viable risk mitigation and recovery strategies, and developing business disaster preparedness plans.

Business impact analysis (BIA): A technique that identifies both tangible and intangible impacts on a business function, service, or program usually over time based on given criticalities. Provides senior management with information to devise a recovery strategy and prioritization.

Technology recovery (TR): The discipline of business continuity planning that provides advanced planning to help ensure the ability to recover and restore critical assets, including IT delivery systems, voice and data networks, business applications, and other critical information technology components within defined RTOs.

Incident management (IM): The discipline of business continuity planning that provides advanced planning to help ensure health and safety of people, limit environmental impacts, and protect company assets. IM includes emergency response, crisis management, and emergency operations.

Recovery point objective (RPO): The term used to identify where in the data stream a company or organization needs to recover lost information. Simply put, how much data can be lost if the primary data storage device is destroyed?

Recovery time objective (RTO): The term used to identify how quickly a company or organization must recover a disrupted business function, service, or program.

So, what should the hospital consider when embarking on a clinical and business continuity program and in what order. The following is an eight-step process that begins with a basic analysis of the present capabilities and an identification of known gaps in capabilities or infrastructure. Then, after the limitations and potential weaknesses in the system are identified, plans can be put in place to improve the issues, train staff, conduct exercises, and evaluate the results. This will lead to a high-quality approach to developing contingency and continuity plans.

1. The organization must commit to an enterprise-wide culture of continuity planning.
2. Know the capabilities and limitations of the institution's mission-critical systems. The first and most critical step is for all department heads and senior hospital management (including clinical leadership) to be aware of these capabilities and limitations. All too often, this understanding is limited to the hospital engineer or facility manager.
3. Ensure that a risk assessment is conducted for all mitigation projects and should be based on the hazard vulnerability analysis, as required by The Joint Commission (TJC). It should ensure that activities to mitigate effects from specified hazards are prioritized. It is also important to consider the impact that the surrounding community may have upon the hospital's critical infrastructure such as proximity to an industrial facility. Once the hospital

administration and clinical department heads are aware of the capabilities and gaps, contingency plans can be developed to minimize risks and sustain operations.

4. Develop mitigation plans and projects to enhance resiliency to ensure continued operations in times of outage. Examples include the capability to provide 100% backup generator power and alternate communications capabilities.
5. Develop contingency plans and redundancies; if mission-critical systems fail, these may include conservation for such things as water, fuel or food shortage, illumination devices in cases of power outages, and downtime procedures such as use of paper records when IT systems are impacted.
6. Assure that service providers are aware of mission-critical systems to prioritize the restoration. Do not assume that hospitals or healthcare agencies receive the highest priority restoration without a discussion and written agreements with providers.
7. Educate leadership, staff, and department heads. Providing emergency management and continuity training so that staff understand their roles and responsibilities so that the plan will be executed in an efficient and timely manner.
8. Exercise plans and the mission-critical systems to fail. Only then will gaps in planning and training be identified and process improvements can take place. Develop a formal process-improvement program that tracks gaps identified in exercises and real events. Update plans accordingly and conduct yearly reviews.

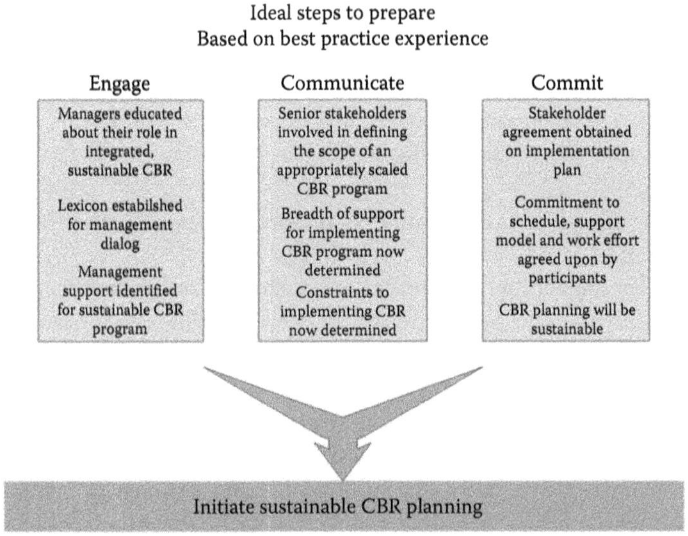

Chapter 10

Communications and Mass Casualty Events

Jeremiah W. Dunlap

Introduction

Even though the twenty-first century is in its early stage, it has already seen some of the greatest tragedies in U.S. history. Be it through the cruel violence of a single individual, well-organized terrorist attack, or the awesome power of Mother Nature herself, we have seen losses of human life on a shockingly large scale. Crisis events such as these, particularly those that claim a significant number of lives, do share common ground even though their cause is extremely different. In each case, the human reaction invariably includes the demand to know how these tragedies could occur or how they could have been mitigated. The answer is simple: Too often, there is a scarcity of communication among those responding to these emergencies.

This chapter explores the factual basis of this assertion to better equip those involved in emergency management (whether by choice or by circumstance) with a deeper understanding of effective emergency communications. Examining several, egregious communication failures will lay the foundation for a look at the components of a well-functioning emergency communications system; the key financial, organizational, and policy constraints will also merit attention, as they will impact the ability of an organization, municipality, state, or other jurisdiction to implement such a network.

Historical Look

No one questions how important being able to communicate is to daily life, particularly if one's survival may rest on the ability to summon assistance. When discussed out of context, any political leader, first responder, or agency head would appear to ascribe incredible value to interdepartmental, interagency communications. As history has proven, the need for a well-structured emergency communications system receives less attention when it is challenged by budgetary constraints or an organizational culture resistant to change. These are unfortunate priorities for any group charged with safeguarding public health and safety, and can lead to system vulnerabilities remaining undetected until disaster strikes.

Failure to Communicate

Tragedy at Virginia Polytechnic Institute

On April 16th, 2007, Virginia Tech student Seung-Hui Cho shot and killed 32 other students and faculty members, before ending his own life (Friedman, 2009, para. 14). In a matter of hours, an entire community was made to question its safety and security. Unfortunately, the school's mishandling of the situation only served to steepen these concerns. Seung-Hui Cho had already shot and killed two students, 2 h before entering Virginia Tech's Norris Hall and opening fire on its occupants. After school police rushed to and assessed the scene, school officials opted not to notify those on campus of what had transpired. Only after 2 h of discussion was an email sent to the students, mere minutes before Cho's rampage began in earnest (Figure 10.1).

Even then the email did not provide sufficient warning—it requested that the students remain observant, instead of instructing them to retreat to a safe location on or off-campus. Swift and appropriately informative communication by school officials or campus police might have saved lives that day; the lack of a single, timely notification is evidence of what can occur in the absence of a well-structured and rehearsed emergency communications plan (Barnett, 2007). Just as individual, organizations and institutions need to address emergency communications in an overarching emergency plan, so do governments at the municipal, state, and federal levels.

Terrorist Attacks of September 11, 2001

During the attacks on 9/11, the catastrophic damage done by the hijacked jetliners overwhelmed many of New York City's (NYC) emergency communication systems. The collapse of the World Trade Center's second tower destroyed an outpost of the New York Fire Department (FDNY), killing several among FDNY's leadership and wiping out the magnetic boards used to track deployed personnel. Off-duty

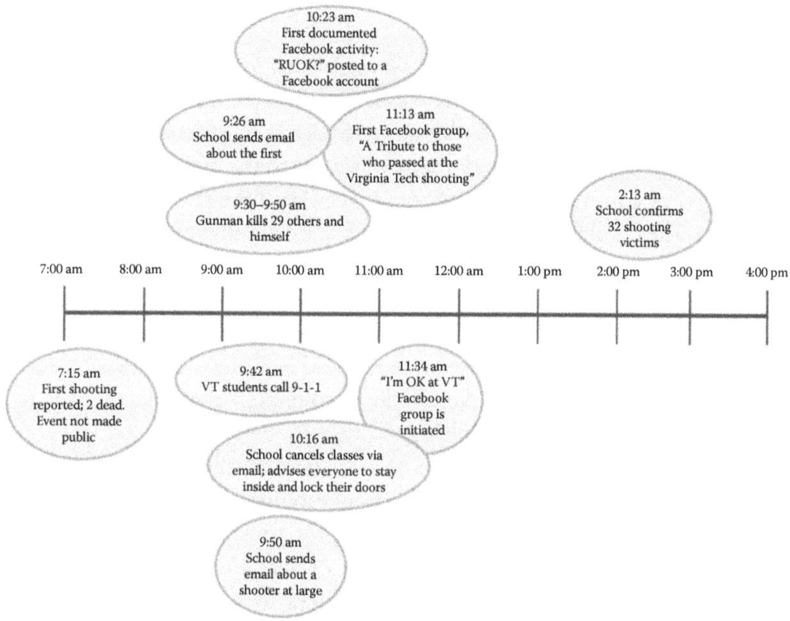

Figure 10.1 Virginia Tech emails students 2 h after the first shooting, minutes before the second and far deadlier attack. The communication through social media, meanwhile, fills the information vacuum. Illustration by Jeremiah Dunlap. (Adapted from Palen, L. et al. 2007. *Social Science Review,* 27(4), 467–480.)

firefighters rallied to assist, but without the means to contact and direct them, organizing off-duty personnel became one more logistical challenge. To make matters even worse, the FDNY's new digital radio system, installed after the 1993 attack on the World Trade Center, was actually less reliable than its old equipment.

The terrorist attacks on NYC claimed the lives of 343 firefighters that day, some of which were in the second tower as it fell, unreachable by the radio equipment in service at the time (Bray, 2011, para. 3, 6, 8, 15–16). Naturally, the attacks on 9/11 have deservedly demanded the attention of those in government, emergency response, and the U.S. citizenry. It is telling, after innumerable analyses of these terrorist attacks, that communication difficulties should attract as much attention as they have. Among the authors of the 9/11 Commission Report, Senator John Rockefeller IV lamented "that the faulty emergency communication on 9/11 was 'probably the greatest killer other than the planes themselves'" ("*Chilling Echoes,*" 2011, para. 7). The collapse of the twin towers created an extremely dangerous environment for emergency first responders, including FDNY; it should not be difficult to believe that the ability to trade information quickly could make the difference between life and death.

Hurricane Katrina

The category five hurricane that struck New Orleans on August 29, 2005 toppled the city's levies, knocked out power, washed out roads, and destroyed countless homes. Over 1000 people died, millions were forced to evacuate and relocate for months, and billions of dollars in damage was done to the city. Here again, an ineffective emergency communications system exacerbated the effects of the disaster. In New Orleans, the communications failure was so severe that many first responders, such as emergency medical technicians (EMTs), were cut off from their departments (Rhoads and Schatz, 2005). The Federal Emergency Management Agency (FEMA) had portable communication facilities, deployable in times of emergency, available, and ready for use. Unfortunately, these facilities were never installed—FEMA had waited for the city of New Orleans to request the units and the request did not come prior to the storm landing.

The incredible strength of Hurricane Katrina and the flood waters disabled both cell phone towers and telephone exchange buildings, leaving a single state-issued satellite phone as one of the few means for communicating to the outside world (Stephan, 2007). As one would expect of a major city, New Orleans had a system in place for emergencies. What it failed to do was build in sufficient redundancies—auxiliary communications capable of picking up the slack for a failed primary system, even if only temporarily. New Orleans had no such option in reserve (Hamblen, 2005). So, when the high-powered hurricane winds blew glass and other detritus into a skyscraper housing the emergency system's transmitter, piercing its radiator and shutting it down in the process, communications in New Orleans were doomed barring swift intervention.

Since the city had also lost electrical power, emergency workers had no way to recharge their communication devices once the batteries went dead. This reality brings two distinct communication failures to light: technological vulnerabilities, such as not having some way to replace the function of a failed or damaged transmitter, and human incompetencies. The lack of effective coordination between FEMA and the local government showcases the need for different parties—responsible for different aspects of the emergency-response process—to collaborate better. An especially notable example of the human communication breakdown involved state police. Technicians had been tasked with the repair of the aforementioned transmitter, but were stopped by state police, who had not been told to let them through (Roane, 2005). Eventually, the technicians were able to bring the transmitter back online, but due to this delay (and others), it was down for more than 3 days.

Improving on the Past: A Retrospective

Each of these cases was chosen for two reasons. First, they are widely recognizable as a major disaster or crisis scenario within the United States. This in turn serves the second reason for their selection: Each of these disasters was extremely costly in terms of lives lost, financial damages, or both. These costs were only worsened through failures in sound emergency communication practices or in the technology employed. Highlighting the past failures in communication offers only a portion of what can be learned from them; the rest lies in trying to understand the corresponding solutions.

While the two shooting sprees at Virginia Tech were hours apart, they caught those on campus by surprise each time. School officials could and should have made use of the campus loud-speaker system to immediately supplement the issued email notifications. Given that many (if not most) college students are equipped with a smart phone, Virginia Tech would have been well served to send emergency notifications via text message, Twitter, or another social media platform. Larger than any other factor, however, is the lack of attention paid to emergency communications. If Virginia Tech had a well-tested emergency response, including campus lockdown scenarios, these timing and technological issues may have been detected before they mattered (Barnett, 2007). Since the shootings on April 16th, the school has adopted some of these very lessons. By December of that same year, at least 4300 students and faculty members had registered for the school's new emergency notification system, capable of sending information via phone, email, text messaging, or instant messaging (Grinberg, 2007).

Certain corrections were made following the terrorist attacks on 9/11. Following the example that focused on the FDNY, better communication technologies were put into place. This included car and subway radio repeaters, and 40-W radios capable of establishing a command post in many different locations. FDNY leadership also changed certain policies, requiring off-duty personnel to report to their firehouse to facilitate a more organized emergency response effort, as one example.

Even so, significant vulnerabilities in NYC's emergency response communications remain. Ten years after the events of 9/11, the FDNY had yet to comply with an international fire code, requiring radio modifications to high-rise buildings (Bray, 2011, para. 8–13, 17–18). An especially pervasive issue, one that goes well beyond the examples cited here, is the ability of different emergency response departments and agencies to communicate and coordinate with one another. It is "called 'interoperability.' Simply put, it means having a structure and technology that allows different agencies in the same area to communicate through a common language and a common system" (Roane, 2004 p. 36). Achieving interoperability may be one of the greatest challenges for those managing emergency response; the progress that many cities, states, and the federal government itself need to make reinforces this nicely.

The handling of Hurricane Katrina is a key example. First responders were quickly isolated not only by dying batteries, but also overwhelmed radio channels. The technology in place was swept aside in a matter of hours and the inability of sister departments to coordinate with one another (e.g., state police not letting technicians into New Orleans to make vital repairs) cost the city and its people dearly (Aziz et al., 2009). While New Orleans has been made aware that its array of cell phone towers, radio transmitters, and telephone-exchange buildings are not invincible, one can only hope that the city and state take lessons learned to heart.

Current Event: Hurricane Sandy

More recently, Hurricane Sandy devastated various locations along the east coast. More than 7 years after Hurricane Katrina, a powerful storm once again cost the United States human lives and billions of dollars in damage. At the time of this chapter's writing, preparations for and responses to Hurricane Sandy have not been fully analyzed. Did the appropriate response agencies take anything away from Hurricane Katrina or was this latest storm a sad (and partially preventable) replay? What commonalities exist between New Orleans in 2005 and the areas Sandy made landfall in late 2012?

As attentions shift from disaster relief to analysis, it will be interesting to determine the involvement of communication in the handling of Hurricane Sandy. Unlike New Orleans during Hurricane Katrina, were an appropriate number of redundancies in place? Did members of different disaster-response agencies coordinate effectively? The degree to which social media was used to warn communities and keep them apprised of the available resources is also worthy of analysis. Social media may have been used effectively by local governments and first responders to warn communities and keep them informed of available resources, or it may have seen significantly greater use by citizens scrambling to fill an information vacuum. The study of Hurricane Sandy stands to be a revealing look at the U.S. ability to adapt disaster communications and should be of great interest to emergency managers.

If the federal government demands that cities bolster interoperability without extending an acceptable amount of financial support, the issue might continue to go ignored (Stephan, 2007). Contending with Hurricane Katrina did crystalize the importance of improved information sharing, but such a consideration ties directly into an organization's culture. Information sharing and interoperability are fine goals, but getting different departments to trust one another can be frustratingly difficult (Raths, 2008). History remains an excellent teacher and clearly has some valuable insights to share regarding communication during dangerous crises. Let us continue by studying some of the people involved in a communication network and the means with which a communication network is built.

Emergency Communications 101

A Local Matter

Part of the difficulty in coordinating emergency responders is that there can be quite a few boots on the ground, especially in the context of a terrorist attack or a major natural disaster. While state and national government agencies respond, a crisis will always begin as a local matter. The nature of the emergency is moot—it will always have a setting, in which one population will have to contend with it first. For this reason, the first responders, police officers, firefighters, and medical personnel require near-instant pipelines of information to save lives and property. Every municipality should have an emergency response plan outlining the individuals in charge, operating procedures, and guidelines in coordinating all the requisite emergency responders (Reilly and Sherd, 2005).

It is crucial that this plan must be tested to calibrate its responsiveness better and to consider any threats deemed most likely to impact that region (Henstra, 2010). Municipal emergency management personnel need the means to alert their community to impending dangers and regularly use sirens, loudspeakers, reverse 9-1-1 calls, television, and radio to do so. Briefing a community on the procedure to follow in one or more crisis scenarios is another way to enhance preparedness and safety, before a problem ever arises. Whether it is a by-product of public education or is driven purely by altruism, it is possible that citizen volunteers will rush to assist those in need during a crisis.

It is then of great importance that a formal structure exists to coordinate volunteers, assuming that the emergency management plan even allows for volunteers. It is incredibly important that efforts must be organized not only to ensure that their efficacy is maximized, but also that volunteers do not become victims themselves (Henstra, 2010). The value of communicating well with the public cannot be overstated and should be a top priority for a municipality's emergency management team (Bruno, 2001). When first responders are unable to coordinate with their colleagues and receive direction, they can become ineffectual and even endangered. However, those in charge of coordinating an emergency response must also keep the populace in the loop or entrust that task to the local government's public information officer (PIO).

Steve Gray, the founder and principal of Denver-based communications firm Rockford Gray, maintained that victims "need to be communicated with first and foremost. Step one is identifying the victims, and step two is communicating with them. People don't like surprises, people take exception to that" (S. Gray, personal communication, April 19, 2012). One can only imagine the fear and frustration of New Orleans residents, fleeing the storm and rising waters, and in many cases, not having any information on where to go. A lack of information is not only stressful to disaster victims, but also it can place them in danger.

Building a Communications Network

Just as there are certain job functions within emergency communications, there are also particular tools. It would be extremely inaccurate to think of emergency communication equipment merely as a collection of loudspeakers, sirens, radios, and other gear. It will take more than a quick stop at an electronics store to build a communications network. While the previously listed equipment has its role to play, a good system has to do a number of things well. The major crises will challenge health providers and emergency medical services (EMSs), and if the death toll is to be minimized, uninterrupted communication via voice, video, or data must be available. If possible, all these means should be made available among health providers, command posts or an Emergency Operations Center (EOC), and sources of support personnel or supplies.

The networks hosting these voice, video, or data exchanges are extremely important to prepare adequately (as history has shown us). Networks need to be reliable, not only covering the necessary land area, but also having the network capacity that would allow a large number of simultaneous users (Qiantori, 2012). Not every municipality will require the hardware that NYC or New Orleans would want to employ. Rather, it is important to objectively assess the risks faced by a specific community. In 2004 and 2006, different areas of Indonesia were struck by a powerful tsunami and earthquake, respectively. Over 130,000 people lost their lives in the two disasters and many thousands were missing or displaced.

During both natural disasters, the destruction impeded or stopped outright efforts to effectively communicate and coordinate. To facilitate better the flow of information to and from command posts, medical centers, and emergency medical staff treating injured victims, a new addition to the existing network was tested. A low-altitude platform (LAP), essentially wireless fidelity (WiFi) equipment resting on a platform suspended by balloons, would be used to relay information. Once deployed in the right area and tethered at the correct height, the LAP system has shown an effectiveness that might save lives the next time Indonesia is beset by Mother Nature (Qiantori, 2012). The take-home lesson is not to extol the virtues of an LAP system for one's town or city, rather, this case is briefly highlighted to reinforce the need for network customization. Different locales will face unique risks and ensuring communication system operability necessitates that it must be taken into account.

Communication Devices and Platforms

Up to this point, the technology utilized to channel information has only been discussed indirectly. Hopefully, the importance of choosing the right tools, capable of disseminating the necessary information even when the system is tested, has been driven home. This section builds on this fundamental understanding, discussing the equipment that an emergency manager might consider for use. Jeff Suggs, the emergency management coordinator of La Porte, Texas, has implemented a highly effective system that offers us two relevant insights. La Porte has been struck by hurricanes Katrina (2005), Rita (2006), and Ike (2008), and in 2007, the town was hit by Tropical Storm Erin.

Preparatory to and during a natural disaster, Mr. Suggs is able to quickly prepare and send custom text or voice messages, which can be received by landline phones, fax machines, mobile phones, and can also be sent to social media sites such as Facebook and Twitter. In 2008, Hurricane Ike buffeted La Porte with 110 mph winds and storm surges over 20 feet, during which over 227,000 messages were sent to direct clean-up efforts, set town curfews, and organize response personnel. To do all this, La Porte had chosen a type of mass message delivery and interactive voice response (IVR) communication system (Klie, 2010). Selecting this technology was wise for two reasons.

First, a massive message delivery option is capable of relaying the same message to many thousands of people and can do so in an instant. Crisis scenarios do not discriminate based on population density and as history has proven, highly populated areas will also face extremely dangerous situations. This leads to the second valuable trait of an IVR system: Its versatility. While mobile phones have become a standard tool for many, not everyone communicates the same way. During an emergency, precious time can be lost if the same information has to be sent repeatedly, each time via a different device. Just as each person is free to choose their preferred mode of communication, there is no set standard for the hardware and software choices from one city to another.

While the particular brand of IVR works well for La Porte, it might not be the preferred platform for everyone. Thankfully, it does not have to be, as an increasing number of technologies are being developed based on the all-important concept of interoperability. Now, there exists communication interfaces capable of receiving radio signals (analog or digital) or phone signals (cellular or landline), delivering them to a router that converts them into an Internet protocol (IP). This digital package can then be converted back into the desired type of signal and can be routed to the corresponding device (Wyllie, 2010, para. 2–3). Even as more traditional means of information exchange are made interchangeable, social media is becoming an increasingly viable method for communicating during emergencies.

In fact, according to a 2010 survey by the American Red Cross, one in five adults (of a total of 1058 respondents) said that they would use a website, email, or social media if 9-1-1 was unavailable (Figure 10.2).

A major crisis, on par with the examples discussed so far, demands that there be multiple communication options. Emergency channels could once again be overwhelmed or collapsed through damage to communication infrastructure. On the same American Red Cross survey, 69% of respondents believed that emergency responders should monitor social media and send assistance based on requests made in that forum, and almost half of the respondents indicated that they believed at least one agency currently does so. Arguably, the most compelling result was that 74% of those surveyed expected help to arrive 1 h after they used Facebook or Twitter to request it (American Red Cross, 2010). The degree to which social media

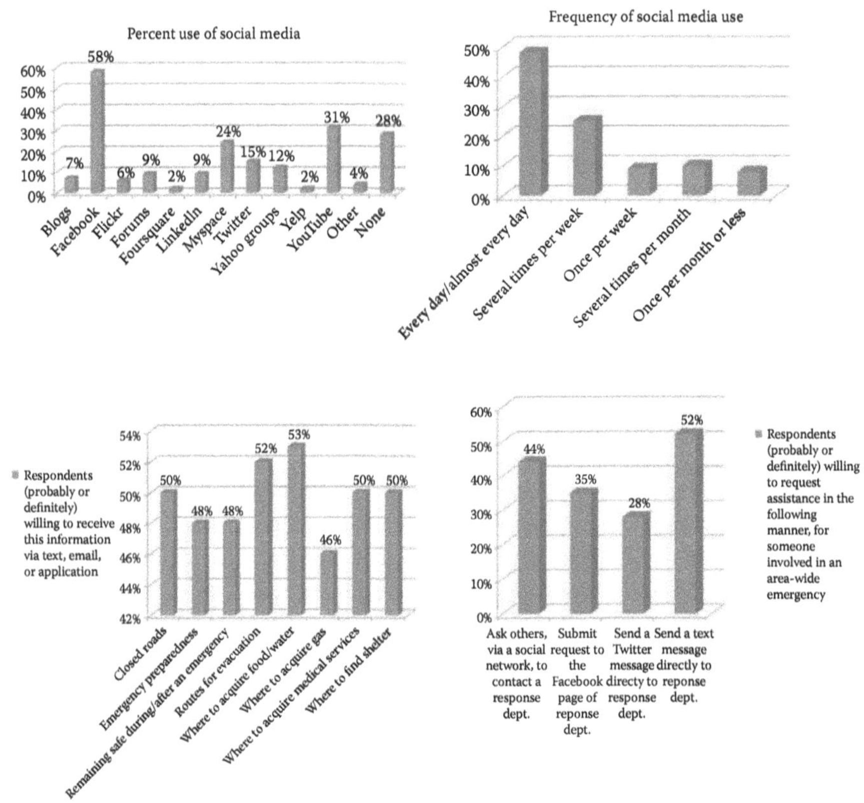

Figure 10.2 Responses to a 2010 American Red Cross survey appear to indicate that social media is widely used and that there is an interest to rely on it during emergencies. Illustrations by Jeremiah Dunlap. (Adapted from American Red Cross. 2010. Web users increasingly rely on social media to seek help in a disaster [releases & statements—2010]. Retrieved from American Red Cross website: http://www.redcross.org/portal/site/en/menuitem.94aae335470e233f6cf911df43181aa0/?vgnextoid=6bb5a96d0a94a210VgnVCM10000089f0870aRCRD.)

pervades Western culture is astounding, particularly when one considers that it is a comparatively young platform.

Information and communication technologies (ICT), including social-networking applications (e.g., Facebook, Flickr, and Twitter) are a new means for anyone, ranging from the victim of a disaster to a helpful passerby, to gather or send important information. During the shootings at Virginia Tech, numerous Facebook groups were created by students or concerned family members to confirm the safety of those on campus. An examination of several of these Facebook groups revealed that through the exchange of information that took place, the 32 shooting victims were all correctly identified before the school released the names (Palen,

2007). Considering social media's popularity and viability as an information collection/dissemination tool, why does not every city have it incorporated into their emergency response plan?

Even now, the communications and emergency response arenas are deliberating on how to best incorporate this medium. The implementation phase is similarly a work in progress. It does appear, however, that the use of social media in this context will necessitate the accrual and maintenance of followers. Organizations, agencies, and political jurisdictions have to campaign to attract more followers and attention, keeping people engaged with updated content (J. Miller, personal communication, April 19, 2012). The need to draw citizens and maintain their digital loyalty may well be one of the growing pains that city, state, and federal departments experience at the outset. If current trends are of any indication, these pains will be well worth it, certainly for any town or city squaring off against a major crisis.

Obstacles

In 2008, 1448 residents of the 20 largest metropolitan areas in the United States were given a survey on their city's emergency notification system. Only 10% indicated that they were aware of their city's system, whereas two out of three respondents were unsure if their city even had one (Emergency communications, 2008). While one can hope that subsequent years saw these numbers improve, it is more likely that U.S. emergency communications still suffer from a sort of identity crisis. For example, there exists a significant variation in the emergency management and response from one community to the next; this may contribute to the difficulty some people have in remembering exactly what comprises their city's emergency notification system.

Misconceptions or outright ignorance is by no means the only hurdle that emergency managers or coordinators face; too often, departments have to compete for limited city or state resources, and have to vie for the attentions of chronically uninformed parties (i.e., politicians and the populace) to obtain them. It can also be difficult for any facet of the emergency response to truly establish its value or measure its effectiveness (Henstra, 2010). The U.S. Government has earned a reputation for being reactive, but it is by no means alone in this. In an idyllic town that has no prior experience with natural or man-made disasters, the accompanying human losses stand to be a tough sell for an emergency manager.

The obvious risk is that critically important issues, such as pushing for greater interoperability of communication technologies, will not be appropriately prioritized until lives are lost in another active shooter scenario, terrorist attack, or natural disaster. Layered onto the difficulties that a tight budget creates, one must also consider the politics and (at points) distrust that exists among different government agencies. This may play into why interoperability, even to promote faster, more reliable, and potentially life-saving communications, does not always receive the attention it deserves (Horan, 2005). Keep in mind that, while cooperation is not always forthcoming, government and agency leaders are not blind to the importance of these matters.

Enhancing interoperability has not merely served as a flaccid sound bite; some specific measures have been proposed to achieve this goal. One such measure was to reallocate a section of frequencies, the 700 MHz D-block spectrum, to services involved in public safety. In February 2012, President Obama signed a law that not only secured these frequencies for public safety, but also secured $7 billion to help construct the nation's broadband safety network (Jackson, 2012, para. 1–3, 8). While a major development, the actual implementation of this law will take considerable time and may even require supplementary funding. Here, we see how a long-term benefit can actually serve as a short-term obstacle: the transition will hinge on how easily the D block can be incorporated into a functional, interoperable, and nationwide communications system.

Rapidly evolving technology is not an obstacle *per se;* it is more of a double-edged sword. If a community is able to afford the fastest and most reliable communication software, for example, then all may be well. Yet, if it cannot afford the next version or latest model, it risks falling behind. Depending on the available budget, an emergency manager, department head, or town leader may find it difficult to keep pace. Properly functioning communication is necessary to keep public safety agencies, EMS, the public, and all the other key parties apprised of situations as they occur. This is by no means a small feat and requires a considerable amount of data processing (integration, analysis) and dissemination.

Figure 10.3 depicts the general function of an information broker, who collects and processes information (a.k.a. data pull) and establishes bidirectional information flow, referred to as communication push (Kuehn, 2011).

Until now this section has focused on the financial resources, organizational will, and political support necessary to provide a top-notch emergency notification and response system. This should not detract from the importance of how the public is engaged during times of emergency, particularly if lives have been lost and the outcry for information is liable to be much greater. Jennifer Miller, a senior consultant of Rockford Gray, explained that those addressing the public may be inclined to speculate on what has transpired, but they must avoid doing so. Anything outside the facts could result in misinformation, which can in turn lead to secondary crises.

Of course, those responsible for briefing the public must also consider that "with social media, everyone can be their own reporter, and if you aren't filling the vacuum, they will" (J. Miller, personal communication, April 19, 2012). Creating a high-quality communications component to local, state, or national emergency response will not always come easily—there are several obstacles that must be negotiated. The effort required is all the more reason to remember what everyone else forgets when emergencies have not occurred in a while: Emergency communication saves lives.

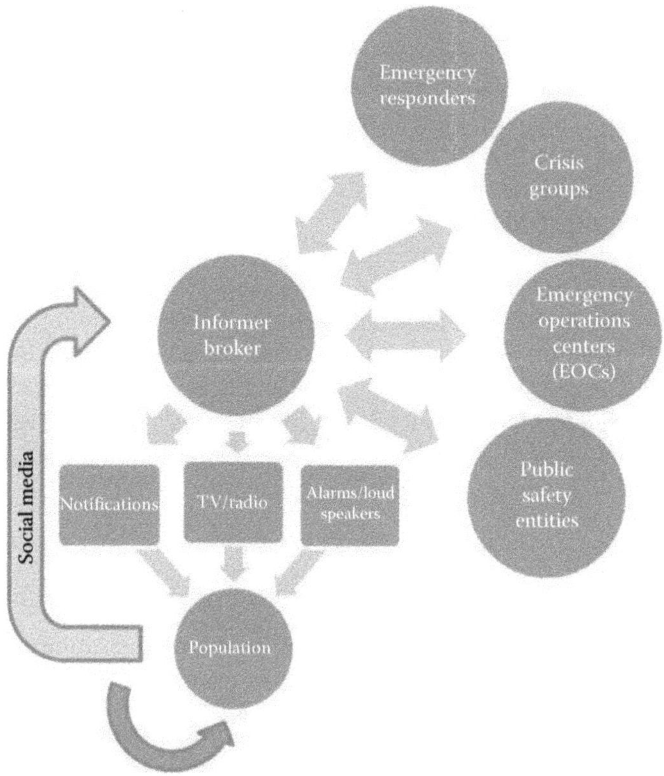

Figure 10.3 Representation of an information broker and the way in which information would be received and distributed. Illustration by Jeremiah Dunlap. (Adapted from Kuehn, A. et al. 2011. *Journal of Theoretical and Applied Electronic Commerce Research,* 6(1), 43–60. http://search.proquest.com/docview/864999934?accountid=12861.)

Conclusion

This chapter attempts to span a divide, beginning with several recent examples of lapses in emergency communication in the United States. In doing so, an uncomfortable question was faced: Could more lives have been saved if greater information exchange and coordination took place? Analyses after the fact agree that great changes must be made for interoperability across the board, ensuring that local, state, and federal agencies are able to talk even when the communications infrastructure is heavily taxed or damaged. The key roles within this effort were discussed, in particular, the public and their information needs.

The advent and continued growth of social media makes it an incredibly germane consideration for emergency managers trying to connect to their community. Social media has leveled the playing field, enabling any individual to report on what they see, hear, and learn. The communications field, particularly with regard to emergency communications, must continue to contend with financial and political resistance to facilitate the inclusion of social media. These platforms are already in heavy use and to ignore them is tantamount to letting the public draw their own conclusions about what is taking place. Conversely, it would be unfortunate to solely trust these platforms: As quick and far reaching as services such as Twitter can be, they are not infallible.

No, social media platforms give emergency managers and coordinators another set of options, but even today the best emergency communication system will include redundancies. What could cause services such as Twitter or Facebook to fail? What would a school or city rely on to disseminate information in that eventuality? Risk assessments and drills can provide direction on the preparations that would best defend communication systems. Ideally, the United States will never see another crisis on par with those discussed in this chapter, but if it does, let us hope that the request for interoperability did not go unheard.

References

American Red Cross. 2010. Web users increasingly rely on social media to seek help in a disaster (releases & statements—2010). Retrieved from American Red Cross website: http://www.redcross.org/portal/site/en/menuitem.94aae335470e233f6cf911df43181aa0/?vgnextoid=6bb5a96d0a94a210VgnVCM10000089f0870aRCRD

Aziz, Z., Feniosky, P.-M., Chen, A., and Lantz, T. 2009. Supporting urban emergency response and recovery using RFID-based building assessment. *Disaster Prevention and Management*, 18(1), 35–48. Doi:10.1108/09653560910938538.

Barnett, N. 2007. The PR response to Virginia Tech and beyond. *Communication World*, 24(4), 14–15.

Bray, C. 2011 . U.S. news—9/11: A decade after: New York prepares for the fire next time—City bolsters communications, sharpens crisis protocol to avoid the kind of carnage that 9/11 visited on first responders. *Wall Street Journal*, A.6–A.6. http://search.proquest.com/docview/887749788?accountid=12861

Bruno, J.L. 2001. New York's emergency response plan—Tested by terrorism. (Counterterrorism). *Spectrum: the Journal of State Government*, 74(4), 7.

Chilling echoes from September 11: There is still no national emergency communications network. (2011, May 23). *New York Times*, pp. A.22–A.22. http://search.proquest.com/docview/867996088?accountid=12861

Emergency Communications Outdated. 2008. *Communications News*, 45(2), 7–7. http://search.proquest.com/docview/202804503?accountid=12861

Friedman, E. 2009. Va. Tech shooter Seung-Hui Cho's mental health records released. *ABC News*. Retrieved April 15, 2012, from http://abcnews.go.com/US/seung-hui-chos-mental-health-records-released/story?id=8278195#.T5uJL-11822

Grinberg, M. and Wade, J. 2007. In the wake of Virginia Tech. *Risk Management*, 54(12), 24–28. http://search.proquest.com/docview/226999490?accountid=12861

Hamblen, M. 2005. Cisco builds system to boost emergency communications. *Computerworld*, 39(43), 5.

Henstra, D. 2010. Evaluating local government emergency management programs: What framework should public managers adopt? *Public Administration Review*, 70(2), 236–246. Doi:10.1111/j.1540-6210.2010.02130.x.

Horan, T. A. and Schooley, B. 2005. Inter-organizational emergency medical services: Case study of rural wireless deployment and management. *Information Systems Frontiers*, 7(2), 155–173. Doi:10.1007/s10796-005-1476-1.

Jackson, D. 2012. *Obama Makes It Official, Signs D Block Legislation. Urgent Communications*. Retrieved from http://urgentcomm.com/policy_and_law/news/obama-signs-dblock-law-20120223/

Klie, L. 2010. Speech in the path of the storm. *Speech Technology*, 15(5), 15–19. http://search.proquest.com/docview/750914042?accountid=12861

Kuehn, A., Kaschewsky, M., Kappeler, A., Spichiger, A., and Riedl, R. 2011. Interoperability and information brokers in public safety: An approach toward seamless emergency communications. *Journal of Theoretical and Applied Electronic Commerce Research*, 6(1), 43–60. http://search.proquest.com/docview/864999934?accountid=12861

Palen, L., Hughes, A.L., Liu, S.B., and Vieweg, S. 2007. Crisis in a networked world: Features of computer-mediated communication in the April 16, 2007, Virginia Tech event. *Social Science Review*, 27(4), 467–480.

Qiantori, A., Hariyanto, H., Ohta, T., Sutiono, A.B., and Suwa, H. 2012. An emergency medical communications system by low altitude platform at the early stages of a natural disaster in Indonesia. *Journal of Medical Systems*, 36(1), 41–52. http://www.springerlink.com/content/p13763j20m156j87/

Raths, D. 2008. Sharing data in a crisis—State and local groups work on interoperability. *KM World*, 17(4), 16–29.

Reilly, J. and Sherd, K. 2005. At MISO, emergency drills mean communications. *Transmission and Distribution World*, 57(5), 33–39. http://search.proquest.com/docview/211146391?accountid=12861

Rhoads, C. and Schatz, A. 2005. In Katrina's wake: Power outages hamstring most emergency communications. *Wall Street Journal*, A.7–A.7. http://search.proquest.com/docview/398947300?accountid=12861

Roane, K. 2004. Excuse me, can we talk? *US News and World Report*, 136(18), 36–37.

Roane, K. 2005. Can't reach out, can't touch. *US News and World Report*, 139(10), 43.

Stephan, K. 2007. We've got to talk: Emergency communications and engineering ethics. *IEEE Technology and Society Magazine*, 42–48. Doi: 10.1109/MTS.2007.906675.

Steve, G. 2012. Rockford Gray. Retrieved April 19, 2012 from http://www.rockfordgray.com/Steve-Gray.html

Wyllie, D. 2010. Leveraging Cisco IPICS and Cisco ISSI for interoperable communications. Retrieved from http://www.policeone.com/police-products/communications/articles/4271685-Leveraging-Cisco-IPICS-and-Cisco-ISSI-for-interoperable-communications/Emergency Communications Outdated. 2008. *Communications News*, 45(2), 7–7. http://search.proquest.com/docview/202804503?accountid=12861

Chapter 11

Emergency Management and the Media

Randall C. Duncan

Introduction

Understanding and working with the media is an important part of an overall emergency management system. This relationship—between the emergency manager and the media—is one that has more opportunity to excel, or fail, than almost any other.

Let us begin our examination of the relationship between emergency management and the media by defining what the media are.

Traditionally, we think of the media as consisting of newspapers, radio, and television. Newspapers have been the media staple since the modern printing press was invented in the mid-fifteenth century by Johannes Gutenberg. Radio and television entered the world of media much more recently, but changed the way media operated and functioned within our society by bringing news on a timelier basis (live reports) and adding the elements of voices (radio) and moving pictures (television).

More recently, the development of the World Wide Web, social network sites, and blogs have led yet another revolution in the way media impacts our lives.

To more fully understand the elements of the relationship between emergency management and the media, it is necessary to understand the characteristics of the various types of media. Let us begin our examination with the traditional media forms of newspapers, radio, and television.

Media

Newspapers

Arguably, the first newspaper in the United States is depicted below and was called Publick Occurrences Both Forreign and Domestick (National Humanities Center 2006). It was published on September 25, 1690 and edited by Benjamin Harris. It only printed one issue, and was banned 4 days after publication by the Governor and Council of Massachusetts (Massachusetts Historical Society 2010). The only surviving copy of the newspaper is in the Public Record Office in London (Library of Congress 2009).

Traditionally, modern newspapers have published on a daily or weekly basis, depending on the size of the reading audience. The circulation of newspapers varies greatly. The newspapers with the three largest weekday circulations in 2012 were the *Wall Street Journal* (circulation 2,293,798), *USA Today* (circulation 1,713,833) and the *LA Times* (circulation 641,369). The newspaper showing the smallest circulation was the Medina (NY) journal *Register* (circulation 1670) (Audit Bureau of Circulation 2012).

PUBLICK
OCCURRENCES

Arguably, the first newspaper in the United States was Publick Occurrences Both Forreign and Domestick (Allentown Art Museum. Publick Occurrrences Both Forreign and Domestick.http://www.renaissanceconnection.org/innovations_science.html)

Newspapers have traditionally been viewed as providing more in-depth coverage than either radio or television because of the amount of space available in which to write the story. Newspapers also provided some of the first coverage of events far removed from the place where they were published by the mechanism of the telegraph (see the section "Radio" for more details). This allowed remote correspondents to send a story from far away back to the newspaper home office, and created a style of journalistic writing known as the "inverted pyramid." The inverted pyramid style of writing called for the correspondent to relay the most important facts first, followed by those of lesser importance in the body of the story (Scanlon 2008).

Based on this information, then, we can anticipate what print organizations want in the way of news.

Item	*Explanation*
Details	Print media wants to paint a picture in the reader's mind with words
Questions	Will be more oriented toward details.
	How many times did the truck roll over?
	How far away from the edge of the road did it come to rest?
	Were there flames? If so, how high?
Background information	History related to an event
	Has this ever happened before?
	History of individuals involved in the event
Deadline	*Traditional:* usually daily. Current policy may be impacted by newspaper website

Radio

It is not possible to talk about the history of radio without mentioning the wired telegraph system. The telegraph was made practical within the United States by Samuel Morse, who did his first public demonstration of the device in 1838 (Smithsonian Institution 2010). In 1843, Congress provided funding to install a telegraph between Baltimore, Maryland and Washington, DC. The Whig Party held its nominating convention in Baltimore on May 1, 1844 and selected Henry Clay as their nominee. This was the first news item relayed by telegraph (Smithsonian Institution 2010). In 1901, Guglielmo Marconi began developing what would become broadcast radio—he sent the Morse code signal for the letter S from a wireless transmitter in Poldhu, Cornwall, England to a wireless receiver in Newfoundland, Canada (Public Broadcasting System 1998a). A few years later—on Christmas Eve, 1906—some wireless telegraph operators on board ships heard the Christmas carol "Silent Night" and a voice reading bible verses interrupt the Morse code they normally heard (Public Broadcasting System 1998b). This marked the first radio broadcast.

From these humble beginnings, radio had an impact on the way we listened to news and found out about other events. We could sit in our living rooms and hear the voices of Presidents, dictators from overseas, and Hollywood stars endorsing commercial products.

Unlike newspapers, radios could bring us the sounds and words of a news event as they happened. Modern radio stations are separated into various interest groups called "formats." Some of the formats in today's radio broadcasting include news/talk stations, music stations, public radio, and non-English radio.

Item	*Explanation*
Details	Radio news utilizes short, concise information in the voice of the newsmaker
	Typically, they are 10–15 s "actualities" or "sound bites"
News/talk	More stories; a little more depth than other radio formats
Public radio	Uses "natural sound." Records background of event happening with open mike
Non-English stations	Help to reach those who speak a language other than English within the community
Music stations	May or may not carry news. If they do, it typically consists of only short news items
Deadline	Hourly—depending on schedule of newscasts

Television

The first authorized broadcast of a television in the United States started on July 2, 1928 in Wheaton, Maryland, a suburb of Washington, DC by C.F. Jenkins (Popular Mechanics 1928). The heyday of television may have come on the evening of March 7, 1955, when one in two Americans watched Mary Martin's portrayal of "Peter Pan" on live television (Bogart 1958, p. 1). Other significant events that marked the impact of television on the way Americans received news included the coverage of such live events as the Kennedy–Nixon debates, mankind's first step on the moon, and more. The addition of live images to go with sound literally brought the world into our homes every night. There are various types of news broadcasts on local television stations. They may range from spot news/breaking news of particular activities currently in progress to the regularly scheduled news programs. In addition, some local television stations may air special investigative reports or programs. Typically, local television stations will have an affiliation with a network, and will present a network-originated program of national and international news.

Item	*Explanation*
Details	Video of the event may determine whether or not there is a story
	A story that otherwise would not make the news may become a story if there is video
	Similarly, a story of real importance may not make the news if there is no video
Types of news	Local news includes spot news, regular local news, and investigative reports. Also, feature news programs (either local or network), national news, and international news
Deadline	Various depending on the type of news that will be broadcast. Major deadlines are typically for the evening news broadcast and the late night news broadcast

Social Network Sites and the World Wide Web

No examination of media would be complete without exploring the impact of the World Wide Web and social media on individuals, as well as the traditional media of newspaper, radio, and television.

The World Wide Web, as we know it today, first became a reality in 1990 with the release of a point-and-click hypertext editor called "World Wide Web" (Berners-Lee 1998). In the ensuing years, the number of websites, their functionality, and the pure amount of information have literally exploded. Of course, with such rapid expansion, there is a need for a "buyer-beware" approach by users. It is as easy to find an academically reputable and accurate source on the World Wide Web as it is the lunatic ravings of fringe elements. It is up to the user to find and place the appropriate value on sources available through this medium.

Aside from the ease of accessing information on the World Wide Web, people soon found that it was becoming a tool for personal communication and networking between friends, leading to the development of social network sites. Social network sites are defined as

> ...web-based services that allow individuals to (1) construct a public or semi-public profile within a bounded system, (2) articulate a list of

> other users with whom they share a connection, and (3) view and traverse their list of connections and those made by others within the system. The nature and nomenclature of these connections may vary from site to site (Boyd and Ellison 2007).

We also think of social network sites as social media—a way to convey information to friends and learn about information from friends and others with opinions. Sites such as Twitter, Facebook, MySpace, blogs, and other sites have become a way to share information, opinions, and even reflect on news events.

These same sites have had an impact on the way traditional media—newspapers, radio, and television—interact with their viewers and listeners. As a result, newspapers now also shoot video of news stories and provide it to their readers through the mechanism of their website. Radio stations do the same thing. Television stations now write news stories and publish them, similar to newspapers, on their websites. This has had a major impact on the traditional deadlines for the various forms of media.

Yet another World Wide Web-based phenomena has emerged recently—the weblog, or blog. This phenomenon has blurred the distinction between traditional press and bloggers.

> In the media world, it used to be clear who was in the news business and who was not. News businesses provided news, non-news businesses did not. Reporters worked for companies who were in the business to provide news. News businesses got paid, usually by advertisers, to collect, package and distribute information of interest to news audiences. Non-news businesses or organizations exist for other purposes—perhaps to deliver public service such as environmental protection, or to produce commercial goods such as fertilizer. Providing news for these businesses is simply not their reason for existence. In an instant news world, that distinction is becoming increasingly fuzzy.
>
> One of the most significant trends to come out of the collection of technologies we call the Internet, is the emergence of citizen journalists. 'Blogging,' from the term 'Weblog,' which used to describe people who would record and publish what they discovered on the Internet, reflects the ease with which almost anyone who writes today can also publish. As mentioned earlier, some of the bloggers have accumulated audiences in the millions and have influence as great as any of the celebrity journalists that used to be staples of our early evening hours at home (Baron 2006, 47).

Because of these factors, the emergency manager or spokesperson for the jurisdiction has to keep in mind that there are now more audiences for the information they prepare than the traditional media. There are the families of those directly impacted by the event or emergency, those in the immediate area of the emergency or disaster, and the "traditional media" along with the citizen journalist.

Dealing with the Media in a Crisis

To begin our discussion about dealing with the media in a crisis, we need to understand some of the basics about communication. We first learn to communicate as babies—before we can even begin to say words. We communicate with gestures and sounds, through a thing called nonverbal behavioral clusters. We will discuss this in more detail shortly.

Communications is extremely complicated for such a seemingly simple thing. The act of communications starts as an idea in our brain. That idea wishes to be expressed or communicated to someone. It must then make its way through the filters of our belief system and perceptions. Then it must be encoded (either in speech or in writing) and then broadcast to a receiver (a reader or listener). The receiver has to get the message, decode it, and run the decoded material through their own filters of belief systems and perceptions to understand the idea we originally wished to communicate to them. An understanding of the complications associated with the process of how we communicate allows for a new appreciation of a statement as seemingly simple as, "Pass the salt, please."

Nonverbal behavioral clusters associated with how we say and express things are more important in conveying meaning than the words we actually say (Blatner 2009). Because these nonverbal behavioral clusters typically convey a larger percentage of our communication than the specific choice of words do, we tend to place more faith in the way the message is expressed. When the message being conveyed to an audience by a speaker's words is in conflict with their nonverbal behavioral clusters, the audience will not believe the speaker. As a simple thought experiment, recall the last time you observed a person on television and your reaction to that person was the thought that you did not believe a word they said. The odds are, you felt that way because there was a conflict between the words of the speaker and their nonverbal behavioral clusters.

The normal process of communications takes a slight detour under a crisis situation. When a crisis is in progress, we need to provide assistance to the elected official or spokesperson to make sure we do not allow circumstances to take away from the messages we need to communicate to the public typically through the media. In other words, we need to avoid media pitfalls. The following tables are adapted from unpublished material from Dr. Vincent Covello, founder and director of the Center for Risk Communication.

Do	*Don't*
Define all technical terms and acronyms (Jargon)	Use language that may not be understood by even a portion of your audience
If you use humor, direct it at yourself	Use humor in relation to safety, health, or environmental issues

(*Continued*)

Do	*Don't*
Refute negative allegations without repeating them	Refer to national problems—"This isn't Love Canal."
Use visuals to emphasize key points	Rely entirely on words
Remain calm. Use the question or allegation as a springboard to say something positive	Let your feelings interfere with your ability to communicate positively
Ask whether you made yourself clear	Assume you have been understood
Use examples, stories, and analogies to establish a common understanding	Talk only in abstractions
Be sensitive to nonverbal messages you are communicating	Allow your body language or your position in the room to be inconsistent with your message
Make them consistent with what you are saying	Dress inconsistently with your message
Attack the *issue*	Attack the person or the organization
Promise only what you can deliver	Make promises you cannot keep
Set, then follow strict deadlines	Fail to follow up on those items you promise to
Emphasize achievements made and ongoing efforts	Say there are no guarantees
Refer to the importance you attach to health, safety, and environmental issues—your moral obligation to public health outweigh financial considerations	Do not refer to the amount of money spent as a representation of your concern
Use personal pronouns (e.g., I and we)	Take on the identity of a large organization
Take responsibility for your share of the problem	Try to shift blame or responsibilities to others
Assume everything you say and do is part of the public record	Make side comments or "confidential" remarks
Discuss risks and benefits in separate communications	Discuss your costs along with risk levels
Use risk comparisons to help put risks in perspective	Compare unrelated risks
Stress that the true risk is between zero and the worst-case estimate	State absolutes or expect laypersons to understand

Do	*Don't*
Emphasize performance, trends and achievements	Mention or repeat large, negative numbers
Focus your remarks on empathy, competence, honesty and dedication	Provide too much detail or get drawn into protracted technical debates
Keep presentation to 15 min total	Ramble or fail to plan the time well
Keep answers to 2 min maximum	Tell people more than they want

In a crisis situation, the media follows certain patterns. Those patterns include

- Searching for background information on the incident
- Dispatching reporters to the scene
- Obtaining access to the scene or the official spokesperson
- Dramatizing the situation
- Expecting a briefing complete with written information
- Expects *you* to panic
- Becomes confused by technical information
- Exhausting resources
- Sharing information among themselves
- Acting professional and expecting the same
- Providing filler for stories if credible information is not available

Public Information Officer

The Public Information Officer (PIO) typically has the responsibility of coordinating the collection, verification, and dissemination of information to the public. These duties may occur as a part of day-to-day organizational operations, or on an emergency basis. Since the focus of deliberations in this chapter is "Emergency Management and the Media," let us concentrate on the roles and responsibilities of a PIO in an emergency.

The PIO has responsibilities to a number of different constituencies. These include

- The public—The largest user group for emergency messages. This implies that the PIO should be aware of any special demographic characteristics of the community being served, and have familiarity with the best media channels to distribute information to those who need it.
- The media—This is one of the most important relationships to establish to make sure information is distributed to those who need it. The PIO will need to understand the traditional and social media outlets within the jurisdiction.
- The agency—This relationship is the basis of trust within the jurisdiction. The PIO has a duty to positively portray the efforts and successes of the agency

they represent. This relationship will be especially important when navigating an agency through the dangerous shoals and reefs of a negative news story.
- The other responding agencies—The PIO needs to have a good working relationship with other agencies responding to the emergency or disaster. These relationships are especially important in helping to avoid conflicting stories or statements. If the incident becomes large enough to engage the joint information system (JIS) or the joint information center (JIC), the PIO needs to be able to function in that environment. In addition, the PIO needs to be aware of the possibility that there will be differing priorities among the agencies responding to the emergency or disaster and how to deal with those differing priorities so that "mixed messages" are not given to the public.

We generally make the statement that the PIO provides public information. It would be helpful to provide a definition of what public information is, with respect to emergencies and disasters. Generally, we can conclude that public information is used by people to save lives, reduce injury and harm, and protect property (FEMA 2009). Given that public information covers such a wide territory, it is understood that almost every piece of information coming from your agency or Emergency Operations Center (EOC) could result in the public taking some type of action to protect themselves or others from the effects of a disaster or emergency. This also emphasizes the criticality of the accuracy and timeliness of your information (FEMA 2009).

We generally expect that information communicated to the public through our PIO will result in action by people, provide information, change behaviors or attitudes, or create a positive view of our agency or EOC within the community. Some examples of public information could include

- The current status of the emergency or disaster
- Agency response actions to the disaster
- Information or warnings as conditions change
- Important locations (i.e., where food, water, and shelters are located)
- Specific evacuation information or directions
- Other pertinent information:
  - What is open or closed
    - Government facilities
    - Stores
    - Roads
    - Schools
  - Status of lifeline systems—electricity, gas, water, and sewer
  - Volunteer recruitment
  - Where people can find aid or assistance
  - Public inquiry telephone numbers

To perform the job of PIO well, the person in that position needs to have a number of qualities. Some of those qualities include

- Knowledge of the organization they represent. This allows them to speak with credibility about the operations of the organization. It demonstrates to the media and the public that the PIO has access to agency leadership. It also provides the media with opportunities for interviews and briefings about the agency.
- A good working relationship with the organization or EOC. This is a necessary quality for the PIO to have access to the information and resources everywhere within the agency or EOC they are representing.
- A certain amount of aggressiveness. The PIO may need to be able to go directly into the organization and get to decision makers and leaders with minimal delays. In addition, the PIO will undoubtedly be called upon to provide advice to leadership—making it necessary that the PIO is in the inner circle of the organization or EOC.
- A high level of trust and ability to strategize. The PIO needs to be able to establish trust within the organization or EOC. Essentially, the PIO becomes an advisor who understands how things will be viewed outside of the agency or EOC, and understand the implications of information to which the public will have access. It will be important to understand what the potential negative consequences of these issues are and how to present them, truthfully, in as positive a light as possible.
- Community relations skills are necessary for the PIO. The PIO needs to understand the demographics of the jurisdiction—who lives and works there and what the prevailing local values, concerns, and interests are. The PIO must also know about organizations within the community and how they work and interact.
- The PIO needs to have good media relations skills. This includes a level of credibility that is usually only developed over time and through hard work.

It is also important for the PIO to have several sets of skills, including

- Writing ability—Organizes clear thoughts in a written format (whether electronic or printed). This includes the ability to develop talking points, guidance, strategy papers, speeches, and general information for management. It is especially important that proper grammar and spelling be utilized in these pieces. This should probably include familiarity with Associated Press Stylebook (http://www.apstylebook.com). Generally, the PIO needs to be able to produce quality documents, whether electronic or paper.
- Other abilities—The PIO should be able to understand the basics of using video as a means of communication. This would include knowledge of the basic elements of photography. In addition, the PIO should be able to clearly communicate and outline ideas in a manner allowing the PIO or other spokesperson to communicate effectively with an audience; the ability to speak effectively and persuasively in front of an audience; and an awareness of the impact of nonverbal behavioral clusters on the delivery of a message.

Joint Information System/Joint Information Center

As you have read previously, the National Incident Management System (NIMS) was developed at the direction of Homeland Security Presidential Decision Directive (HSPD)-5. The original NIMS document was developed in 2004, updated in 2006, and updated again in 2008 (FEMA 2008). Public information, and the process of establishing the system to collect, integrate, and coordinate it, is defined as a part of NIMS Component IV—Command and Management (FEMA 2008, pp. 70–74). The overall system is called the JIS and the specific place where this process happens is called the JIC.

Typically, this process involved a PIO, who supports the incident command structure, and who is also a member of the command staff. The responsibilities of this position typically include

- Responding to inquiries from the media, public, and elected officials
- Supervising the process of collecting, integrating, and coordinating information for emergency public information
- Supervising the process of collecting, integrating, and coordinating information for warning information
- Monitoring for rumors and responding to them
- Relations with the media
- Creating coordinated and consistent messages through
  - Identifying key information to be communicated to the public
  - Creating the message that provides the key information in a clear and easily understood method
  - Prioritizing messages so the most important gets out first and that the public is not overwhelmed with the amount of information
  - Verifying the accuracy of information
  - Making sure the message gets out through the most effective means available

The overall JIS provides the means to coordinate the messages being released to the public by all elements of government involved in the disaster response—whether multiple local jurisdictions, or local, state, and federal governments; multiple disciplines involved in the response; nongovernmental organizations (NGOs) involved with the response; and the private sector. This coordination is particularly important because all the voices involved in the disaster should be providing substantially the same message.

One of the important elements to keep in mind when dealing with a large emergency response situation is that different disciplines may have their own spokesperson or PIO present and different jurisdictions may have their own spokesperson or PIO present as well. It is possible that these PIOs may serve as the basis for the JIC staff. Remember that each of these officials will have the primary responsibility for making sure the story—as it relates to their agency, discipline, or jurisdiction—gets out to the media. But there is no reason they cannot work together and collaborate to establish the JIS and provide the personnel for the JIC.

The JIC is basically an instrument to help facilitate the processes that take place within the JIS; much like the EOC is an instrument to help facilitate the processes that take place within the Local Emergency Operations Plan (LEOP). As a result, there are other parallels between these two elements of public information. The elements of the JIS must be worked out well in advance of the occurrence of a disaster. The system plans and processes, much like the roles and responsibilities within the LEOP, must be worked out and understood in advance of their application.

There should also be consideration given as to what kind of triggers might initiate the activation of the JIC. Some suggestions might include

- The creation of a standard operating procedure or guideline (SOP/SOG) that defines the opening of the facility. This SOP/SOG could be modeled after the existing document for the activation of the Emergency Operations Center.
- An analysis of the potential impact of the incident.
- An analysis of the potential media interest in the incident.
- The potential duration or the response and recovery phases of the emergency or disaster.

Other items noteworthy about the JIS and JIC are outlined in the below table.

Joint Information System Provides a Structure and System for	*Joint Information Center Provides a Place to*
Developing and delivering coordinated interagency messages	Centrally facilitate operation of the JIS during and after an incident
Creating, recommending, and executing public information plans and strategies	Increase information coordination
Advising incident commander about incident-relevant public affairs issues	Reduce misinformation
Monitoring and correcting erroneous information circulating among the media or the public	Maximize resources for dealing with the public and the media
Be adaptable to the size and scale of the incident—from three PIOs at the scene to 150 PIOs at a major disaster from multiple locations	Provides "one stop shopping" for the media

The below table presents a number of different types of JICs. It is adapted from the information found in FEMA G290 Basic Public Information Officer Course (FEMA 2009).

JIC Type	*Description*
Incident	Typically, an incident-specific JIC is established at a single, on-scene location in coordination with federal, state, tribal, and local agencies or at the national level, if the situation warrants. It provides easy media access, which is paramount to success. This is a typical JIC.
Virtual	A virtual JIC is established when a physical colocation is not feasible. It connects PIOs through e-mail, cell/landline phones, faxes, video teleconferencing, web-based information systems, and so on. For a pandemic incident where PIOs at different locations communicate and coordinate public information electronically, it may be appropriate to establish a virtual JIC.
Satellite	A satellite JIC is smaller in scale than other JICs. It is established primarily to support the incident JIC and to operate under its direction. These are subordinate JICs, which are typically located closer to the scene.
Area	An area JIC supports multiple-incident ICS structures that are spread over a wide geographic area. It is typically located near the largest media market and can be established on a local, state, or multistate basis. Multiple states experiencing storm damage may participate in an area JIC.
Support	A support JIC is established to supplement the efforts of several incident JICs in multiple states. It offers additional staff and resources outside of the disaster area.
National	A national JIC is established when an incident requires federal coordination and is expected to be of long duration (weeks or months) or when the incident affects a large area of the country. A national JIC is staffed by numerous federal department and/or agencies, as well as state agencies and NGOs.

JIC Readiness Assessment (FEMA 2009, A39–A42)

Plans		
Do you have systems and procedures for:	Yes	No
✓ Developing an emergency response or crisis communications plan for public information and media relations?	☐	☐

Does your emergency response or crisis communications plan have systems and procedures for:	Yes	No
✓ Designating and assigning line and staff responsibilities for the public information team?	☐	☐
✓ Identifying and updating current contact numbers for PIO staff and other public information partners in your plan?	☐	☐
✓ Identifying and updating current contact numbers for regional and local news media (including after-hours news desks)?	☐	☐
✓ Establishing the JIC at the Emergency Operations Center (if activated)?	☐	☐
✓ Securing needed resources (space, equipment, people) to conduct the public information operation during an incident 24 h a day, using such mechanisms as memorandums of understanding, contracts, and so on?	☐	☐
✓ Creating messages for the news media and the public under severe time constraints, including methods to clear these messages within the emergency response operations of your organization (including multijurisdiction and/or agency cross-clearance)?	☐	☐
✓ Disseminating information to news media, the public, and partners (e.g., website capability 24/7, listservs, broadcast fax, printed news releases, and door-to-door leaflets)?	☐	☐
✓ Verifying and clearing/approving information prior to its release to the news media and the public?	☐	☐
✓ Operating a public inquiry hotline with trained staff available to answer questions from the public and control rumors?	☐	☐
✓ Activating the Emergency Alert System, including the use of prescripted messages?	☐	☐
✓ Coordinating your public information systems planning activities with other response organizations?	☐	☐
✓ Testing the plan through drills and exercises with other response team partners?	☐	☐
✓ Updating the plan as a result of lessons learned through drills, exercises, and incidents?	☐	☐

(*Continued*)

People		
Do you have systems and procedures for:	Yes	No
✓ Identifying staffing capabilities needed to maintain public information operations for 24 h per day for at least several days? (*Note:* Staff may include regular full- and part-time staff as well as PIOs from other agencies or departments, disaster employees, volunteers, etc.)	☐	☐
✓ Establishing and maintaining agreements for acquiring or borrowing temporary staff? (*Note:* Such agreements may be mutual aid arrangements or memorandums of understanding.)	☐	☐
✓ Granting emergency authority to hire or call up temporary staff or those on loan from other organizations?	☐	☐
✓ Establishing and maintaining job descriptions and qualifications for individuals serving as your organization's PIO and other roles during an incident?	☐	☐
✓ Assigning a staff member and at least one alternate person the role and responsibilities of PIO.	☐	☐
✓ Determining if the assigned PIO(s) is qualified? Sample qualifications include o Experience and skills in providing general and emergency public information. o Ability to represent your organization professionally (can articulate public information messages well when dealing with the media and the public, and can handle on-camera interviews). o Written and technical communication skills (writing/editing, photography, graphics, and Internet/web design proficiency). o Management and supervision experience and skills needed to run a JIC.	☐	☐
✓ Establishing and maintaining a list of language translators available to assist with public information? (*Note:* Such network should include sign language interpreters and individuals capable of writing and speaking the non-English language(s) used by individuals in your jurisdiction.)	☐	☐
✓ Establishing and maintaining working relationships with PIO partners from other organizations that you might need to work with during an incident (e.g., PIOs from other jurisdictions, other government agencies or departments, NGO, and private entities)?	☐	☐

✓ Developing and maintaining working relationships with your local and regional media, and established procedures for providing information to those media entities effectively and efficiently during incidents?	☐	☐
Logistics		
Do you have a go-kit for PIO use during an incident, including:	Yes	No
✓ Laptop computer capable of linking to the Internet/e-mail?	☐	☐
✓ Cell or satellite phone, pager, and/or PDA/palm computer with wireless e-mail capability?	☐	☐
✓ Digital camera, photo storage media, and charger/backup batteries?	☐	☐
✓ Flash drives, CDs, and/or disks containing the elements of the crisis communication plan (including news media contact lists, PIO contact lists, and information materials such as topic-specific fact sheets, backgrounders, talking points, and news release templates)? *Remember:* Redundancy is important in case the computer you are using does not have a USB port, CD, or floppy drive?	☐	☐
✓ Office supplies such as paper, pens, self-stick notes, and so on?	☐	☐
✓ Manuals and background information necessary to provide information to the media and the public (e.g., your Smart Book)? (*Note:* A Smart Book is a compilation of factual information assembled about your jurisdiction, such as population, number of schools and hospitals, size and description of geographic or infrastructure features, etc.)?	☐	☐
✓ Hard copies of all critical information?	☐	☐
Do you have systems for:	Yes	No
✓ Acquiring and maintaining go-kits with a funding mechanism (e.g., credit card) that can be used to purchase operational resources? (*Note:* A go-kit is a mobile response kit that allows PIOs to maintain communications in the event that they are working outside of their normal place of operation.)	☐	☐
✓ Ensuring PIOs can access the go-kit when serving at an incident?	☐	☐

(*Continued*)

✓ Acquiring and maintaining portable communications equipment, critical up-to-date information, and supplies?	☐	☐
✓ Acquiring and maintaining essential media production equipment (cameras, digital storage, laptops, etc.)?	☐	☐
✓ Acquiring and maintaining a Smart Book (or equivalent technologies) to assist PIOs in accurately informing the media and the public during an incident?	☐	☐
✓ Identifying a dedicated location to house the JIC? (*Note:* The location selected must be wired for telephone, Internet access, cable, etc.)	☐	☐
✓ Securing and maintaining the necessary JIC equipment and supplies to allow information to be disseminated to the media and the public?	☐	☐
✓ Inventorying and restocking the PIO go-kit after an incident?	☐	☐
✓ Inventorying and restocking JIC equipment and supplies after an incident?	☐	☐
✓ Periodically updating your Smart Book with current information?	☐	☐
Logistics		
Do you have equipment and supplies needed for a JIC, including:	Yes	No
✓ Computers on a LAN with Internet access and e-mail listservs designated for news media and partner entities?	☐	☐
✓ Laptop computers?	☐	☐
✓ Electric and manual typewriter(s) in case of power outage or other problems interfere with computer/printer usage?	☐	☐
✓ Fax machine preprogrammed for broadcasting fax releases to news media and partner entities?	☐	☐
✓ Printers and copy machines, with supplies such as toner and paper?	☐	☐
✓ Paper shredder and trash bags?	☐	☐
✓ Televisions with access to cable hookups and VHS VCRs or other recording media?	☐	☐

✓ Cell or satellite phones, pagers, and/or PDAs/palm computers with wireless e-mail capability?	☐	☐
✓ Digital camera, photo storage media, and charger/backup batteries?	☐	☐
✓ Audio recorder and batteries?	☐	☐
✓ Flash drives, CDs, and/or disks containing the elements of the crisis communication plan (including media contact lists, PIO contact lists, and information materials such as topic-specific fact sheets, backgrounders, talking points, and news release templates)?	☐	☐
✓ Office furniture/accessories such as desks, chairs, file cabinets, bulletin boards, white boards, trash cans, lights, in/out baskets, landline phones, clocks, large calendars, and so on?	☐	☐
✓ Audio equipment and furniture necessary for conducting news conferences (e.g., wireless microphones, lectern, and multibox)?	☐	☐
✓ Office supplies (e.g., white and colored paper, pens, self-stick notes, folders, blank tapes, binders, overnight mail supplies, tape, poster board, erasable and permanent markers, chart paper, easels, staplers and staples, press kit folders, binders, computer disks/CDs, hole punch, organization logo on stickers, letterhead, and postage stamps)?	☐	☐
✓ Manuals, directories, and background information necessary to provide information to the media and the public (e.g., your Smart Book)?	☐	☐
✓ Hard copies of all critical information?	☐	☐

References

Allentown Art Museum. *Publick Occurrences Both Forreign and Domestick.* http://www.renaissanceconnection.org/innovations_science.html.

Audit Bureau of Circulation. 2012. *US Newspapers—Search Results.* Retrieved April 15, 2010, from Audit Bureau of Circulation: http://abcas3.accessabc.com/ecirc/newstitlesearchus.asp

Baron, G. R. 2006. *Now Is Too Late2: Survival in an Era of Instant News.* Bellingham, WA: Edens Veil Media.

Berners-Lee, T. 1998, May 7. *The World Wide Web: A Very Short Personal History*. Retrieved December 27, 2012, from World Wide Web Consortium (W3C): http://www.w3.org/People/Berners-Lee/ShortHistory.html

Blatner, A. 2009, June 29. *About Nonverbal Communications*. Retrieved December 27, 2012, from Adam Blatner's website: http://www.blatner.com/adam/level2/nverb1.htm

Bogart, L. 1958. *The Age of Television*. New York: F. Unger Publishing Company.

Boyd, D. M. and Ellison, N. B. 2007. *Social Network Sites: Definition, History, and Scholarship*. Retrieved July 16, 2013, from *Journal of Computer-Mediated Communication*: http://jcmc.indiana.edu/vol13/issue1/boyd.ellison.html

FEMA. 2008. *National Incident Management System*. Retrieved December 27, 2012, from Federal Emergency Management Agency: http://www.fema.gov/pdf/emergency/nims/NIMS_core.pdf

FEMA. 2009. *G290 Basic Public Information Officer Training*. Washington DC: Federal Emergency Management Agency.

Library of Congress. 2009. *Eighteenth-Century American Newspapers in the Library of Congress*. Retrieved December 27, 2012, from Library of Congress: http://www.loc.gov/rr/news/18th/200.html

Massachusetts Historical Society. 2010. *The Boston Newsletter, number 1*. Retrieved July 16, 2013, from Massachusetts Historical Society: http://masshist.org/database/186

National Humanities Center. 2006. *Publick Occurrences Both Forreign and Domestick*. Retrieved December 27, 2012, from National Humanities Center: http://nationalhumanitiescenter.org/pds/amerbegin/power/text5/PublickOccurrences.pdf

Popular Mechanics. 1928. What Television Offers You. *50* (5), 820–824.

Public Broadcasting System. 1998a. *Marconi Receives Radio Signal over Atlantic 1901*. Retrieved December 27, 2012, from A Science Odyssey: People and Discoveries: http://www.pbs.org/wgbh/aso/databank/entries/dt01ma.html

Public Broadcasting System. 1998b. *KDKA Begins to Broadcast 1920*. Retrieved December 27, 2012, from A Science Odyssey: People and Discoveries: http://www.pbs.org/wgbh/aso/databank/entries/dt20ra.html

Scanlon, C. 2008. *The Inverted Pyramid Structure*. Retrieved December 27, 2012, from Purdue Online Writing Lab: http://owl.english.purdue.edu/owl/resource/735/04/

Smithsonian Institution. 2010. *History Wired: A Few of Our Favorite Things*. Retrieved December 27, 2012, from National Museum of American History, Smithsonian Institution: http://historywired.si.edu/detail.cfm?ID=324

The AP Style Book. The Associated Press. http://www.apstylebook.com.

Chapter 12

Volunteer Management

Mark Chambers

Background

Since the beginning of man's time on this earth, situations have developed where one man must turn to another and seek assistance. Lending aid to our fellow man seems inherent to being human and there are numerous examples exemplified through volunteerism. The intricacies of seeking, finding, and utilizing this assistance are at the heart of emergency management and often lie within the Emergency Operations Center (EOC). Admittedly, each jurisdiction retains and traditionally uses the ability to determine the most effective and efficient way to manage volunteer assistance, so there are numerous successful methods from which to choose. In times of need, especially in disaster situations, it is vital for emergency management to be prepared to utilize assets and resources sufficient to mitigate the incident and initiate recovery as otherwise needed. The one resource man has called upon since the beginning of disaster response is perhaps the most complex and diverse of all current available resources. Although our fellow man remains the greatest resource in any disaster, we sometimes struggle with how best to utilize this resource.

Managing volunteers can arguably be one of the most challenging tasks for emergency management and EOCs. Failure to manage volunteers appropriately can and will have detrimental effects on the success of various missions as well as the final outcome and associated recovery from disasters. This effect is magnified in large-scale events. The effective management of volunteers can be directly attributed to the overall success of many missions. Volunteers bring a host of positive substance to disaster mitigation, preparedness, response, and recovery operations. We will explore many of the pros and cons to various volunteer management tactics throughout this chapter. We will also examine some of the pitfalls of managing volunteers while taking a look at best practices and successful techniques. There is no right or wrong way to manage volunteers, but experience and history can often point us in the right direction and lead us on a path with a high potential for success.

EOCs as well as the plans that drive them can be very diverse and complex. You must choose appropriate tactics or methods to manage volunteers that will be successful given the existing plan or EOC structure. In some cases, the plans must be changed to facilitate the best tactics to manage volunteers. EOC managers and emergency management staff must constantly evaluate the effectiveness of their plans for managing volunteers and be willing to change them as needed. The Federal Emergency Management Agency (FEMA) published the *Comprehensive Preparedness Guide (CPG) 101* to help emergency planners and plan authors standardize plans. This guide does not stipulate the planning structure or framework to the point where no options are available to the authors, but it does provide a framework that planners can utilize along with their own choices of strategy, operations, and tactics.

FEMA breaks down planning into three types:

1. Strategic—policy, objectives, and overall guidance
2. Operational—roles and responsibilities, tasks, integration, and actions
3. Tactical—personnel, equipment, and resource management

The most successful volunteer management programs address all three planning types and continuously update their guidance documents for consistent current information. CPG 101 then allows several planning approaches such as scenario-based, functional, and capability-based planning.[1]

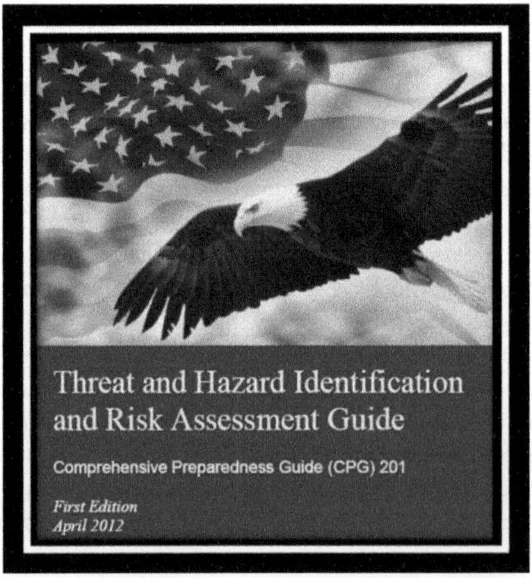

It is easy to see that planners retain many options for how to manage volunteers. Volunteer managers should be familiar with planning principles to devise and articulate management tactics appropriately. This is especially true if these tactics are to be integrated into existing planning structures or drafted to stand alone.

Since volunteers are most often considered resources, a great deal of volunteer planning occurs at the tactical level. Volunteer groups may sometimes fulfill tactical and some operational-level planning goals and objectives, leaving the volunteer managers to integrate these plans with appropriate strategies. Regardless of the planning preference, volunteer management must also integrate into the jurisdiction's EOC. The EOC may also be set up in any number of ways, none of which would be considered wrong as the preference of the jurisdiction dictates the layout and responsibilities of any established sections or groups. A National Incident Management System (NIMS)-compliant EOC will mirror the command and general staff functions of the Incident Command System (ICS). It is important to study both the plans and the EOC design to achieve the highest degree of functionality in managing volunteers from the EOC level. Volunteer managers should consult with EOC officials and take advantage of any opportunities to secure appropriate supplies, equipment, space, and access to relevant EOC programs. As an example, some EOCs provide space for call center operations specifically for volunteer management. Another successful technique involves combining volunteer and donations management functions, so similar management assets such as call centers and logistical support can be shared and perhaps more effectively utilized.

EOCs are often set up in a similar manner to the emergency operations plan that guides them. An emergency operations plan that utilizes emergency support functions will likely reflect its parallel in the design of that EOC. Likewise, an emergency operations plan that utilizes functional annexes instead of emergency support functions could also be used by an EOC that utilizes more traditional strict functional sections. In rare cases, volunteer management duties fall outside of emergency management. Depending on the jurisdiction, senior officials may appoint volunteer coordinators outside of emergency management and simply direct them to coordinate volunteers with emergency management officials. In any case, managing volunteers is not always about the volunteers themselves. The emergency management system of plans and EOCs must be an integrated part of volunteer management as volunteer management must be integrated into these same plans and facilities designated to support emergency management missions.

Volunteer Types

The word "volunteer" is often used to describe any number of separate and distinct resources. It seems that the only common denominator is financial in nature. Volunteers are generally a pro bono asset and thus have a profound positive characteristic as compared to other more costly resources. Emergency managers that successfully recruit and utilize volunteer resources often accomplish many significant tasks at little to no cost.

The most obvious classification of volunteers is individuals and groups. Individuals, when used and managed correctly, are excellent resources for shoring up other assets and resources where manpower is weak. Depending on the individual's specific skill set, individuals can also serve in key roles such as command and general staff. The higher the degree of skill, the more valuable an individual volunteer can be. A good example would be public health and medicine. Highly trained medical professionals can be specifically and individually placed in key positions, thereby acting as a force multiplier for various health and medical groups and teams. This concept has been used across various other disciplines. Individual volunteers rarely respond prepared to manage their own logistical needs, so the EOC that plans on utilizing individual volunteers should consider any logistical support that may be needed.

Groups of volunteers tend to be useful in that they can perform any number of missions depending on their makeup and capabilities. Organized groups often respond with some level of logistical support. Volunteer groups may specialize in one area of expertise or simply provide general manpower and support to an existing mission or ongoing operation. As stated earlier for specific disciplines, volunteers can also be used as the force multiplier, taking on the nonspecific support roles for other professionals. In many cases, highly credentialed professionals with unique skills are tasked in support roles because there are not enough support staff to properly utilize these individuals.

One successful approach has been to maintain a pool of deployable and versatile staff to augment existing teams of professionals. These staff may be subcategorized as needed but for the most part they remain flexible and available for any nonspecialized assignments. Some volunteer groups contain various areas of expertise through its individual members but very little capability in any of those areas due to the small numbers of those individuals. In this instance, and if possible, the tactic of group splitting may be of use. Group splitting occurs when specific expertise is removed from one group to either augment strength or strengthen a weakness of another group. The donor group is essentially split apart to utilize its pieces because the whole did not fit a needed capability. A more simple explanation may be placing the group's individual members into the volunteer pool. This technique can be difficult and will require a high level of coordination and understanding from the donor group. There have been many instances where groups have been turned away because their capabilities as a group did not fit any existing missions or their logistical needs outweighed their usefulness as a uniform group. Group splitting is one method of utilizing volunteers in an existing system while strengthening existing teams and avoiding prolonged staging or lack of mission assignment all together. Again, this method requires a high level of cooperation as reassignment requires logistical alterations as well.

It is important to maintain a significant amount of knowledge regarding the skill sets of individuals in the labor pool so as not to delay their assignment and to ensure their appropriate assignment. Volunteers tend to become frustrated if they are not utilized in a reasonable amount of time. The crucial mistake of undervaluing and underutilizing volunteers may be financially costly and even jeopardize the timely success of the overall mission.

The EOC should be prepared to match both individuals and groups with the appropriate missions to be successful. Since volunteers come in many shapes and sizes, this task often proves tremendously difficult. In many cases, emergency management may not have sufficient information or practiced and proven techniques in place to properly categorize volunteers much less match them to the appropriate missions. Volunteers for disasters often have other commitments in their lives that preclude them from extended missions. Therefore, volunteer time on mission should be considered when assigning individuals or groups to a task.

The specifics of managing volunteers make the process detail sensitive and intensive. Ad hoc methods are rarely successful, so emergency management authorities have developed many techniques to aid in gathering, analyzing, and using information to better manage volunteers. There are numerous systems that include databases, spreadsheets, and software to help keep track of this information but all those systems require this data to be gathered, entered, and kept current for the system to be useful. The system should also be user friendly. Often times, volunteer management systems with extensive capabilities are purchased only to discover that they are too complicated to operate or require extensive constant training to utilize.

Although it seems obvious to the public when help is needed, emergency management should not forgo offering direction to would-be volunteers. Self-deployed volunteers come with the potential to make a positive impact but more often than not, they end up making a negative impact. Where volunteers are concerned, being unprepared to manage the self-deployed versions can be just as detrimental as not having a system to integrate them into at all. The EOC should be prepared to give direction to those who are considering volunteerism or have self-deployed at the time of disaster. Successful methods of engaging this type of volunteer include public information message maps for news opportunities prepared before the event as well as various social media outlets. Specific news interviews regarding volunteering have also been successful, but you must rely on the news outlets to grant this opportunity. The occasion will only be offered under the guidance of the news media as well.

In recent times, social media such as Twitter and Facebook have shown tremendous potential in communicating and updating the public on volunteering as well as many other topics. Self-deployed volunteers often have no information regarding the true needs of the community or the common operating picture for the event. They lack situational awareness and are not plugged in to the intelligence that authorized official responders are privy to know. For these reasons and others, self-deployed volunteers can hamper existing operations and offset the balance of needed services and provided services in a disaster-struck community. Examples include a group of volunteers beginning debris removal before search and rescue is complete or perhaps setting up a medical clinic where medical resources are adequate, while another area goes without needed medical assistance. Self-deployed volunteers often do not have the logistical support to sustain their operation and therefore burden the community they are trying to help.

Emergency management and its partners should screen volunteer groups to determine what, if any, logistical support is needed upon activation. Volunteer groups that are not expected or part of the emergency management system can rapidly drain resources and hamper response and recovery efforts across the spectrum. Self-deployed groups that request and need support from emergency management may be denied that support if resources are not plentiful. Emergency management and EOCs may find it very difficult to predict the number of these volunteers or their impact on an incident. It can be assumed that they will exist and that there must be a plan to communicate and manage the self-deployed as well as the official volunteers.

Volunteer Assessment

As you can see, managing volunteers is no easy task. The larger the event or incident, the more difficult it becomes. Volunteer managers must have a high degree of knowledge concerning the volunteers to be used in a disaster deployment. To obtain and keep this knowledge, various assessments can and should be done. There is no formal model for assessing volunteers, but a generalized information relative to the corresponding operations plan should be readily available and current. Volunteer managers should acquire this information and update it on a schedule that keeps it current within reason.

In addition to knowing the plans, EOC structure, and types of volunteers, volunteer managers must understand the various laws that govern volunteers. There are federal and state laws that directly govern the use of volunteers. There are two specific areas of law that often hamper the volunteer effort and the management of volunteers. The first one is the Fair Labor Standards Act (FLSA), which is monitored by the U.S. Department of Labor. The FLSA covers many wage-related issues, but most often causes the need for concern when professionals volunteer their professional services. Specific instances of volunteering should be examined for FLSA and other legal compliance by the entity authorizing the volunteer action.

The other common legal hurdle involves one of the most complex pieces of volunteer management. That hurdle is credentialing. Credentialing is a vital part of a successful volunteer management program, especially where professional licenses and certifications are being used. Credentialing is generally combined with the initiation of resource tracking. Depending on whether the EOC is charged with this duty, credentialing can be carried out through logistics and monitored by the planning section. Credentialing is an important part of any response system, particularly those with volunteers and especially those where licenses and certifications are required for the individual tasks assigned to volunteers. Emergency management officials should be confident that individuals and groups assigned to certain tasks retain the appropriate credentials to do those tasks.

There are many approaches to accomplishing credentialing. Recently, one of the most common tools used to aid the credentialing process is electronic software systems. Electronic systems can be integrated with various boards and credentialing agencies to verify credentials. These systems may be costly to implement and require extensive support to remain functional and current. When signing contracts or entering agreements with volunteer agencies, emergency management officials should include language that advocates credentialing and compliance with all applicable credentialing bodies. Volunteers that are not part of an established and trusted group that is contracted with the authority having jurisdiction will need to be monitored by an appropriate authority to ensure that proper credentials are present. This may or may not be the ultimate responsibility of emergency management and the EOC.

The medical profession offers a common example. Doctors, nurses, and emergency medical technicians are often credentialed by separate authorities, and these authorities vary from state to state. Furthermore, there are various levels of credential in each of these professions and none of them enjoy universal reciprocity. Various credentialing laws and rules may be relaxed with the stipulation of certain declarations. The volunteer manager may need to be familiar with these circumstances and their limitations.

An unfortunate common occurrence in disaster response and volunteer management is the presumption of relaxed rules and laws. Incorrect assumptions regarding these rules and regulations may result in unnecessary legal liabilities down the road. Along those same lines, the legal status of liability insurance, malpractice, and workers compensation are details often overlooked that later cause immense legal difficulties during or after the response. The health and medical functions of state-level emergency management often utilize a federally subsidized program called the Emergency Systems for Advanced Registry of Volunteer Healthcare Professionals (ESAR-VHP). This system can typically verify credentials for most healthcare professions in many states. Depending on the software used for the system, even this robust program may have limitations. In many cases, out-of-state credentials are not included in the system's database.

Another form of credentialing may be more simplistic, but no less important. Any individuals or groups operating under appropriate authorities should be readily identifiable. For paid-response organizations, this is a standard practice. For volunteers, emergency management may need to take extra steps to secure appropriate identification for the credentialing of volunteers. As perimeters are set up and the disaster area is defined and secured, disaster relief workers who are authorized to work in the protected area will need appropriate levels of identification. The use of standardized identification is a valuable tool in managing volunteers and will help steer nonsanctioned self-deployed groups toward integrating with the existing emergency management-authorized volunteer management system (Figure 12.1).

Successful tactics for credentialing include the establishment of a centralized check-in and staging point that has the capability of issuing official credentials (Figure 12.2). Tactics such as these may also be under the direction of an EOC. Credentialing support is commonly a joint effort in the incident command structure and in most state and local plans. Jurisdictions that strictly follow the model of the National Response Framework (NRF) and its Emergency Support Function (ESF) structure tend to assign credentialing to ESF-7, which are logistics management and resource support. Other models have the credentialing function in several other ESFs, and in some cases, ESFs are added for volunteer and donations management to include credentialing.

The EOC or its assigned partner agency should establish the capability to catalog and type known volunteers and volunteer groups. Just as we discussed this need for individuals and self-deployed volunteers, it is also needed for known volunteer groups. To assign volunteers to an appropriate mission, the authority with the responsibility of assignments should have knowledge of volunteer capabilities. Volunteer groups come with numerous capabilities and levels of those capabilities. Vague descriptions of these capabilities and limited knowledge of the volunteer group can be a recipe for shortcomings and other inadequate resource assignments.

Figure 12.1 Disaster relief volunteers take a break from recovery cleanup duty. (Image provided by Faith-Based Volunteers, http://www.fema.gov/photolibrary/photo_details.do?id=55666.)

Figure 12.2 It is important to document volunteer check-in and check-out information for both accountability and various other reasons for response and recovery information analysis. (Photo by Mark Chambers.)

Volunteer coordinators should make a specific effort to develop and cultivate relationships with established volunteer groups. There are many national volunteer groups with local chapters that are well versed in the details of collaborating with emergency management and jurisdictional volunteer coordinators. Some of these groups are larger than others, but many have developed their own niche and do specific work rather well. These groups are typically nonprofit nongovernmental organizations or faith-based organizations. Examples include the well-recognized Red Cross, the National Voluntary Organizations Active in Disaster (VOAD), and the faith-based services of various Southern Baptist organizations.[2]

The National VOAD, for example, has a mission of sharing knowledge and resources throughout the disaster cycle—preparation, response, and recovery—to help disaster survivors and their communities. Members of National VOAD form a coalition of nonprofit organizations that respond to disasters as part of their overall mission.[3] There are various Southern Baptist and Baptist organizations that are known nationally for services that range from meal provisions to child care. Almost every well-established faith-based organization participates in disaster relief to some degree.

Established volunteer organizations have proven to be valuable pieces in the formula for disaster response success. If you have a trusted relationship and current visibility on these resources, they can ensure success where traditional response agencies often fall short. While cultivating these relationships, it is also important to maintain consistent communications. An occasional call or visit can yield a number of useful results. Volunteer agencies, even the large ones, may have a high degree of turnover or unstable means to make good on capabilities. It is important to maintain good communications to account for a change in capabilities or their ability to support themselves from a logistical standpoint. One example would be food services. In some cases, a volunteer group may retain the capability to establish a food kitchen and feed disaster victims or even responders for a preestablished amount of time (Figure 12.3). This capability may be dependent on donations, so when donations wane, volunteer managers will need to make adjustments for logistical support or mission assignments. These actions can be undertaken during emergency, but it is far easier to maintain an accurate capability picture than to manage large amounts of contingency actions.

Another successful technique in making the most of your volunteer resources is training. Although there are countless programs already out there, such as Citizens Emergency Response Teams (CERT) or Medical Reserve Corps (MRC), there are just as many organized and unorganized groups and individuals who simply need some direction to be a successful part of the response. After examining existing plans and studying historical disaster operations, the volunteer manager may want to develop just in time (JIT) training for specific duties. JIT has proven useful in many disaster responses. During Hurricane Katrina, thousands of self-deployed volunteers were given JIT from hundreds of groups, which allowed them to be a useful part of the response.

Many emergency managers and volunteer managers underestimate the positive and the negative economical and functional impact of volunteer groups. Volunteer operations can infuse a disaster area with much-needed spending. Volunteer groups

Figure 12.3 Volunteers mobilize to provide feeding support. (Image provided by Faith-Based Volunteers, http://www.fema.gov/photolibrary/photo_details.do?id=55594.)

that help restore either normal operations of local businesses or the ability of patrons to reestablish business with them can prove vital to the recovery effort. Nontraditional operations such as home cleanup, small debris removal, child care, food services, and even security can rapidly jumpstart recovery operations far in advance of the work that traditional emergency services provides. Volunteer services often provide the initial actions needed for people to return to their homes and business after a disaster strikes. However, these services can outlast their need as well. The provision of free services in place of paid services that preexisted the disaster should serve as temporary bridge to recovery and not intentionally or unintentionally supplant the services that were in place prior to the disaster. The art of removing free or volunteer services at the appropriate time is difficult to master. Volunteer managers and emergency managers should develop assessment techniques and establish triggers to pull back free services so as not to permanently cripple businesses in their own process of recovery. Large-scale disasters such as hurricanes may require long-term response and recovery services in the disaster area. In several instances, local businesses have been forced out of business by identical services for free provided by volunteers. This is especially true in the small retail and medical fields.

The demobilization of volunteers is an important and intuitive process that is often overlooked by emergency management officials. Admittedly, emergency management may have little to no authority over volunteer organizations but good communication and guidance from emergency management officials with regard to the economic and functional stability of a community may go a long way in protecting that community from being overserviced. A good volunteer management plan should contain triggers for mobilization as well as demobilization. These triggers should account for long-term effects of volunteer services that may have profound economic impact.

The management of volunteers is a complex and vital service that is essential in all disasters. As with any asset or resource, preparing to receive them and integrate them into a uniform and organized response can be a formula for profound success in small- and large-scale events. Emergency management and volunteer management officials should maintain a current common operating picture of volunteer services that is cataloged and easily used for mission tasking. Officials should also prepare to manage self-deployed volunteers and constantly examine the short- and long-term effects of volunteer services. Finally, knowing when and how to demobilize volunteer services may prove just as useful in the long term as knowing how to use them in the first place.

References

1. Federal Emergency Management Agency. 2010. FEMA Comprehensive Preparedness Guide 101, http://www.fema.gov/pdf/about/divisions/npd/CPG_101_V2.pdf. Last accessed August 31, 2013.

2. U.S. Department of Health and Human Services. Emergency System for the Advanced Registry of Volunteer Healthcare Professionals, http://www.phe.gov/esarvhp/pages/default.aspx. Last accessed August 31, 2013.
3. The National Voluntary Organizations Active in Disaster. National VOAD, http://www.nvoad.org/index.php?option=com_content&view=article&id=53&Itemid=188. Last accessed August 31, 2013.
4. Federal Emergency Management Agency. Faith-Based Volunteers, http://www.fema.gov/photolibrary/photo_details.do?id=55666. Last accessed August 31, 2013.
5. Federal Emergency Management Agency. Faith-Based Volunteers, http://www.fema.gov/photolibrary/photo_details.do?id=55594. Last accessed August 31, 2013.

Chapter 13

Legal Considerations in Threat Response Management

Ernest P. Chiodo

Introduction

There are a number of legal issues that must be considered by the threat response manager. The threat response manager must be aware of legal issues that are likely to arise in a crisis. These issues include hoarding of vital supplies, including medical supplies as well as prescriptions; provision of medical services by physicians distant to the local patient; the unique qualifications required for a physician to make public health decisions as opposed to medical decision; and the police powers possessed by public health organizations as opposed to medical organizations. While there are a wide range of legal issues that may be of interest to the threat response manager, the issues discussed in this chapter are not likely to be readily accessible in the general legal literature and the insights are crucial in a crisis. The legal issues will be considered from a U.S. federal, state, local, and when appropriate international prospective.

Legal Counsel

It is important to note that this is not a "how to" chapter. Any plans, decisions, or actions should be made in conjunction with the advice of a qualified lawyer. This chapter is not legal advice. The legal counsel should be utilized throughout the threat response management process. Law is a discipline that is outside the scope of expertise of most emergency response managers. The prudent emergency response manager will recognize that the legal counsel is of the utmost need in charting the water of the legal pitfalls that can arise in emergency circumstances. In particular, a qualified legal counsel should be involved in the formulation and review of all threat response management plans and procedures. The legal counsel should be a part of the management team. It is important that a relationship be established with the legal counsel where the legal counsel is available on an emergency basis to provide legal advice as an emergency response circumstance arises and progresses. This may require the payment of a retainer to the legal counsel for the purpose of being available immediately if the need arises. This is analogous to a physician being "on call" to deal with medical issues. You will also need a lawyer to be "on call" to deal with legal issues. Of course, there is the need to have contact names and telephone numbers readily on hand for the "on call" legal counsel. Emergency plans need to be periodically reviewed and revised. The legal counsel should be involved in the review and revisal process of emergency response plans.

Hoarding

During a crisis, there may be concerns about hoarding of vital supplies and food stuffs by individuals and corporations. During times of crisis in the past, there have been efforts by individuals and corporations to store large quantities of supplies.

Federal Legal Issues

The federal government implemented a program of rationing of foods and consumer goods during the World War II. Rationing has not occurred to any extent during any military conflict involving the United States since the World War II. The lack of need for rationing is a manifestation of the large surplus production capacity of the United States compared to the needs of the military during the time of active conflict. However, the collapse of the Soviet Union in the early 1990s occurred at a time of a major economic expansion in the United States. At that time, there were thoughts that one of the reasons for the major economic expansion was the "war dividend" due to the lack of further expenditures to fight the "Cold War."

Consequently, there is some evidence that the ability of the United States to maintain production at a level to counter a military treat has some limits even during times of normal economic functioning. The disruptions that would be anticipated to occur during a major crisis would be likely to place tremendous stresses upon the supply of foods and consumer goods. The plans of individuals and corporations to stock-up on food and other supplies may run afoul of the plans of the U.S. federal government. However, it should be noted that the

Center for Disease control website does provide the same recommendations for the general public.

It should be noted that the U.S. Food and Drug Administration (FDA) has authority over foods as well as drugs sold within the United States. The earliest federal legislation that dealt with food and drugs was the Drug Importation Act of 1848, which was passed by the U.S. Congress to require the U.S. Customs Service to conduct inspections so as to stop the importation of adulterated drugs from overseas (*Legal Medicine*, 2001, p. 560). The first comprehensive Food and Drugs Act was enacted by the U.S. Congress under the Theodore Roosevelt administration on June 30, 1906 (*Legal Medicine*, 2001, p. 561). In 1938, the U.S. Congress passed the Federal Food, Drug, and Cosmetic Act (FDCA) (21 United States Code [U.S.C.] 321 to 394) (*Legal Medicine*, 2001, p. 561).

State Legal Issues

Each individual state within the United States may have concerns about stresses placed upon local sale and supply of food and consumer goods that may mirror the concerns of the federal government.

Local Legal Issues

The local governments such as counties and municipalities may have concerns about disruption of supplies that mirror the federal and state concerns.

International Legal Issues

Large corporations have employees in many countries throughout the world. The governments of each nation with employees of a corporation may have concerns about hoarding of food and consumer supplies. The hoarding of food and consumer goods in second- and third-world countries may place great stress upon the resources of these nations. Consequently, corporations that recommend that their employees store supplies of food and materials may be giving recommendations contrary to the mandates of the various nations in which the corporation has employees. In fact, during times of great national crisis, many nations have made hoarding of food and consumer goods a criminal act. During the World War II, persons who were involved in hoarding were often considered to be "black marketers."

> "Black market" was the term given to the illegal trade in consumer goods, manufactured products and raw materials without regard to rationing or price fixing statutes, practiced because of the scarcity of goods. The constantly escalating game of bidding and the risks that the traders ran pushed prices up to unbelievable levels. (*The Historical Encyclopedia of World War II*, 1980, p. 57)

In Britain and the United States, black marketeers were subject to imprisonment. These hoarding activities were a capital offense in totalitarian regimes such as Nazi Germany and the Soviet Union.

Supplies of Prescription Drugs

Obtaining adequate supplies of prescription drugs is likely to become difficult during a crisis. Even obtaining the prescriptions may become very difficult since licensed physicians are likely to be scarce during a crisis.

Federal Legal Issues

Medical licensure within the United States is conducted by the various states of the union. Consequently, a prescription written by a physician in one state may not be deemed valid in another state. The various states in the United States have the right to protect the health of their citizens by means of controlling licensure of physicians to practice medicine within the borders of the state.

> The right of a physician to toil in his profession ... with all its sanctity and safeguards is not absolute. It must yield to the paramount right of government to protect the public health by any rational means. (*Lawrence v. Board of Registration in medicine, 239 Mass. 424, 428, 132 N.E. 174 (1921)*)

This may result in difficulties in filling of prescriptions in states other than the state where the prescription was written. The licensure statutes of the states had the original propose of ensuring that physicians had the minimum scientific knowledge to safety practice (*Legal Medicine*, 2001, p. 70). However, in the case of *Hawke v. New York, 170 U.S. 189, 194 (1898)*, the U.S. Supreme Court ruled that states have the right to use standards of behavior and ethics as factors in granting or removing physician licensure. Consequently, the various states have the power to control the practice of medicine within their borders through medical licensure. The writing of prescriptions in one state to be filled by the patient in another state may be deemed a clear violation of the right of the state where the prescription is filled to control the practice of medicine in that state.

In addition, during the time of an epidemic, there may be migration of persons from colder areas of the United States to warmer areas in the mistaken brief that there is less risk of becoming infected with influenza. As a result, there may be abnormal stresses placed upon the supplies of noninfluenza-related medications that may be problematic for health delivery. For example, a state such as Florida may experience an abnormally large demand for medication needed to treat diabetes such as insulin. This may result in a lack of necessary medications for diabetics who usually live in Florida. Conversely, there may be an oversupply of insulin in a cold northern state such as Michigan where there may be a lower prevalence of diabetes due to a younger age demographic. The federal government may wish to control the supply of vital medical supplies across state borders. This control of the flow of medical supplies across state borders may be controlled by the federal government through the operation of the Interstate Commerce Act. Interstate and foreign commerce are defined as follows:

> Commerce between a point in one State and a point in another State, between points in the same State, through another State or through a foreign country, between points in a foreign country or countries through the United States and a point in a foreign country or in a Territory or possession of the United states, but only insofar as such commerce takes place in the United States. The term "United States" means all the States and the District of Columbia. 18 U.S.C.A. § 831. (*Black's Law Dictionary*, 1979, p. 735)

Interstate commerce was defined in *Gibbons v. Ogden, 22 U.S. (9 Wheat.) 1, 6 L.Ed. 23* as follows:

> Traffic, intercourse, commercial trading, or the transportation of persons or property between or among the several states of the Union, or from or between points in one state and points in another state; commerce between two states, or between places lying in different states. (*Black's Law Dictionary*, 1979, p. 735)

Furst v. Brewster, 282 U.S. 493, 51 S.Ct. 295, 296, 75 L.Ed. 478 further defines interstate commerce as "It comprehends all the component parts of commercial intercourse between different states" (*Black's Law Dictionary*, 1979, p. 735).

The Interstate Commerce Act provides for the following:

> The act of congress of February 4, 1887 (49 U.S.C.A. § 1 et seq.), designed to regulate commerce between the states, and particularly the transportation of persons and property, by carriers, between interstate points, prescribing that charges for such transportation shall be reasonable and just, prohibiting unjust discrimination, rebates, draw-backs, preferences, poling of freights, etc., requiring schedules of rates to be published, establishing a commission to carry out the measures enacted, and prescribing the powers and duties of such commission and the procedure before it. (*Black's Law Dictionary*, 1979, p. 735)

Clearly, the federal government has the power to have some degree of control over the flow of medical supplies and pharmaceuticals across state borders.

However, the federal government may not be able to control the flow of persons across state borders due to the privileges and immunity clause of the U.S. Constitution. Perhaps more importantly, the various states may not normally control the flow of U.S. citizens into their state from other states. The Privileges and Immunities Clause of the U.S. Constitution provides the following protections:

> There are two Privileges and Immunities Clauses in the federal Constitution and Amendments, the first being found in Art. IV, and the second in the 14th Amendment, § 1, second clause, clause 1. The provision in Art. IV states that "The Citizens of each State shall be entitled to all Privileges and immunities of Citizens in the several States," while the 14th Amendment provides that "No state shall make or enforce any law which shall abridge the privileges or immunities of citizens of the United States."
>
> The purpose of these Clauses is to place the citizens of each State upon the same footing with citizens of other states, so far as the advantages resulting from citizenship in those states is concerned. (*Black's Law Dictionary*, 1979, p. 1079)

Consequently, there is likely to be a bar against actions by the governments of the various states in attempting to prevent the entry of U.S. citizens into their state from another state. This bar may be effective even during the time of a crisis such as an avian influenza epidemic.

State Legal Issues

The various states within the United States regulate the prescriptive authority of the physicians, dentists, and nurse practitioners within each state. Consequently, one state may have a different law concerning prescriptive authority than another state. For example, one state may allow a nurse practitioner to write prescriptions while another state may not allow nurse practitioners to write prescriptions. As a result, a prescription written by a nurse practitioner in a state granting prescriptive authority to nurse practitioners will almost certainly not be honored in a state that does not grant nurse practitioners prescriptive authority. In addition, even if a practitioner such as a physician is of a profession that is granted prescriptive authority in two different states, the writing of a prescription by a physician licensed in one state with knowledge that the patient intends to fill the prescription in another state in which the physician is not licensed may be considered the unauthorized practice of medicine. It should be noted that the unauthorized practice of medicine is usually considered a felony. Consequently, prescriptions written by a physician in one state where the physician knows that during a crisis the prescription will be filled and dispensed in a foreign state may present issues concerning the unauthorized practice of medicine.

The onset of an epidemic may lead people to attempt to obtain medical advice and treatment in means other than through direct contact. There are likely to be attempts to obtain medical advice and treatment via electronic devices such as the telephone or the Internet. The reasons for the desire to obtain medical treatment over a distance through electronic means are likely to be threefold. First, there will be reluctance by persons to attend physician offices where there will be an increased likelihood of coming into contact with sick persons who may be infected with an infection disease that is causing the crisis such as avian influenza. There may also be a reduction in the availability of physician services within the patient's immediate geographic area. This will cause patients to seek medical advice and treatment at a distance. The providers of this medical advice and treatment may be located across state borders. In addition, patients may seek the advice of specialists with particular expertise in the diagnosis and treatment of disease caused by the infectious agent of concern such as avian influenza. These considerations will encourage patients to seek medical care from physicians who may not be licensed to practice medicine within the state of residence of the patient. Physicians may also be drawn to provide care to patients at a distance who may not be located within the same state as the physician and who may have never been seen by the physician. The factors that will draw physicians into this telemedicine activity are also threefold. First, there may be commercial factors that may draw physicians into areas of practice in great demand. Owing to the demand, these areas of specialized practice may be very lucrative. Physicians may also wish to provide medical care without the risk to their personal health due to infection that would arise with direct patient contract. In addition, there will be the admirable desire on the part of physicians to provide needed services to persons that may not have access to medical services in their own communities due to remote geography or other factors limiting access to care. The issue of telemedicine raises a number of legal concerns.

The first legal consideration concerning telemedicine services involves legal liability for medical malpractice. The case of *International Shoe v. State of Washington*, 325 U.S. 310, 66 S.Ct. 154 (1945) established the principle that for a state to have legal jurisdiction over a person, that person must have some minimal contacts with the state seeking the jurisdiction. Within the realm of medical malpractice, it generally considered that the physician can only be sued within the state where he provided the medical service even if the patient has a place of residence in another state (*Legal Medicine*, 2001, p. 238). However, in *Bullion v. Gillespie*, 895 F. 2d 213 (5th Cir. 1990), a physician was held to be subject to the jurisdiction of another state where the patient lived when it was shown that the physician attracted patients to his practice through nationally distributed marketing literature. In *Kenndy v. Freeman*, 919 F. 2d 126 (10th Cir. 1990), a physician was found to be subject to the jurisdiction of another state when he was involved in the reading of biopsy specimen sent by physicians located outside of his own state. It should also be noted that several states have forbidden the prescription of medications for patients where the only contact with the patient was through telemedicine (*Legal Medicine*, 2001, p. 238). In *re B.T. Taylor, M.D., Action report, Medical Board of California (Oct 1999)*, the Board of Medicine of the State of California reprimanded a California physician when it received notification of a disciplinary action by the State of Colorado due to the physician having prescribed medications to patients in Colorado with whom the physician only had contact via the Internet.

Local Legal Issues

Local governments may have concerns about stresses placed upon the healthcare delivery network in their areas. Public healthcare delivery services such as municipal and county health departments may be overwhelmed by the demands placed upon them by persons seeking prescriptions.

International Legal Issues

Each national government has the power and right to regulate the practice of medicine within its national borders. Prescriptions written in foreign countries are not likely to be honored in a nation other than the country of origin of the prescription. The public policy issues in this matter are similar to the concerns of the various states of the United States in preventing the unauthorized practice of medicine. It should be noted that the U.S. Supreme Court only ruled in 1973 that a requirement that a person be a citizen of the United States to obtain medical licensure is an unconstitutional discrimination (*Legal Medicine*, 2001, p. 71). It should not be surprising if foreign nations may consider the medical advice to its citizens or residents by a physician that is not a citizen of that nation is an unauthorized practice of medicine.

Hoarding of Nonprescription Drugs and Other Health Supplies

During a crisis, individuals and corporations may wish to stockpile nonprescription drugs and other health supplies that may be needed during the crisis. These supplies include toiletries, and products for personal and dental hygiene such as soap, deodorants, and tooth paste. Fluids with electrolytes may also be important in cases of dehydration due to diarrhea.

Federal Legal Issues

The public policy issues in this matter are essentially the same as in the recommendations for food and consumer goods since the issues for prescriptive authority do not apply.

State Legal Issues

The public policy issues in this matter at the state level are essentially the same in the case of the recommendations concerning food and consumer goods since the recommendation concerns nonprescription medications.

Local Legal Issues

Local authorities may have concerns about local distribution of the personal hygiene products because of the risk of disease spread with declines in hygiene. There may be a need for an increase in disease surveillance on the part of local health departments.

Autonomy and Direction of Care

During a crisis, question will likely arise as to who will direct the care of person who becomes ill during the crisis.

Federal Legal Issues

In Western democracies, including the United States, there is the understanding that persons have the right to make their own decisions about matters influencing their health. This concept is termed *autonomy* from the Greek *auto nomos*, which means self-rule. This means that in a medical setting a patient has freedom of choice (Legal Medicine, 2001, p. 291). In *Schoendorf v. Society of New York Hospital,* 1914, 105 N.E. 92 (N.Y.C.A.), Justice Cordozo stated that "every human being of adult years and sound mind has a right to determine what shall be done with his own body."

International Legal Issues

In 1948, there was a Universal Declaration of Human Rights that was ultimately adopted by the World Medical Association and the World Health Organization. These measures were designed to prevent some of the great evils that occurred during World War II. The focus was to prevent human experimentation of the type that occurred during the Nazi regime (*Legal Medicine*, 2001, p. 291).

Qualifications of Physicians Making Public Health Decisions

During a crisis, there will be a need for crucial public health policy decisions to be made by qualified physicians. However, the issue of what are the necessary qualifications for a physician to make public health policy decisions is often not known by most physicians. While the qualifications needed by a physician to properly determine public health policy and procedures is not strictly a legal matter, the failure of threat response managers to recognize what type of physician is qualified to provide public health guidance raises obvious liability concerns.

Federal, State, Local, and International Legal Issues

There is specialized training required for the effective functioning of a physician as a public health official.

For example, the fundamental tools of public health officials are not part of the routine medical or nursing training. Epidemiology and biostatistics are crucial tools in the arsenal of any public health official. However, these disciplines are either not taught or only provided only superficial coverage in most medical and nursing schools. In addition, course work in public health management is virtually nonexistent in medical and nursing school curriculum. Public health-related disciplines such as epidemiology and biostatistics are taught primarily in schools of public health. In addition, subject matter focused upon public health issues and identification and control of disease in populations of persons is not part of the specialty training of most physicians. Issues concerning public health are covered in training in the specialties under the American Board of Preventive Medicine. The American Board of Preventive Medicine requires that a physician have documented formal education in the following areas to qualify for the board certification examinations in the preventive medicine specialties.

> In addition to the knowledge of basic and clinical sciences and the skills common to all physicians, the distinctive aspects of preventive medicine include knowledge and competence in:
>
> 1. Biostatistics,
> 2. Epidemiology,
> 3. Administration, including planning, organization, management, financing, and evaluation of health programs,
> 4. Environmental and occupational health,
> 5. Application of social and behavioral factors in health and disease,
> 6. Application of primary, secondary, and tertiary prevention measures within clinical medicine.
>
> (*The American College of Preventive Medicine Directory of Preventive Medicine Residency Programs in the United States and Canada*, 1988, p. 1)

The American College of Preventive Medicine defines the roles of preventive medicine specialists as follows:

The roles of preventive medicine specialists include serving as:

1. Health planners and administrators,
2. teachers of preventive medicine,
3. Researchers in preventive medicine,
4. Clinicians applying preventive medicine in health care.

(*The American College of Preventive Medicine Directory of Preventive Medicine Residency Programs in the United States and Canada*, 1988, p. 1)

The American College of Preventive Medicine describes that setting in which preventive medicine is practiced as follows:

> The setting in which preventive medicine specialists may be found include:
>
> 1. Government organizations such as local, state, national, and international public health departments and other military and civilian agencies concerned with the health of populations,
> 2. Educational institutions such as schools of medicine, public health, and allied health,
> 3. Organized medical care programs in industry, other employment settings, and in the community, where clinical practice involves prevention and health maintenance,
> 4. Voluntary health agencies, professional health organizations, and related organizations.
>
> (*The American College of Preventive Medicine Directory of Preventive Medicine Residency Programs in the United States and Canada*, 1988, p. 1)

The American Board of Preventive Medicine provides board certification to physicians in three areas of specialization. The American Board of Preventive Medicine specialties are *occupational medicine*, *aerospace medicine*, and *public health and general preventive medicine*.

Occupational medicine is the medical specialty that is primarily focused upon the diagnosis, prevention, and treatment of disease in persons in a work setting. The practice of *occupational medicine* requires skills that allow the detection of disease in populations such as the work force of a company. An outbreak of an occupational disease that can arise due to poor industrial hygiene practices such as occupational asthma may not be apparent to a physician not trained in *occupational medicine*. Consequently, the medical directors of most major industrial corporations are board certified in *occupational medicine*.

Aerospace medicine is the area of board certification held by physicians involved in the medical aspects of air and space flight. The physicians working for NASA are usually board certified in *aerospace medicine*. The public health skills required to detect disease in populations rather than solely in individuals are crucial for the functioning of physicians working in *aerospace medicine*.

Public health and general preventive medicine is the medical specialty that is focused upon detection, prevention, and control of disease in large populations such as a large city, county, state, or nation. The United States Centers for Disease Control and Prevention (CDC) "hot zone" doctors and the medical directors of major city and state health department are usually required to be board certified in *public health and general preventive medicine*.

In addition to additional training required for physicians to function as public health officials, the healthcare system and public health systems differ in that public health systems have "police powers" while healthcare systems do not have "police powers."

Police powers are those powers that are necessary to function in a police capacity. The police powers of local public health officials in the State of Michigan in the event of an epidemic, emergency, or case where quarantine is required are described in the Public Health Code of the State of Michigan (Act 368 of 1978, as amended) as follows:

> 333.2453 Epidemic; emergency order and procedures; involuntary detention and treatment.
>
> Sec. 2453. (1) If a local health officer determines that control of an epidemic is necessary to protect the public health, the local health officer may issue an emergency order to prohibit the gathering of people for any purpose and may establish procedures to be followed by persons, including a local governmental entity, during the epidemic to insure continuation of essential public health services and enforcement of health laws. Emergency procedures shall not be limited to this code.
>
> (2) A local health department or the department may provide for the involuntary detention and treatment of individuals with hazardous communicable disease in the manner prescribed in sections 5201 to 5238.

The difference between a public health department and a hospital as part of a healthcare system is illustrated by what happens in the case of a person with a serious infectious disease such as tuberculosis. A hospital that has a patient with infectious tuberculosis who refuses to take antitubercular medication cannot force the person to take the medication. If the person with infectious tuberculosis wishes to leave the hospital and enter into the general community with the risk of spread of tuberculosis, the hospital must allow the infectious person to leave since the hospital is a nonpublic organization without police powers. In this situation, the hospital must contact the local public health department, which then exercises its police powers to detain the infectious individual and compels the person to take antitubercular medications.

The essence of the above dissertation is that the healthcare and public health systems are fundamentally different. The focus of the healthcare systems is on provision of healthcare services to individuals. The focus of the public health systems is detection and control of disease in populations. Public health systems have police powers. Healthcare systems do not have police powers. There is a fundamental difference in the training and therefore the mind set of physicians working in public health compared to physicians working in health care.

In summary, the threat response manager must be aware of legal issues that are likely to arise in a crisis. These issues include hoarding of vital supplies, including medical supplies as well as prescriptions; provision of medical services by physicians distant to the local patient; the unique qualifications required for a physician to make public health decisions as opposed to medical decision; and the police powers possessed by public health organizations as opposed to medical organizations.

References

Black's Law Dictionary, 1979. Fifth Edition. West Publishing Co., St. Paul, Minnesota. ISBN 0-8299-2041-2.

Legal Medicine, 2001. Fifth Edition. Mosby, Inc., St. Louis, Missouri.

The American College of Preventive Medicine Directory of Preventive Medicine Residency Programs in the United States and Canada, 1988. Fifth Edition. American College of Preventive Medicine, Washington, DC.

The Historical Encyclopedia of World War II, 1980. Facts on File Inc. Greenwich House, New York, ISBN 0-517-431491.

Chapter 14

Sport Venue Emergency Planning

Stacey Hall

Introduction

Providing a safe and secure environment is a priority for sport venue operators due to the potential for all-hazard emergencies and subsequent legal implications. The education and training of all key security personnel is critical to ensure that effective security measures are implemented and that an all-hazards approach to emergency planning is employed (Sauter and Carafano, 2005). An effective sports event security management system requires the involvement and commitment of many agencies and individuals, including professionals, volunteers, public agencies, and outsourced contractors.

Emergency Management

An emergency situation is defined as "any incident, situation, or occurrence that could affect the safety and security of occupants, cause damage to the facility, equipment and its contents, or disrupt activities of the facility" (Center for Venue Management Studies, 2002, p. 1). Examples of emergency situations at sport venues include bomb threats/explosions, medical emergencies, fire hazards, natural disasters/inclement weather, power or equipment failure, crowd control issues, hazardous material release, domestic or international terrorism, and evacuations (partial, full, and shelter in place) (Hall et al., 2012). Emergency planning is imperative for several reasons:

- Sport facility operators must adhere to professional industry standards and expectations to provide a safe environment for patrons.
- Previous incidents at sporting events have demonstrated that planning, training, and implementation of an emergency plan have reduced the potential damage and loss associated with emergencies.
- Sport facility operators must adhere to legal public safety requirements, such as fires safety codes and American Disability Act (ADA) requirements.
- Documentation of an emergency plan will minimize the liability to facility, owners, and operators when an emergency occurs.
- Good planning practices will minimize negative media exposure and enhance a facility's positive image with patrons and local community stakeholders.

Emergency management practices attempt to avoid or reduce potential losses from all hazards, and ensure appropriate assistance to achieve rapid and effective recovery after an emergency (Warfield, 2008). Sport venue managers must focus on all components of this interrelated process (see Figure 14.1). The sport venue command group (CG) must work closely with external agencies involved in the planning, preparedness, response, and recovery operations at venues and events.

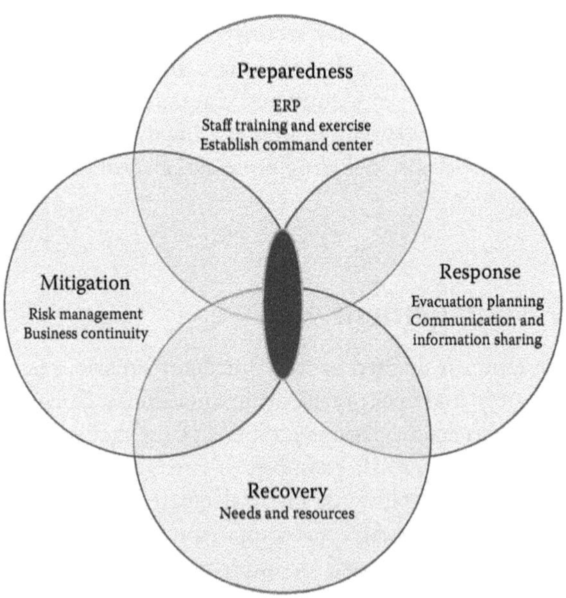

Figure 14.1 Emergency management components.

Sport Venue Command Group

A sport venue CG is composed of key personnel from multiagencies that collaborate to provide a safe and secure sporting environment for spectators, athletes, officials, sponsors, and community stakeholders. The CG is composed of representatives from the following key areas: sport venue management, law enforcement, emergency management, fire/hazardous materials (HAZMAT), and emergency medical services (EMS). Other key entities and individuals that may play a role in security operations include state and federal government security, public health and safety, media/public relations, public utilities, contractors, vendors, and temporary employees (i.e., volunteer groups). The CG serves as the core functional group for all security efforts, including incident management, risk management, staff training, developing and implementing plans and protective measures, conducting exercises, and coordinating response, recovery, and continuity efforts. It is important to establish this multidiscipline team prior to the sport season (or event day), and it is critical for them to continue to meet on a regular basis (Hall et al., 2012).

Preparedness

Preparedness is a key measure to reduce the impact of an emergency by taking actions before an incident occurs. It is impossible to ensure a risk-free environment; therefore, planning and preparedness activities are critical to build and sustain capabilities to protect against, respond to, and recover from all hazards. Preparedness is a continuous improvement process and involves all key stakeholders in the sport venue CG, as well as local, state, and federal agencies as needed. Preparedness activities may include developing an emergency response plan (ERP), training staff and conducting exercises, and establishing command and control.

Emergency Response Plan

The ERP describes how the sport venue will do business in an emergency situation. The emergency plan highlights the steps and actions to be taken by the sport venue staff and response agencies to minimize or eliminate potential harm to facility patrons and/or damage to the venue (Hall et al., 2012). The ERP's ultimate goal is to reduce risk exposure and potential consequences through prevention, detection, communication, damage control, and recovery efforts (Center for Venue Management Studies, 2002). The sport venue CG becomes the emergency planning team (EPT) responsible for developing the sport venue ERP in coordination with local and state authorities. The EPT have working knowledge of the venue (including its systems and resources) and are decision makers that have the ability to implement the ERP. The plan should be developed with specific actions and responses to any possible emergency.

Staff Training and Exercise

Once an ERP has been developed, key staff and facility personnel responsible for implementation should be trained and tested. The three primary levels of staff training include the multiagency team (CG), supervisors, and event staff (i.e., ticket takers, ushers, checkers, parking attendants, etc.). The following list highlights possible training techniques (Hall et al., 2012):

- Training sessions (seminars/workshops)
- Practice drills to test the staff's knowledge of the venue ERP
- Prepost event briefings
- Hand-held cards provided to event staff containing information pertinent to emergencies and specific roles and responsibilities
- Include external public agencies involved in responding to venue emergencies (i.e., police, fire, public health, media relations) in training sessions
- Produce training videos to orient new employees on the ERP

Once sport venue managers have assessed threats and risks, developed plans, and trained staff members, they should consider testing their operational plans to assess their level of preparedness. The CG should conduct exercises to test plans and promote awareness of roles, responsibilities, and position assignment during an incident scenario. "An exercise is a focused practice activity that places the participants in a simulated situation requiring them to function in the capacity that would be expected of them in a real event" (Federal Emergency Management Agency, 2008, p. 2). Exercises improve readiness by evaluating operational capabilities, reinforcing the concept of teamwork, and addressing identified gaps. Exercises help venue managers to (1) clarify roles and responsibilities, (2) improve interagency coordination and communication, (3) reveal resource gaps, (4) develop individual performance, (5) identify opportunities for improvement, and (6) gain program recognition and support of management (Federal Emergency Management Agency, 2009).

The U.S. Department of Homeland Security Exercise and Evaluation Program (HSEEP) promotes seven different types of exercises, categorized as either discussion based or operation based. Discussion-based exercises familiarize participants with current plans and policies, and may be used to develop new plans, policies, and procedures. Discussion-based exercises include seminars, workshops, table-top exercises, or game simulations. Operation-based exercises are more complex than discussion-based exercises. Operation-based exercises validate plans and policies, clarify roles and responsibilities, and identify resource gaps in security operations and capabilities. Operation-based exercises normally involve the deployment of resources and personnel; these include drills, functional exercises, and full-scale exercises (HSEEP, 2007).

Establishing a Command Center

Most major sport venues include a command center (command post) with communication capabilities for security forces to monitor activities inside and outside the venue perimeter. The command center controls the security functions for the event and is normally staffed with the facility security director, venue operations manager, police, fire, EMS, private security, and media representatives. Copies of security and facility plans, phone directories, and backup technology systems are located at this site. The command center has reliable communications and the capability to access the venue public announcement system, fire alarm system, voice activation system, turnstile system, and access control systems. Command center capabilities include (Hall et al., 2012, p. 51)

- Coordinate internal response to all minor incidents
- Refer support requests to external agencies for major incidents
- Manage all event communications
- Document venue incidents
- Manage event timeline
- Maintain a safe, orderly environment
- Direct and manage venue evacuation
- Expand or contract based on the incident

The command post is usually established prior to event day and is activated in the event of an incident. The command post signifies the location of the tactical-level, on-scene incident command and management organization. The sport venue CG must decide where the command post will be established, and ensure the location offers enough space for multiagency personnel. The CG must consider what other government officials should be notified and included. The identification and availability of additional emergency resources and amount of time to access these services should be estimated. In addition to possessing adequate resources, certain facilities should be designated for emergency use during a crisis, for example, shelters to house displaced victims, distribution centers for food, water, and emergency supplies, and storage areas for equipment (The Spectrum of Incident Management Actions, 2006). A general guidelines checklist for emergency preparedness is presented in Appendix A.

Response

According to Lindell et al. (2007), emergency response has three distinct goals: (1) protect people, (2) limit damage from primary impact, and (3) minimize damage from secondary impacts. Response efforts begin when an emergency is imminent or immediately after an incident occurs. Response includes tasks and activities that address the direct (short-term) effects of an incident. Response activities may include execution of the ERP and evacuation procedures, increasing security operations, and implementing emergency communication systems.

Evacuation Planning

Planning for an evacuation at sporting events requires coordination, communication, and cooperation by venue operators and the response community (federal, state, local, and private). An evacuation plan should be an essential component of the facility's ERP and should take into consideration all potential hazards for a particular venue. Making the decision to evacuate, shelter in place, or relocate during an incident is complicated and requires input from various entities knowledgeable about the structure of the stadium, the size and distribution of the spectators and participants, the hazard involved, and the anticipated response to that hazard (U.S. Department of Homeland Security, 2008). The incident commander (i.e., police chief/emergency management director) determines whether to conduct a full evacuation, partial evacuation, or shelter in place. A full evacuation may be conducted due to a major structural failure, earthquake, explosion, chemical spill, or severe storm. A partial evacuation may be conducted due to a minor fire, small explosion, bomb threat, minor structural damage, or fan violence. In some instances, it may be necessary to shelter in place by keeping patrons inside the facility, for example, when an event has occurred outside the facility such as inclement weather. A stadium evacuation plan template is provided in Appendix B. Venue managers should consider the following items when developing evacuation procedures (Center for Venue Management Studies, 2002, p. 13):

- Venue staff must direct patrons to a safe area to reduce panic and chaos that may ensue.
- Identify a chain of command, including the person of authority who will make an evacuation decision (normally incident commander).
- Determine alternate points of egress from every point in the facility in advance.
- When an incident occurs and is localized (i.e., fire emergencies), consider staying in place or evacuating to another area of the facility rather than evacuating large crowds to the outside.
- Adequately train supervisory and event staff through preevent briefings and issuing handheld information cards on specific positions, locations, and evacuation procedures.
- Prepare and display evacuation announcements when an evacuation is deemed necessary.
- Develop a plan on what to do with the patrons once they have been evacuated outside the venue to ensure they are out of danger and the emergency is under control.

If venue management chooses to shelter in place rather than evacuate, the CG must consider several items, such as communication procedures with patrons to provide updates on the emergency situation. This can be achieved through prescripted and prerecorded public address announcements, scoreboard signage, video screens, and verbal commands between event staff and patrons. Patrons should also be allowed the opportunity to contact family and/or friends if communication capabilities exist. Complimentary food and beverages may be offered to patrons if they are required to remain in the venue for an extended period of time. Furthermore, the sport organization should develop partnerships (i.e., mutual aid agreements) prior to the sport season with local emergency response agencies to obtain emergency supplies if needed (Hall et al., 2012).

Communication and Information Sharing

Normal day-to-day communications and emergency communications differ. Emergency information is vitally important and can mean the difference between life and death, or it can provide reassurance to those affected that response and recovery efforts are underway. Unfortunately, people find it difficult to hear messages during an emergency because of stress or change of routine (Federal Emergency Management Agency, 2005). Additionally, information shared with the public should be consistent and relay the same intended message since many parties respond to an emergency. When communicating in a crisis, the speaker should (1) word the message precisely, (2) avoid jargon, codes, or acronyms, (3) use common names for all personnel and facilities, (4) omit unnecessary information, (5) speak in sync with other agencies, and (6) keep messages consistent across all media outlets. The types of communication methods are highlighted below (Hall et al., 2012, p. 35):

- Emergency alert system: An established communication system that warns of impending dangers. Individuals should be aware of warning tones, messages across TV screens, cable TV override, and National Oceanic and Atmospheric Administration (NOAA) weather radio.
- Oral communication: Phone conversations, briefings, public speeches, TV interviews, radio announcements, and public service announcements.
- Print communication: Fax, email, public notice, flier, press release, or feature article.
- Social media: Utilization of social media outlets such as Facebook, Twitter, or YouTube.

The most effective communication tool is one that reaches the intended target audience in a timely manner, delivers the message reliably, enhances comprehension of the message content, and can be accessed within resource limitations. The right technology can support and enhance communication capabilities. Most often, a combination of methods is used to deliver a message. The CG should consider (1) how they issue emergency communications, (2) what areas of emergency communications can be improved, (3) what steps need to be taken to improve communications, and (4) whether they should collaborate with other agencies in this effort (Federal Emergency Management Agency, 2005).

The ways of sharing critical information must be agreed upon prior to any incident, for example: What is the chain of command for communication? Does information flow through a preset protocol (during day-to-day activities) or through an all-channel network to preestablished personnel (during an incident response)? Emergency events in the past have highlighted the inability of incident command teams to communicate in a time of crisis because of a lack of basic skills, or incompatible equipment. Venue management should obtain satellite phones to allow continued communication in case hard-wire phone lines into the venue are out of service. All stakeholders involved in response and recovery efforts should focus on alleviating this problem. In times of crisis, all parties must possess high levels of communication skills so that they can send, receive, and understand information (U.S. Department of Homeland Security, 2009).

The Department of Homeland Security (DHS) Office of Intelligence and Analysis has created state and local fusion centers to share information and intelligence within their communities. As of July 2009, there were 72 designated fusion centers in the United States. Sport organizations in the United States should become familiar with their state or local fusion center representatives. This type of support can assist in gathering information pertaining to potential threats or issues at upcoming sporting events. This support is tailored to the unique needs of the locality and serves to (1) help the flow of classified and unclassified information, (2) provide expertise, (3) coordinate with local law enforcement and other agencies, and (4) provide local awareness and access (U.S. Department of Homeland Security, 2009).

Recovery

The goal of recovery is to ensure the sport organizations operations return to normal as soon as possible. Recovery efforts are unique to each incident, dependent on the extent of damage caused and the resources available. This includes (1) identification of needs and resources, (2) addressing long-term care of affected persons, (3) implementing measures for organizational and venue restoration, and (4) identifying lessons learned and incorporating mitigation measures to lessen effects of future incidents (The Spectrum of Incident Management Actions, 2006). Short-term recovery begins immediately postincident and is an extension of the response efforts to restore basic services and functions. Long-term recovery is the restoration of personal lives and livelihood of the sport organization and surrounding community.

The CG should collect critical data that is needed to recover from a disaster, that is, personnel listings with telephone numbers, inventories of equipment, list of vendors, storage locations, data backup files, and important contracts. The CG should also assess current capabilities by reviewing existing plans and policies, that is, evacuation plan and mutual aid agreements. Additionally, the CG should review the availability of internal assets to respond and recover from an incident, such as emergency medical teams, public relation representatives, fire equipment, communication equipment, warning systems, emergency power, or shelter areas. Consultation with external groups will also provide insight to the availability of external resources available in coordination with agencies such as local emergency management, fire and police departments, ambulance services, public works, hospitals, and volunteer organizations such as the Red Cross. When the CG has decided what measures and procedures will be taken pre-, during, and postincident, they will document their efforts into a comprehensive plan. Supporting documents may also be included, for example, building and site maps indicating utility and shutoff locations, escape routes, location of emergency equipment, and hazardous materials (Hall et al., 2012).

Recovery is primarily the responsibility of local government, but by presidential declaration of a disaster in the United States, a number of assistance programs may be available under the Stafford Act. The Stafford Act defines "emergency" and "major disaster" declarations. An emergency is defined as "any occasion or instance for which, in the determination of the President, Federal assistance is needed to supplement State and local efforts and capabilities to save lives and to protect property and public health and safety, or lessen or avert the threat of a catastrophe in any part of the United States." A presidential declaration of an emergency provides assistance that (1) is beyond state and local capabilities, (2) serves as supplementary emergency assistance, and (3) does not exceed $5 million of federal assistance. The governor of an affected state must request a presidential declaration for an emergency within 5 days of the incident (Fundamentals of Emergency Management, 2010).

A major disaster is defined as "any natural catastrophe or, regardless of cause, any fire, flood, or explosion, in any part of the United States, which in the determination of the President causes damage of sufficient severity and magnitude to warrant major disaster assistance under this chapter to supplement the efforts and available resources of States, local governments, and disaster relief organizations in alleviating the damage, loss, hardship, or suffering caused thereby." Major disasters may be caused by natural events such as floods, hurricanes, and earthquakes. Disasters may include fires, floods, or explosions that the president believes are of sufficient magnitude to warrant federal assistance. Although the types of incidents that may qualify as a major disaster are limited, the federal assistance available for major disasters is broader than that available for emergencies. A presidential disaster declaration provides assistance that (1) is beyond state and local capabilities and (2) supplements available resources of state and local governments, disaster relief organizations, and insurance. The governor of an affected state must request a presidential declaration for a major disaster within 30 days of the incident (Fundamentals of Emergency Management, 2010).

The two major categories of federal aid are public assistance and individual assistance. Public assistance is for repair of infrastructure and public facilities and removal of debris. Individual assistance is for damage to residences and businesses or for personal property losses. Recovery from a disaster is unique to each sport or event business and local community depending on the amount and kind of damage caused by the disaster and the resources that the community has available or has access to (Federal Emergency Management Agency, 2010). The chapter case study provides an example of a sporting business' response and recovery efforts after a natural disaster.

Mitigation

Mitigation activities try to prevent disasters or lessen the damage of unavoidable disasters and emergencies (Hall et al., 2008). Mitigation activities may be implemented prior to, during, or after an incident and are incorporated from lessons learned from prior incidents. Mitigation measures are identified in conjunction with a threat/risk analysis that identifies potential hazards to the sport venue, the probability that an event will occur, and the potential consequences, such as loss of life, destruction of property, disruption of critical services, and economic impact of recovery. The sport or event business' mitigation strategy must consider ways to reduce risks associated with all hazards and potential losses. According to FEMA's Fundamentals of Emergency Management (2010), an effective mitigation strategy is based on several factors:

- Prevention measures: to prevent existing risks from becoming worse or implementing new prevention measures in areas that have not been developed or are in an early phase of development, that is, physical protection security features in building design.
- Property protection measures: to modify buildings or their surroundings to reduce the risk of damage from a known hazard, that is, raising generators to prevent damage from flooding.
- Emergency services measures: to protect people before and after an incident occurs, that is, warning notifications, response tasks, and protective measures for critical facilities.
- Structural projects: to protect people and/or property through the construction of man-made structures to control the damage from a known hazard (i.e., bollards to prevent a vehicle-borne improvised explosive device).
- Public information: to inform and remind people about potential hazards and measures that should be taken to avoid damage or injury. Public information measures may include outreach projects, technical assistance, or educational programs (i.e., basic security awareness workshops hosted by the sport organization).

The mitigation strategy developed must consider the risks faced, the potential damage or impact, and the overall needs of the organization and venue. The mitigation measures must be consistent with the strategy and considered as part of the larger emergency management cycle.

Risk Management

According to Decker (2001), "risk management is a systematic and analytical process to consider the likelihood that a threat will endanger an asset, individual, or function and to identify actions to reduce the risk and mitigate the consequences of an attack" (p. 1). To determine threats and vulnerabilities, a sport organization must conduct a risk assessment. Input from the CG during the risk assessment process is critical to ensure all information and intelligence sharing about threats and vulnerabilities inside and outside of the sporting venue. Venue management can significantly reduce liability exposure by successfully managing risks and foreseeable actions that lead to injuries (Schwarz et al., 2010). An all-hazard risk management approach is critical for sport organizations to protect physical and human assets against potential threats.

The DHS considers major sport stadia as critical infrastructure in the Commercial Facilities Sector. There are many benefits of conducting an assessment and sport venue operators must embrace risk management processes. The risk assessment will indicate the current security profile of the sport venue and highlight areas in which improved security is required. It will also help to develop a justification of cost-effective countermeasures and increase security awareness by reporting strengths and weaknesses in security processes to all staff members, including management (Broder, 2006).

Sport leagues, teams, and venues must prepare for a wide range of possible threats at their venues (Hurst et al., 2007). "A threat is a product of intention and capability of an adversary, both man-made and natural, to undertake an action which would be detrimental to an asset" (Vulnerability Assessment Report, 2003, p. 11). Sport venue managers can obtain threat information from various local, state, and federal sources. Local sources of threat information are obtained from the venue security director, local law enforcement, and state or regional law enforcement. Many states also have threat and investigation working groups such as the Joint Terrorism Task Force (JTTF) (Biringer et al., 2007). Timely and accurate information is available from the Federal Bureau of Investigation (FBI) and the DHS. The DHS provides information on current threat levels through the Homeland Security Advisory System (HSAS), including Homeland Security Bulletins. The FBI operates the National Threat Center and maintains the National Security Threat List (NSTL) (2007). Other techniques used to gather information relevant to specific threats include brainstorming and table-top exercises, modeling and simulation, walk-through of the facility preevent and during the event, knowledge transfer from industry peers, and security surveys (Stevens, 2007). The threat data collected assists the CG to determine who or what the threats are, their capabilities, and the potential and severity of the threat. Sport facility threats include terrorism, natural catastrophes, crowd control, vandalism, theft, fire, fraud, personal assault, traffic incidents, facility intrusion, and technological problems (Stevens, 2007). Terrorism has been cited as one of the most common threats to sport venues, and DHS has issued warnings indicating that sport stadia as critical infrastructure are vulnerable targets. International, domestic, and lone-wolf terrorists have considered stadia as targets since such facilities present open access and opportunity to achieve objectives of mass casualties, economic damage, and social/psychological impact. Scenarios of particular concern to DHS and FBI are explosive devices and the use of aircraft and chemical weapons to attack stadiums and arenas (Office of Intelligence and Analysis, 2009).

Vulnerabilities expose the asset to a threat and eventual loss. The General Security Risk Assessment Guideline (2003) defines vulnerability as "an exploitable capability; an exploitable security weakness or deficiency at a facility, entity, venue, or of a person" (p. 5). Hall et al. (2007) identified vulnerabilities at major sport venues relative to emergency preparedness, perimeter control, physical protection systems, access control, credentialing, training, and communication:

- Lack of emergency response and evacuation plans specific to their facility
- Inadequate searching of the facility prior to an event
- Inadequate lock-down procedures
- Inadequate searches of fans and their belongings
- Unsecure concession areas
- Inadequate signage concerning searches and restricted items
- Lack of closed circuit television (CCTV) coverage of the sport facility or surrounding areas
- Storage of dangerous chemicals inside the sport facility
- Lack of accountability for vendors and their vehicles
- Lack of security notification system for fans, players, staff, and so on
- Inadequate training of staff members
- Inadequate communication capabilities among responding agencies

Countermeasure improvements (also referred to as risk reduction strategies) are recommendations provided to management based on venue assessments to address security weaknesses and enhance emergency planning and recovery efforts. Countermeasures are any actions involving physical, technical, and administrative measures taken to reduce the probability and severity of risks and enhance decision-making abilities (Long and Renfroe, 1999). Possible security measures to reduce the likelihood of an undesired event include increasing security protection system effectiveness through physical upgrades, good personnel practices, information security, staff training, and preventative facility maintenance (Ammon et al., 2010). Physical upgrades include detection, delay, and response strategies; for example, detection methods can include intrusion sensors, identity check access control, alarm communication, and CCTV. Delay mechanisms may include the use of locks and security personnel stationed at restricted access areas of the venue (Biringer et al., 2007). Some measures are designed to be implemented on a permanent basis to serve as a routine protection for a facility; these are referred to as baseline countermeasures. Additional measures can be implemented or increased in their application during times of heightened alert (Department of Homeland Security, 2008). Responding to threat elements requires intelligence and information sharing at the local, state, and federal level. The DHS created an HSAS with corresponding alert levels.

Business Continuity

"Business continuity involves developing measures and safeguards that will allow an organization to continue to produce or deliver goods and services under adverse conditions" (Sauter and Carafano, 2005, p. 333). Continuity planning makes good business sense and is important for numerous reasons, such as reducing the cost of downtime, cost of rebuilding, cost of reconstructing lost critical data, and loss of revenue. The costs associated with the aftermath of a large-scale incident may severely damage the sport organizations operations or even inhibit their ability to fully recover (Broder, 2006). Sport organizations should therefore develop a business continuity plan to ensure operations are maintained. According to Broder (2006), "A business continuity plan is a comprehensive statement of consistent action taken before, during, and after a disaster or outage" (p. 179). The business continuity planning process is also a training exercise for the sport security CG as they must think through contingencies and be familiar with actions required to recover from all types of incidents (Broder, 2006).

Examples of issues to consider include relocation of athletes, use of alternate facilities, enrollment of athletes in other institutions, and/or rescheduling of games or events. This all occurred in the aftermath of Hurricane Katrina and was, in large measure, not planned for. Hurricane Katrina affected many professional and college sports programs in the Gulf Coast region and New Orleans area. Contracts should be in place for immediate restoration and secondary locations identified to hold event bookings in case of an incident. Mutual aid agreements should be included as part of both the ERP and recovery plans. This assures prearranged resources and services are available and provided, if needed. Sauter and Carafano (2005) identified the following key planning steps in developing a business continuity plan: (1) obtain management commitment, (2) establish a planning committee, (3) perform a risk assessment, (4) establish operational priorities, (5) determine continuity and recovery options, (6) develop a contingency plan, and (7) implement the plan.

There are several common flaws in continuity plans. First, a one-size-fits-all approach to continuity planning is not effective. Each plan should be customized to a specific venue, as each venue is unique and offers different resources and capabilities. Training venue staff and exercising the plan is important. Not testing your plan could result in failure to notice significant gaps in the plan that may be exposed during a crisis. When a plan has been developed, it should be updated on a regular basis as business regulations or standards may change within the sport industry. A lack of senior management (venue owner/operator) support could prohibit a successful planning process. The CG will need the support of management to develop and test the plan, especially if any financial support needed. In conclusion, the belief that something bad or wrong will never happen to me, or my organization, may result in complacency and a lack of urgency to plan and be prepared for all-hazard emergencies (Hall et al., 2012).

Appendix A: General Guidelines Checklist for Emergency Preparedness*

Facility Preparedness

- Emergency power to lighting and publics address system, that is, emergency generators.
- An evacuation plan outlining responsibilities and duties of event staff.
- Proper fire equipment available and personnel trained in operating use.
- Adequate signage for exits and general directions are displayed in and around the facility.
- Adequate ingress and egress points and checking emergency exits are accessible in case of an emergency.
- Prepared evacuation messages are activated in event of an emergency.
- A public address system that has the ability to broadcast outside the facility perimeter.
- Cell phone available within facility in case hard-wire phone lines are out of service.
- Copies of local emergency phone numbers are readily available, that is, bomb squad, local utility companies, police, fire, and EMS.

Documentation and Record System

- Predeveloped forms on file to be completed after an incident.
- Facility policy requiring an incident to be reported on every type of situation involving a patron or employee.
- Copies of reports completed by external agencies, that is, police/medical, are filed with the facility's incident report.
- Insure proper documentation and backup information are gathered for legal and insurance purposes.

---

* Adapted from Center for Venue Management Studies (2002).

- Maintain records of all facility emergency preparedness and response efforts, that is, training, policies, standards, and so on.
- Designate an event incident report person to complete all necessary paperwork.
- Use video to record an incident when appropriate.

Emergency Medical

- Ensure appropriate number of emergency medical staff are present at major events.
- Maintain constant communication with EM staff.
- Position EM staff in a predetermined location with adequate supplies/equipment.
- Exercise EM response to certain scenarios.

Bomb Threat

- Have a recording device on phone lines receiving calls during events.
- Individual receiving the call should take note of as much information as possible and keep the caller on the line for as long as possible.
- Take every threat seriously.
- Complete a visual search of the facility.
- Contact local law enforcement, fire department, and facility management.
- Contact event manager about the possibility of evacuating patrons.
- If exploding device is found, immediately evacuate area using security staff and prepared announcements. Refrain from using radio communication as some explosive devices can be triggered by radio waves.
- If nothing is found, the facility manager and police representative should decide whether an evacuation is still necessary.

Fire

- Local fire inspector should conduct periodic inspections of facility.
- Implement facility policies regarding materials that can be taken into the facility.
- A licensed pyrotechnician should be utilized when necessary.
- Contact fire department in event of fire, regardless of size.
- Evacuate facility or threatened area if necessary.
- Fire and smoke alarms should be checked frequently and inspections and certifications should be documented.

Appendix B: Evacuation Plan Template for Stadiums

Introduction

Events at *(Insert Name of Stadium)* are considered premier events hosted in *(Insert Name of State)*. As such *(Insert Name of Stadium)*, needs to be prepared for any eventuality where it may become necessary to evacuate, shelter in place, or relocate spectators, participants, and staff from within the stadium, or to redirect traffic around the stadium. Assessing risk, reducing vulnerabilities, and increasing the level of preparedness will help to minimize potential threats and consequences. It is essential, therefore, that key security personnel at *(Insert Name of Stadium)* are well trained in risk factors, planning an appropriate response, informing the public, and implementing the evacuation plan. This evacuation plan is a supplement to the *(Insert Name of Stadium)* emergency plan (ERP).

Purpose

This evacuation plan provides instructions and guidance to effectively address the safety of all individuals in attendance at *(Insert Name of Stadium)* with regard to evacuation, sheltering in place, or relocation. The emergency plan describes procedures for responding to an emergency or critical incident at the *(Insert Name of Stadium)*. The evacuation plan provides guidance for developing and implementing procedures to evacuate, shelter in place, or relocate in response to an emergency or critical incident.

This evacuation plan was prepared by *(Insert Name)*, *(Insert Name of Stadium)*, Security/Safety Director and *(Insert Name)*, *(Insert Name of County/City)*, Emergency Management Director on X_X_/_X_X_/_X_X_. This document was prepared in coordination and cooperation with the following, and they have signed off with their concurrence:

Chief of Police ________________, & Staff ________________ Police Department
Fire Chief ____________________, & Staff ____________________ Fire & Rescue
Sheriff ___________________, & Staff ____________________ Co. Sheriff's Office
Emergency Management Director ________________________________
Emergency Medical Services Director ____________________________
State Highway Patrol Captain __________________, & Staff __________________
State Bureau of Investigation __________________, & Staff __________________
FBI Special Agent in Charge __________________, & Staff __________________
Bureau of Alcohol Tobacco and Firearms ____________________________
Area Substance Abuse Council ____________________________________
Federal Aviation Administration, Flight Standard Office __________________
Other—if additional or different people, continue to list __________________

Relevant Plans

This section provides an overview of the plans, policies, and guidance documents that are applicable to the *(Insert Name of Stadium)*. Plans may be maintained by the county or city where the stadium resides.

A. Owner's security and safety guideline reference manual
B. Emergency action plan
C. Security and safety plan
D. Other (as appropriate)

Command Structure/Response Organization

The command structure/response organization for evacuation, sheltering in place, and relocation activities should mirror the normal command structure, as found in Section *(Insert Section Number)* of the emergency plan.

A. Jurisdiction and liability
   - Identify laws, ordinances, and authorities that affect evacuation activities.
   - Identify any issues of liability associated with evacuation activities.

B. Evacuation team—Roles and responsibilities
   - Define for each entity, designate, and identify key personnel.

C. Direction and control—Roles and responsibilities
   - Define for each entity, designate, and identify key personnel.

D. Local, state, and federal assistance—Roles and responsibilities
   - Define for each entity, designate, and identify key personnel.

E. Surrounding industry/private sector assistance—Roles and responsibilities
   - Define for each entity, designate, and identify key personnel.

F. Local transportation structure—Roles and responsibilities
   - Define for each entity, designate, and identify key personnel.

Pre-event Planning Considerations

Preevent planning considerations need to be considered prior to a scheduled event at the *(Insert Name of Stadium)*. This section of the evacuation plan provides further information on the types of potential hazards/scenarios that could occur at the stadium and the number and makeup of the spectators and participants of the stadium.

Potential Hazards/Scenarios

Table 14.1 includes the potential hazards that the *(Insert Name of Stadium)* can expect. The table also illustrates the likelihood of the hazard and whether evacuation, sheltering in place, or relocation, would be the appropriate response for each hazard.

Table 14.1 (*Insert Name of Stadium*) Hazards

Hazard/Scenario	*Likelihood of Hazard High/Medium/Low*	*Evacuation, Shelter in Place or Relocation Decision*
Weather		
• Rain		
• Lightning		
• Tornado		
• Heat		
• Severe thunderstorm/heavy rain/flooding		
• High winds		
• Hurricane		
• Heavy snow		
Accidental release (chemical, biological, radiological)		
IED or bomb threat		
Active shooter situation		
Mass casualty event		
Civil disturbance		
Food-borne illnesses		
Fire		
HAZMAT		
Structural collapse		
Terrorism—WMD, explosion, chemical, or biological event, dirty bomb		

Source: Evacuation Planning Guide. 2008. The U.S. Department of Homeland Security, Washington, DC. Retrieved from http://www.dhs.gov/publication/evacuation-planning-guides.

References

Ammon, R., Southall, R., and Nagel, M. 2010. *Sport Facility Management: Organizing Events and Mitigating Risks.* Morgantown, WV: Fitness Information Technology.

Biringer, B.E., Matalucci, R.V., and O'Connor, S.L. 2007. *Security Risk Assessment and Management.* Hoboken, NJ: Wiley.

Broder, J.F. 2006. *Risk Analysis and the Security Survey.* Third Edition. Oxford, United Kingdom: Butterworth-Heinemann Business Books.

Center for Venue Management Studies. 2002. *Best Practices Planning Guide Emergency Preparedness.* IAVM Safety and Security Task Force. TX: International Association of Venue Managers.

Decker, R.J. 2001. *Key Elements of a Risk Management Approach.* Washington, DC: U.S. General Accounting Office. Retrieved from www.gao.gov/new.items/d02150t.pdf

Evacuation Planning Guide. 2008. The U.S. Department of Homeland Security, Washington, DC. Retrieved from http://www.dhs.gov/publication/evacuation-planning-guides

Federal Emergency Management Agency, Emergency Management Institute. 2005. Communicating in an emergency. In *Effective Communication: Independent Study.*

Federal Emergency Management Agency, Emergency Management Institute. 2008. Introduction to Exercise Design. In *Exercise Design: IS-139,* Washington, DC.

Federal Emergency Management Agency, Emergency Management Institute. 2009. Exercise basics. In *An Introduction To Exercises: IS-120.A,* Washington, DC.

Federal Emergency Management Agency, Emergency Management Institute. 2010, January 14. *Fundamentals of Emergency Management: IS-230.A.* Retrieved from http://training.fema.gov/EMIWeb/IS/IS230a.asp

General Security Risk Assessment Guideline. 2003. ASIS International. [Online]. Available: http://www.asisonline.org/guidelines/guidelinesgsra.pdf

Hall, S., Cooper, W.E., Marciani, L., and McGee, J.A. 2012. *Security Management for Sports and Special Events—An Interagency Approach.* Champaign, IL: Human Kinetics.

Hall, S., Marciani, L., and Cooper, W.E. 2008. Sport venue security: Planning and preparedness for terrorist-related incidents. *Sport Management and Related Topics Journal, 4*(2), 6–15.

Hall, S., Marciani, Cooper, W.E., and Rolen, R. 2007. Introducing a risk assessment model for sport venues. *The Sport Journal, 10*(2), 1–6.

HSEEP. 2007. Homeland Security Exercise and Evaluation Program. The U.S. Department of Homeland Security.

Hurst, R., Zoubek, P., and Pratsinakis, C. 2007. *American Sports as a Target of Terrorism: The Duty of Care after September 11th.* Retrieved from www.mmwr.com/_uploads/UploadDocs/publications/American%20Sports%20As%20A%20Target%20Of%20Terrorism.pdf

Lindell, M.K., Prater, C., and Perry. R.W. 2007. *Introduction to Emergency Management.* New Jersey: John Wiley & Sons.

Long, L.E. and Renfroe, N.A. 1999. *A New Automation Tool for Risk Assessment.* 15th Annual NDIA Security Technology Symposium. Session: Risk and Threat Assessment Techniques. Retrieved from www.dtic.mil/ndia/technology/smith.pdf

Office of Intelligence and Analysis. 2009, January 26. *Threats to College Sports and Entertainment Venues and Surrounding Areas.* U.S. Department of Homeland Security.

Sauter, M.A. and Carafano, J.J. 2005. *Homeland Security: A Complete Guide to Understanding, Preventing, and Surviving Terrorism.* New York: McGraw-Hill.

Schwarz, E.C., Hall, S., and Shibli, S. 2010. *Sport Facility Operations Management: A Global Perspective*. Oxford, United Kingdom: Butterworth-Heinemann Business Books.

Stevens, A. 2007. *Sports Security and Safety: Evolving Strategies for a Changing World*. London, England: Sport Business Group.

The Spectrum of Incident Management Actions. 2006. In *Principles of Emergency Management: Independent Study*. Emergency Management Institute, Federal Emergency Management Agency (FEMA).

U.S. Department of Homeland Security. 2008. *Evacuation Planning Guide for Stadiums*. Retrieved from www.dhs.gov/xlibrary/assets/ip_cikr_stadium_evac_guide.pdf

U.S. Department of Homeland Security. 2009. *State and Local Fusion Centers*. Retrieved from www.dhs.gov/files/programs/gc_1156877184684.shtm

Vulnerability Assessment Report. July 2003. Office of Domestic Preparedness, U.S. Department of Homeland Security. Retrieved from http://www.ojp.usdoj.gov/odp/docs/vamreport.pdf

Warfield, C. 2008. *The Disaster Management Cycle*. Retrieved from http://gdrc.org/uem/disasters/1-dm_cycle.html

Chapter 15

Pandemic Preparedness

Douglas E. Himberger

Pandemic preparedness is a key topic in the realm of overall preparedness and is relevant to nearly all citizens at one time or another. While public awareness of pandemics may be high, preparedness for them may not be. Much has been written on this subject, including a chapter in a previous textbook in this series.[1] This chapter presents highlights of that earlier work, along with new information on pandemic preparedness.

Despite great sums spent by both public and private entities on pandemic preparedness in recent years, our preparedness for these infectious disease events has changed little. This state of affairs is attributed to a combination of effects: the nature of pandemics as compared with other crises; the unique preparedness requirements of pandemics; and the challenging pandemic preparedness planning activities. This chapter discusses each of these effects, along with activities during and after a pandemic, and the way ahead in pandemic preparedness planning.

Nature of Pandemics

Health Concerns of Pandemics

A pandemic is defined as "… an epidemic (a sudden outbreak) that becomes very widespread and affects a whole region, a continent, or the world."[2] The *Collaborative International Dictionary of English* describes a pandemic as an "everywhere epidemic." Pandemics are often thought of in terms of notable historical events, such as the first (of as many as seven, to date[3]) Asiatic cholera pandemic of 1817; the "Spanish Flu" or "Great Pandemic" of 1918–1819; and the severe acute respiratory syndrome (SARS) pandemic of 2002–2003 (probably more appropriately deemed an "epidemic" because of its relatively limited geographic scope). Pandemics span an enormous time frame, ranging from the first recorded Greek bubonic plague in 430 BC to the recent 2003–2012 "avian influenza" (type A/H5N1 variant) and 2009–2010 "swine influenza" (type A/H1N1 variant) pandemics.[4]

Pandemic crises are recurring on average three times per century, and preparing for "the next one" should be continuous. Many experts believe that another epidemic is inevitable and may happen soon; if true, preparedness is imperative.

In pandemics, the most devastating impact is in the area of health care. The scores of people sickened and killed by the infectious diseases affected families, communities, and even entire countries or global regions. The diseases placed "sudden and intense demands on health systems,"[5] and overarching community structures also were often broken.

The illness and death caused by these pandemics resulted in societal disruption. The "psyche" of populations was so severely degraded that the outlook of those citizens was not simply depressed—the pandemics colored the very future of their societies.

Pandemics have killed staggering numbers of people worldwide; the worst single pandemic, the Spanish Flu pandemic* of 1918–1919, killed an estimated 20 million to 40 million people globally (with some estimates as high as 50 million to 100 million deaths). This impact takes a heavy toll on affected populations, not only in terms of treating the sick and dying but also in terms of dealing with mass burials in a culturally acceptable way and tending to the associated mental health impact.

In absolute terms of sickness and death (or morbidity and mortality), the recent "avian flu" potential pandemic associated with the type A/H5N1 virus was relatively minor: only 606 cases resulted in a relatively high mortality of 357 deaths.[6] However, the effect on society worldwide was chilling: the "worried well" tapped valuable health resources, and the fear generated by the disease spread across entire continents, disproportionally affecting behavior. Compare these effects with those caused by the 2009–2010 "swine flu" (type A/H1N1 influenza virus) pandemic: the virulence of the disease when it had run its course was less than some of the more profound historical pandemics, but the resultant fear and anxiety affected

* The pandemic was referred to as the "Spanish Flu" pandemic, although the disease is believed to have possibly started at a military base in Kansas (note that other research indicates that it might have begun in the Far East or Austria) and been carried by World War I U.S. troops to other parts of the world. Because Spain was a neutral country during the conflict and had no wartime censorship in place, it was one of the few news outlets reporting on the pandemic (particularly when it moved from France to Spain)—hence, the term "Spanish Flu" might have been coined as a result. The variant of the influenza virus dominant in the Spanish Flu pandemic was H1N1.

many aspects of society. For example, Vice President Biden at one point said that he would advise his family not to travel on subways or airlines, and many citizens followed suit.

The most troubling characteristic of these viruses may be their unpredictability. Public health academic Philip Alcabes, author of *Dread*, says that instead of looking to physicians to predict epidemics, "... we should leave the job of seeing the future to the mystics, prophets, and fortunetellers." This level of uncertainty breeds anxiety.

Nonetheless, there are certainties in pandemics: pandemics will recur, and they will have great impact on society. In a prescient statement in 2005, a Congressional Budget Office (CBO) report stated, "... [the H5N1 virus] could evolve in a way that rendered it harmless, and a pandemic could arise from an entirely different virus subtype."[7,8] Although the H5N1 avian flu virus could not be deemed "harmless," certainly the type A/H1N1 swine flu virus became a public pandemic concern. As Professor Frederick Hayden of the University of Virginia School of Medicine said recently, "What is clear is that it is *when*, not *if*."

Community Continuity Concerns of Pandemics

The impact of pandemics goes beyond health implications. The entire socioeconomic system is deeply affected, with effects being felt well beyond the health community. "[Pandemics] expose existing weaknesses in these systems and, in addition to their morbidity and mortality, can disrupt economic activity and development."[9] Nearly all facets of a community, or even a nation, can be affected gravely, from mundane daily tasks to broad strategic operations.

These operations are disrupted on many levels. The people required for any level of community activities—governance, education, health care, transportation, and food distribution—are the parts of the infrastructure that are affected profoundly. Not only are there fewer people to conduct these vital activities, but the people with appropriate skills are also affected in unpredictable ways. Although most experts forecast high absenteeism, the specific absentees themselves are not predictable. As much as 20–40% of a population may be unable to function because of their own or a family member's sickness or death[10]; there is no way to predict exactly which individuals would be affected. Critical functions might be affected to even a greater degree, and overall continuity of operations could be compromised severely.

Communities need a pandemic preparedness plan, but how do they know when the plan must be put into practice? Several tools exist for measuring the pandemic threat. Two of the best known are the World Health Organization (WHO) pandemic alert level tool,[11] and the U.S. Federal Government (Department of Health and Human Safety [HHS]) response stage tool.[12] Figure 15.1 depicts these tools.

For each WHO phase or HHS stage, there are "triggers" that lead the organization to elevate the levels and also inform communities and populations about

<table>
<tr><th colspan="2">WHO Phases</th><th colspan="2">Federal Government Response Stages</th></tr>
<tr><td colspan="4">Interpandemic period</td></tr>
<tr><td>1</td><td>No new influenza virus subtypes have been detected in humans. An influenza virus subtype that has caused human infection may be present in animals. If present in animals, the risk of human disease is considered to be low.</td><td rowspan="2">0</td><td rowspan="2">New domestic animal outbreak in at-risk country</td></tr>
<tr><td>2</td><td>No new influenza virus subtypes have been detected in humans. However, a circulating animal influenza virus subtype poses a substantial risk of human disease.</td></tr>
<tr><td colspan="4">Pandemic alert period</td></tr>
<tr><td rowspan="2">3</td><td rowspan="2">Human infection(s) with a new subtype, but no human-to-human spread, or at most rare instances of spread to a close contact.</td><td>0</td><td>New domestic animal outbreak in at-risk country</td></tr>
<tr><td>1</td><td>Suspected human outbreak overseas</td></tr>
<tr><td>4</td><td>Small cluster(s) with limited human-to-human transmission but spread is highly localized, suggesting that the virus is not well adapted to humans.</td><td rowspan="2">2</td><td rowspan="2">Confirmed human outbreak overseas</td></tr>
<tr><td>5</td><td>Larger cluster(s) but human-to-human spread still localized, suggesting that the virus is becoming increasingly better adapted to humans, but may not yet be fully transmissible (substantial pandemic risk).</td></tr>
<tr><td colspan="4">Pandemic period</td></tr>
<tr><td rowspan="4">6</td><td rowspan="4">Pandemic phase: increased and sustained transmission in general population.</td><td>3</td><td>Widespread human outbreaks in multiple locations overseas</td></tr>
<tr><td>4</td><td>First human case in North America</td></tr>
<tr><td>5</td><td>Spread throughout United States</td></tr>
<tr><td>6</td><td>Recovery and preparation for subsequent waves</td></tr>
</table>

Figure 15.1 WHO and federal pandemic response levels. (From U.S. Department of Health and Human Services, 2009. www.PandemicFlu.gov.)

changes in the pandemic. As Figure 15.1 illustrates, WHO and HHS monitor these triggers, but entities at all levels (state and regional public health organizations, local community health authorities, and even businesses) should also monitor the triggers. The more aware these organizations are, the more prepared they will be to implement preparedness actions.

The WHO phases measure the transmissibility of a virus, not the severity of a resultant pandemic. This is a key difference because many pandemic plans are based on severity of the illness. Although the WHO might measure a pandemic as being at a high level (e.g., levels 5 or 6), this means that triggers show the virus to be transmissible at the levels indicated in the phase chart (e.g., "widespread human infection" for levels 5 and 6), not that resulting illnesses are necessarily severe

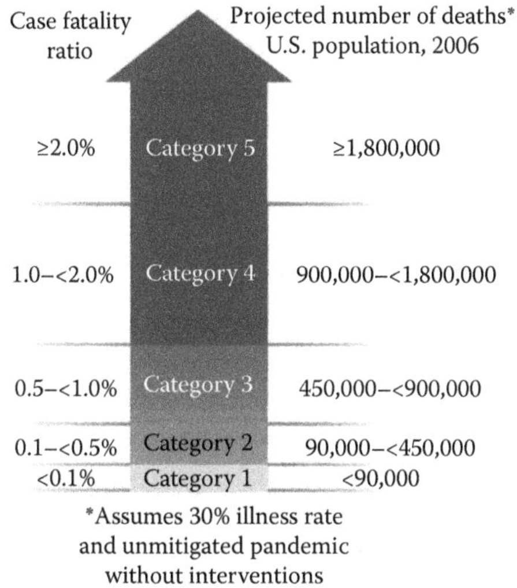

Figure 15.2 Pandemic severity index. (From U.S. Department of Health and Human Services, Community Strategy for Pandemic Influenza Mitigation, February 2007.)

(although that might also be true). The WHO made clarifications to their phase structure in 2009,* but the phases themselves still relate to virus transmissibility.

With respect to severity, at this writing, the WHO has not yet developed a system for measuring this characteristic of a pandemic. "Severity" refers to not only the degree of virulence of a virus (e.g., number of severe illnesses and deaths, contagiousness of the virus, age distribution of cases, prevalence of chronic health problems and malnutrition of the population, viral mutations, number of waves of illness, and quality of health services) but also the overall socioeconomic impact of the outbreak.[13]

Though WHO does not yet measure severity,[14] HHS has such a severity index in place, based on hurricane classifications (see Figure 15.2).[15] The HHS Pandemic Severity Index is part of the agency's guidance for combating a pandemic. The index is based on case fatality rates (CFR), with a CFR of 2% or greater signaling the most severe pandemic (Category 5). The outbreaks of 1957 and 1968 would be rated as Category 2 events, with CFRs between 0.1% and 0.5%.[16] There is potential

* The World Health Organization (WHO), Current WHO global phase of pandemic alert: Avian Influenza A(H5N1), http://www.who.int/influenza/preparedness/pandemic/h5n1phase/en/ (accessed June 1, 2012). Of note, "…the grouping and description of pandemic phases have been revised to make them easier to understand, more precise, and based upon observable phenomena. Phases 1–3 correlate with preparedness, including capacity development and response planning activities, while Phases 4–6 clearly signal the need for response and mitigation efforts. Furthermore, periods after the first pandemic wave are elaborated to facilitate postpandemic recovery activities."

for public confusion when using this index in combination with the WHO pandemic phases; this may have led to WHO's delay in presenting a similar index. HHS has presented numerous cross-references between their index and WHO's and has included characteristics and interventions for each severity level.

The WHO and HHS tools depend on faithful and timely reporting of relevant triggers. During the 2009–2010 swine flu pandemic, it was reassuring to see that no apparent cover-up existed for any nation with respect to reporting the illness (as some had believed might have happened in reporting the 2002–2003 SARS event). However, there was concern not only that some time had passed—perhaps three weeks or more[14]—between the first cases in Mexico and the first WHO warnings, but also that some time had passed before the Centers for Disease Control and Prevention (CDC) identified the novel variant of the virus. Going forward, delays must be minimized so that pandemic preparedness plans can be implemented when they can have the greatest impact.

Psychosocial Concerns of Pandemics

As stated earlier, the effects of a pandemic go well beyond health concerns. The cyclical nature of pandemics, as well as the rapid transmission of the diseases, generates fear. The "worried well" overwhelms an already besieged healthcare infrastructure. Individuals whose family members or friends have been sickened or have died suffer from mental health issues that could disable even the strongest among us. Other practical issues result; during the Spanish Flu pandemic of 1918–1919, Philadelphia suffered such losses of adult citizens that the Bureau of Child Hygiene determined that it was unable to care for a large number of orphans.[18] Even worse, the city was unable to cope with the numbers of dead. If a pandemic of similar virulence were to occur today, there could be "1.7 million deaths in the United States and 180 million to 360 million deaths globally,"[19] clearly overloading the healthcare system, as well as the funereal communities, and taking these essential services beyond the breaking point.

One does not need to look to horrific pandemic data to sense the fragility of a community's healthcare infrastructure. In 2006, a two-week heat wave in California caused as many as 141 deaths, leading "coroners…[to deal with] the large jump in the number of bodies that were stuffed, some piled on top of others, into the freezers at the Fresno County morgue."[20]

Another consideration for planning for pandemics is the very nature of human behavior. Though such behavior is nonlinear, and therefore seemingly unpredictable, behavior can be predicted with some fidelity if sufficient information is known about a community's culture, history, and motivations. However, a deep understanding of all these factors is seldom possible, particularly after a pandemic has started. Therefore, the more a community knows about itself and its underlying structure, and the more that knowledge is factored into preparedness planning, the more effective the plans.

When developing a plan that fits a community, it is important to consider all aspects of that population. There are five major aspects: (1) culture of the group; (2) policies, strategies, and plans in place; (3) economics, management, and budgeting of the entities involved; (4) governance and operations; and (5) technology implemented to deal with a crisis (e.g., information databases and communication networks).[21] Planners must thoroughly understand all five aspects. For instance, if the culture of a group is to be prepared, but the policies for preparedness are not in place, or the technologies to enable the group to be thoroughly prepared are not implemented, preparedness might not be possible. Similarly, developing plans for such preparedness must take into account all these elements, or the plans will not work.

Economic Impacts of Pandemics

As already described, pandemic effects of go well beyond health, specifically taking an economic toll that is far-reaching. These costs occur at both ends of a pandemic. Costs to prepare for pandemics can be enormous, as can costs to respond to and recover from the crisis itself. For example, within days of the first outbreaks of the swine flu in early 2009, the World Bank alone responded to the pandemic with "fast-disbursing funds" ($25 million for drugs and supplies, and $180 million for epidemiologic, regulatory, institutional, and operational activities[22]), and many other agencies and institutions followed course. These amounts were only a portion of the total funding applied; the details of those funds are unknown at this writing, but the costs of a pandemic were and will be high.

As another example, in late 2005, the Bush administration requested $7.1 billion in emergency funds for pandemic preparedness[23] (though less was ultimately made available to planners and responders, the majority of this funding request was allocated). The funded activities included plan development, antiviral stockpiling, vaccine production, and similar key actions. Long-term infrastructure was strengthened, including source control and surveillance, vaccine research and development, antiviral drug research and development, and healthcare system readiness.[24]

The last of these is of particular importance. For pandemics, our healthcare infrastructure is fragile and insufficient. During a severe pandemic, there would be far more demand on the U.S. healthcare system than the system could accommodate (as many as 5 million to 10 million sick individuals,[25] exceeding the roughly 970,000 staffed hospital beds and 100,000 ventilators,[26] with 75% of those in use at any given time under normal, nonpandemic situations). Consequently, building a stronger healthcare infrastructure before a pandemic would be key. Even a mild pandemic such as the 2009–2010 swine flu pandemic (considered mild in severity, although transmissible at a relatively high level) demonstrated to communities that the stress on the health infrastructure could be enormous. A sizable percentage (about 25%) of the federal stockpile of antiviral drugs was distributed quickly to states and localities; however, it was clear that local health departments would have difficulty distributing them rapidly to the population as a result of local budget cuts and layoffs. Treating a sick population would have been difficult or even impossible. Supporting the recovery of those who had been ill would add another level of stress to the healthcare system.

The indirect costs (e.g., decreased supply from a shrinking workforce, and dramatic decline in demand for goods and services due to avoidance of shopping malls, restaurants, and other public places) during and after a pandemic can be even greater. These indirect costs of recent pandemics, even relatively minor ones such as the 2002–2003 SARS pandemic, have been enormous. Some estimate that the costs of disruption of activity in business, civil, and governmental domains related to the SARS event were as much as $30–100 billion (again, these figures continue to be debated). In 2005, the World Bank estimated that a 2% loss of global gross domestic product (GDP) would result from a pandemic of similar severity to 1918–1919, which translates into $800 billion in losses for a year.[27] Recent estimates raise this expected impact to nearly 5% of global GDP, or about $3 trillion.[28] This impact is "greater than in recent recessions and roughly the same size as the average postwar recession."[29]

Unique Preparedness Requirements of Pandemics

It is clear that we must prepare for pandemics. But with what goal? The debate continues, but the CBO (drawing on WHO and others) has identified the following goals for the federal government[30]:

- Support for the efforts of governments of other countries and international organizations to contain [virus] strains and control their evolution to diseases that are transmitted easily from person to person.
- Building stockpiles of vaccines, improving antiviral drugs, and putting in place new technologies that allow effective vaccines to be produced more rapidly and in large quantities.
- Improving the capacity of the health system to care for many people in all parts of the country who are sick simultaneously.
- Lower-level plans (states, regions, localities, and individual organizations and entities) have parallel goals appropriate for their needs, but the last goal typically is the driving one for planners.

Pandemics at Hand—Pandemic Influenzas: Avian and Swine

A pandemic of great recent concern is avian flu. This flu infects birds, such as domestic poultry (e.g., chickens) and wild fowl (e.g., ducks). "Avian flu. . .is caused by influenza viruses that occur naturally among wild birds. Low pathogenic [avian flu] is common in birds and causes few problems [to humans]. Highly pathogenic H5N1 is deadly to domestic fowl, can be transmitted from birds to humans, and is deadly to humans. There is virtually no human immunity, and human vaccine availability is very limited."[31] In fact, "viruses of the H5 subtype are not known to have ever circulated among the human population, which means that there would be little immunity to it."[32] Therefore, it is this highly pathogenic version of avian flu that is a risk to communities and has the potential for pandemic impact. The avian flu pandemic was of great concern from 2005 to 2009, but in reality, it began as early as 2003, and continues into 2012.

Interest in the avian flu peaked again in 2012 when researchers at Erasmus Medical Center in Rotterdam found that "without the help of another virus, the deadly avian flu (H5N1) could easily mutate in mammals to become transmissible through the air, like a true pandemic strain, through a sneeze or a cough. And it might need as few as five mutations to make that leap."[33] This research was controversial, with publication of the findings in scientific journals being held up until a review by other researchers and the National Science Advisory Board for Biosecurity cleared the way for such publication (in both *Science* and *Nature*).

Similarly, swine flu is caused by influenza viruses that occur in swine. However, although the immunity and human vaccine availability is like that of avian flu, the 2009–2010 swine flu pandemic appeared to be a variant that was a mutation (or "reassortment") of avian flu, swine flu, and human flu.* The resultant virus appeared to be transmitted with some efficiency from human to human, leading the WHO to raise the pandemic alert phase to 5 in April 2009. This phase carried with it the fact that the virus was "characterized by human-to-human spread...into at least two countries in one WHO region. Although most countries [would] not be affected at this stage, the declaration of Phase 5 [was] a strong signal that a pandemic [was] imminent and that the time to finalize the organization, communication, and implementation of the planned mitigation measures [was] short."[34] A significant number of cases had been reported worldwide in 2009, with numbers rising and at levels that many considered alarming. ("As of 27 December 2009, worldwide, more than 208 countries and overseas territories or communities...reported laboratory confirmed cases of pandemic influenza H1N1, including at least 12,220 deaths."†)

* An animated depiction of the genetic reassortment of four different influenza virus strains, including human influenza gene segments, swine influenza from North America and Eurasia, and avian gene segments from North America, can be found at the Food and Agricultural Organization (FAO) website (2009), Evolution of Swine Influenza Virus in North America, www.fao.org/ag/againfo/programmes/en/empres/AH1N1/ Background.html (accessed June 1, 2012).

† U.S. Centers for Disease Control and Prevention (CDC), *Pandemic (H1N1) 2009—Update 81*, http://www.who.int/csr/don/2009_12_30/en/index.html. As of May 1, 2009, 331 cases were reported in 11 countries, with 10 resultant deaths. By May 10, 2009, only nine days later, the reported cases had risen dramatically to 4379, with 49 deaths (World Health Organization [WHO], http://www.who.int/csr/don/2009_05_10/en/index.html).

The WHO subsequently raised the outbreak to a full pandemic level, Phase 6, indicating a global pandemic was underway.

Note that every community already suffers significant impact from seasonal flu, a common respiratory illness transmitted from person to person. Although many people have some immunity to this illness, and vaccines are available annually, each year in the United States alone, an average of 36,000 deaths occur as a result of seasonal influenza. These deaths, along with more than 200,000 U.S. hospitalizations and more than $10 billion in U.S. economic cost,[35] accompany more than one-quarter to one-half million deaths worldwide annually.[36] This toll should lead to a heightened awareness of hygiene principles—principles that would be effective for not only seasonal flu but also a pandemic. Unfortunately, our culture has adopted a *laissez-faire* approach to this disease and to measures effective in preventing its transmission.

There are encouraging developments, however. Recently, public service announcements (PSAs) and other media communications have been emphasizing the importance of hygiene for limiting transmission of infectious disease (e.g., suggesting to citizens that they wash their hands regularly, and taking the washing time necessary to sing the tune "Happy Birthday" twice). Although it has been suggested that the public suffered from "information fatigue" as a result of extensive coverage of the avian flu pandemic, and that subsequently media reports waned, subsequent coverage again rose, particularly as a result of the swine flu pandemic. This coverage can be helpful in preparing the public for pandemics, but only if the messages are clear and easily understood.

Numerous interesting tools and techniques related to pandemics are being introduced—not only to the healthcare community and emergency preparedness professionals but also to the public. As an example of novel technologies available to the healthcare community, Brown University researchers "have created a reliable and fast flu-detection test that can be carried in a first-aid kit. …The technology could lead to real-time tracking of influenza."[37] In the public domain, keying on Internet searches, Google's Flu Trends looks at the "relative popularity of a slew of flu-related search terms to determine where in the U.S. flu outbreaks may be occurring." "What's exciting about Flu Trends is that it lets anybody—epidemiologists, health officials, moms with sick children—learn about the current flu activity level in their own state based on data [from any given] week."* Given the popularity of web tools, including Google Earth (now being used to track avian flu outbreaks and

* Landau, Elizabeth (2008), *Google Tool Uses Search Terms to Detect Flu Outbreaks*, http://edition.cnn.com/2008/HEALTH/conditions/11/11/google.flu.trends/index.html. As further stated in the article, "The Centers for Disease Control and Prevention collaborated with Google on the project, helping validate and refine the model, and has provided flu tracking data over a five-year period." Furthermore, "In the 2007 and 2008 flu season, Google accurately estimated current flu levels one to two weeks faster than published CDC reports in each of the nine U.S. surveillance regions, Google said in a statement." Google's Flu Trends can be found at www.google.org/flutrends.

mutations worldwide) and mobile applications,* the public will have techniques never before available to aid them in their preparedness. Plans must include these tools, and planners must use the tools in exercises.

Persistence and Pervasiveness of Pandemics

Pandemics are different from other crises. Events such as floods, tornadoes, hurricanes, and wildfires are often limited to relatively small areas. By definition, pandemics are crises that affect areas often covering entire countries, continents, or even the globe.

This geographic coverage greatly complicates preparedness planning. For most emergencies, planners can depend on other resources and aid from areas not affected by the crisis. Firefighters often travel to serious fires; first responders are often brought to disaster scenes such as floods or hurricanes; and food and other resources are typically flown to stricken areas. Planners for pandemics must assume that the pandemic will similarly hit all other areas, and people in other areas will be unable to assist. Pandemic plans must build on locally available staff, food, water, power, transportation, and other resources. The plans must call on the affected areas to weather the pandemic organically. As is sometimes said, the citizens must be "their own first responders."

Temporal Requirements of Pandemic Preparedness

Pandemic planners must not only assume that little or no outside help will exist but also accommodate another unique aspect of pandemics: typically, pandemics are not bounded by a short, easily defined timeframe. Rather, they often occur in "waves" (each of several weeks in duration) separated by some time period. These waves can be individually devastating; collectively, they can sap a population's energy and resources as response, recovery, and planning phases begin to overlap one another.

These waves can be separated by many years, as has been the case for the global cholera pandemics (e.g., seven of which have occurred from 1817 to the present). Pandemics separated by such long periods are more manageable in some ways. Preparedness for a successive wave can build on lessons learned from a previous wave, with sufficient time elapsing for replenishment of vital medical and other resources. However, if pandemic waves are separated by shorter periods (measured in weeks or months), the collective impact can be demoralizing and overwhelming. For example, the Spanish Flu pandemic of 1918–1919 attacked in three waves: the first occurring in the spring/summer of 1918; the second, in the fall of 1918; and

* Mobile "apps" are becoming popular and easily acquired. An example of this is the "Fight the Flu" app, developed by Genetech, the makers of the antiviral Tamiflu. The app "helps you to: track flu activity in your area; receive free email alerts; check the symptoms of flu; get flu prevention tips; and learn what to do if you've been exposed to flu." (http://itunes.apple.com/us/app/fight-the-flu/id398369706?mt=8#).

the third and final, in the spring of 1919. Future influenza pandemics are expected to follow a similar pattern, causing great concern to planners and responders alike.

Pandemic Preparedness Planning

The concept of pandemic planning is not only a notion of practical need but also one of policy. The federal government has developed new broad strategic guidance—the National Security Strategy (May 2010)[38]—noting that "...pandemic disease threaten[s] the security of regions and the health and safety of the American people." Other federal planning guidance addressing all crisis preparedness included the Department of Homeland Security's (DHS's) National Response Plan (NRP) (issued in December 2004), which was superseded by the National Response Framework (NRF) (issued on March 22, 2008). The NRP (like the NRF) is an "all-discipline, all-hazards plan intended to establish a single, comprehensive framework for managing domestic incidents."[39] The NRF describes federal support to be implemented through activation of 15 emergency support functions (ESFs), including several relevant to pandemic planning: Mass Care, Emergency Assistance, Housing, and Human Services (ESF #6); Public Health and Medical Services (ESF #8); Public Safety and Security (ESF #13); and Long-Term Community Recovery (ESF #14). The 15 ESFs, including those relevant to pandemics, address crisis planning, implementation, training, and exercising.[40]

Federal guidance also provides specific details with respect to pandemics—the National Strategy for Pandemic Influenza[41]—and was motivated largely by the threat of the type A/H5N1 virus (avian flu). The National Strategy addresses preparation for and response to any pandemic, and it is fully consistent with higher-level policy and planning documents such as the NRP and the subsequent NRF. The National Strategy "guides our preparedness and response to an influenza pandemic, with the intent of (1) stopping, slowing, or otherwise limiting the spread of a pandemic to the United States; (2) limiting the domestic spread of a pandemic, and mitigating disease suffering and death; and (3) sustaining infrastructure and mitigating impact to the economy and the functioning of society."[42] There are three "pillars" in the National Strategy: (1) preparedness and communication, (2) surveillance and detection, and (3) response and containment. Each has considerable detailed guidance for communities and the general population in business, civil, and governmental situations.

Following the National Strategy, the National Strategy for Pandemic Influenza Implementation Plan was released.[43] This plan presents specific actions for more than 300 federal departments and agencies (e.g., Department of Defense, HHS, and Veteran's Administration), and agency implementation plans followed.

This section addresses the process for developing pandemic preparedness plans, and focuses on essential elements of not only the plans themselves but also the implementation of those plans.

Developing a Pandemic Preparedness Plan

The National Strategy includes key overarching goals (described previously) that are elaborated in a U.S. Homeland Security Council publication.[44] Any lower-level pandemic preparedness plan must be in consonance with these high-level goals. Such lower-level plans may well be developed carefully, and are almost certainly well intentioned, but many are "not legally or logistically feasible." Furthermore, it has been said that "lessons from simulations had not been drawn on to revise plans."[45] In short, the plans must be realistic, appropriate to the entity or entities being addressed, and exercised thoroughly. These exercises must be carried out repeatedly so that participants will become trained and plans can be updated, based on exercise results. Situations for an affected population may change, and the plan must be agile enough to accommodate these changes, but these can be appreciated only in the context of realistic exercises. Often, a preparedness plan (for pandemics or for other crises) becomes "shelf-ware"—developed with care, but then relegated to a shelf until a crisis hits. At this point, it is too late to update the plan or exercise the population.

Numerous tools are available for pandemic preparedness, including the following: (1) vaccines, (2) antiviral medications, (3) infection control measures, and (4) community mitigation measures. Each is complex and could potentially have a broad effect on a pandemic; however, none is likely to completely address the severe impact of a pandemic. Most likely, all these tools (and perhaps others) will be needed. Lisa Koonin, Senior Advisor of the Influenza Coordination Unit and Lead for Pandemic Medical Care and Countermeasures at CDC, likens this issue to "Swiss cheese." A single mitigation tool is like a single slice of cheese—full of holes. However, when one slice is laid over another, and then another, and then another, we find that the holes may be partially—or even completely—filled. Similarly, these preparedness techniques are additive; as many as possible must be used to achieve the maximum mitigating effect.

Vaccines are a unique element of a pandemic preparedness plan. Typically, vaccines can be quite effective, providing a measure of individual immunity from the virus. They are in widespread use for other viruses, including seasonal influenza, where they have been shown to be effective. Typically, pandemic preparedness plans specifically address the availability, distribution, and use of vaccines.

However, a vaccine must be matched precisely to the virus in circulation. The laborious process for developing this "match" typically takes months.* This time lag is crucial; if a highly transmissible virus were to cause a pandemic, many thousands (even millions) of people could be infected, many of whom could die

* There is hope for great improvement in this area. New vaccine development techniques promise dramatic decreases in development time. For example, the National Institute of Standards and Technology and the University of Queensland have "successfully demonstrated...findings that could reduce the time...to produce a vaccine from months to weeks..." (National Institute of Standards and Technology, *NIST Tech Beat*, December 8, 2008).

before a vaccine was available. The time lag is not the only issue concerning vaccines. It is possible or even likely that vaccines, once developed, would not be available in sufficient quantities to be distributed to all affected populations. Therefore, prioritization of recipients will be required (e.g., healthcare workers and emergency management responders could be first on such a list). Practical and ethical issues are certain to arise. For example, should those in failing health (unrelated to the influenza) be administered the vaccine rather than those who are younger and more vital? Should a vaccine be administered to "key" staff before those deemed less critical for operations? Government and healthcare officials would be expected to resolve these endless issues. Other issues would also need to be considered, including topics such as use and efficacy of prepandemic vaccines (ones that might be available before a pandemic, but not perfect matches to the circulating virus).

Antivirals do not function like vaccines. Antivirals do not provide immunity, but they may make the illness less severe or shorten the course of the illness. They are not specific to a virus in circulation (although considerable differences exist in effectiveness and the like for each variation of antiviral), and large stockpiles (e.g., tens of millions of regimens in the United States alone) already have been developed and stored. Antivirals have demonstrated some effectiveness in prophylactic use, but they are not intended to immunize a recipient to a virus.

However, antivirals also have shortcomings. Because they can be used prophylactically, there is fear that they will be overused or cause the virus strain to develop resistance. Resistance, whether a result of overuse or not, can be significant. For the 2008 and 2009 seasonal flu, virtually all flu cases have shown resistance to Oseltamivir,* a popular antiviral. However, the prescribing of antivirals, while correlated to flu (and pandemic) activity, also showed heed to CDC guidance (e.g., instructing physicians to change from one antiviral to another) that lessened the resultant resistances.[46] Furthermore, although large quantities of antivirals are stockpiled, there still might be a need to prioritize who gets them and when.

Another tool is infection control. Many control measures are commonplace and intuitive, but others are less so. Encouraging standard hygiene processes (e.g., hand washing, and covering coughs and sneezes) has a strong impact on limiting transmission. Other personal protective equipment (PPE) such as surgical masks and other respiratory devices (e.g., N95 masks and respirators) has been shown to be effective, although such equipment can be burdensome, expensive, possibly misused, and effective only in certain situations. The key issue for all these techniques is compliance; without constant and vigilant use, these tools are ineffective.

---

* *L.A. Times*, Tamiflu No Longer Works for Dominant Flu Strain, February 4, 2009, http://articles.latimes.com/2009/feb/07/science/sci-flu7. Although this news report states specifically that only the H1N1 virus is showing this resistance (thus avoiding such resistance from the H5N1 Avian Flu variant), this caused further concern over the H1N1 Influenza A (Swine Flu) variant that posed a global threat in 2009.

As discussed earlier, training is a key element in developing compliance; the more individuals are trained, the more likely that they will comply with these infection control approaches.

Finally, community mitigation techniques are an important part of pandemic preparedness planning. A key mitigation approach is social distancing, requiring that people maintain a minimum distance between one another to "reduce the duration and/or intimacy of social contacts and thereby limit the transmission of influenza,"[47] either through shifts in operation times and/or locations* or individual behavior modification. The CDC and U.S. Office of Personnel Management (OPM) published guidance about such behavior modification for the 2009–2010 swine flu crisis: maintain a distance of 6 feet from each other unless PPE masks are used. Other guidance for mitigation included using flexible work schedules to reduce face-to-face interactions.[49] Other approaches could be used: minimizing mass gatherings (e.g., schools, churches, and malls); practicing isolation (for people who are sick or tending to the sick); or even quarantining (a method not typically expected to be used extensively in a modern pandemic, but might be used in severe circumstances). These tools are available to planners and responders, and pandemic plans will almost certainly include several of these techniques.

Many excellent examples of pandemic preparedness plans have been developed, most of which are available online at state emergency management sites (an example of such a site is one from Virginia Emergency Management, http://www.vaemergency.com/em-community/plans/coveop), as well as others specific to businesses and civil entities. Each plan typically includes guidance regarding assumptions, scenario descriptions, roles and responsibilities, a concept of operations, incident management actions, and maintenance of the plan itself.

Of note, the WHO released a checklist for pandemic influenza preparedness planning[50] based on their guiding document, "Pandemic Influenza Preparedness and Response."[51] This guidance draws on "extensive practical experience...gained from responding to outbreaks of highly pathogenic avian influenza A (H5N1) virus infection in poultry and humans, and from conducting pandemic preparedness and response exercises in many countries. There is greater understanding that pandemic preparedness requires the involvement of not only the health sector, but also the whole of society."[52] Furthermore, "WHO decided...to update its guidance to enable countries to be better prepared for the next pandemic."[53]

Training for and Exercising Pandemic Preparedness

Training is critical to the success of pandemic plans. It is said that the military is successful largely because they "train like they fight, and fight like they train."

* These shifts can notably include "telework." OPM announced a new government-wide telework policy in April 2009, which combined components of the Telework Improvements Act (H.R. 1722) and the Telework Enhancement Act (S. 707).[48]

Pandemic plans may call for significant changes in behavior (e.g., social distancing described previously); these changes will come about only with frequent training and exercising, particularly if behaviors are cultural in nature. An example is the common practice of shaking of hands when meeting. If not practiced often, *not* shaking hands might be perceived as rude rather than a way to lessen the transmission of infectious diseases.

Training and exercising of preparedness plans is becoming more commonplace, even for pandemic plans. Setting the tone for such exercises was the series of federal exercises, Top Officials (TOPOFF), and the subsequent National Exercise Program (NEP).* These exercises address emergencies such as the release of biological agents, radiological dispersal devices, and chemical agents. TOPOFF was a "congressionally mandated, national...exercise that was designed to identify vulnerabilities in the nation's domestic incident management capability by exercising the plans, policies, procedures, systems, and facilities of federal, state, and local response organizations against a series of integrated terrorist threats and acts in separate regions of the country." Top officials were engaged in the decision-making processes they would face in a real-world disaster, from public health concerns to communications challenges. "The purpose of the open exercise design was to enhance the learning and preparedness value of the exercise through 'building-blocks', and to enable participants to develop and strengthen relationships in the national response community. Participants at all levels stated that this approach has been of enormous value to their domestic preparedness strategies."[54]

In particular, lessons learned from the exercises have helped pandemic planners design exercises at federal, regional, state, or local levels for pandemic scenarios: PANEX 07, a Federal Emergency Management Agency–hosted, joint federal–state exercise; and FBIIC/FSSCC Pandemic Flu Exercise of 2007, a Department of the Treasury–hosted exercise involving the banking and financial services sectors. The latter example has clear goals: "enhance the understanding of systemic risks...[on] sector[s]; provide an opportunity for firms to test their pandemic plans; and examine how the effect of a pandemic flu on other critical infrastructures will impact... sector[s]."[55] These should be the goals for any exercise intended to validate a pandemic preparedness plan.

Though specific tools and approaches that are components of pandemic plans must be trained and exercised, often, compliance with these techniques is key to their effectiveness. For example, although social distancing appears to be an

* "[NLEs are] designated as a Tier I National Level Exercises. Tier I exercises (formerly known as the Top Officials exercise series or TOPOFF) are conducted annually in accordance with the National Exercise Program (NEP), which serves as the nation's overarching exercise program for planning, organizing, conducting and evaluating national level exercises. The NEP was established to provide the U.S. government, at all levels, exercise opportunities to prepare for catastrophic crises ranging from terrorism to natural disasters." (U.S. Department of Homeland Security, National Level Exercise 2011 (NLE 2011) Fact Sheet, http://www.fema.gov/media/fact_sheets/ nle2011_fs.shtm).

effective mitigation technique, it is neither easy nor foolproof. A recent study[56] showed that the process can be described, a plan for implementation developed, and a study group tested—but that the overall effectiveness of social distancing depends on overall compliance of the subjects, so clearly diligent training and exercising is critical.

Other guidance for developing exercises intended to validate pandemic plans also is widely available. It is noted that "any plan requiring coordinated action by a number of stakeholders, which has not been validated through a process of practice, or 'simulation exercise,' is simply a collection of ideas and concepts waiting for translation into action."[57] The leadership and advice that these documents provide will increase the effectiveness of our pandemic plans, but only if they are used to thoroughly and frequently exercise the plans themselves, and involve those who will be affected by a pandemic.

Dynamically Replanning for Pandemic Preparedness

Any plan must be revisited regularly and adjusted for changes in the threat, environment, and participants. Therefore, plans must be reviewed and exercised not only periodically but also when circumstances necessitate review, such as a change in guiding policy and a change in the nature of a virus.

During and after a Pandemic

Even with diligent prepandemic planning, some issues need to be addressed when a population is in the midst of a pandemic. Pandemics may come in waves, and successive planning, taking into account lessons learned from a previous wave, can drastically reduce the socioeconomic impacts of subsequent waves. This section presents some issues that might be faced.

Responding to Pandemic Infection

One question that will be relevant to any population is whether individuals are infected and who these people are. There is considerable risk in overreacting to this question (ranging from social "shunning" to possibly illegal quarantining); therefore, screening, detection, and response will be necessary, and all should be considered and included in a holistic pandemic plan. Of interest will be a set of techniques for this screening and testing. In early 2009, CDC developed a rapid diagnostic test kit to detect the type A/H1N1 virus and distributed these kits to all 50 U.S. states, the District of Columbia, Puerto Rico, and internationally. Although of lower sensitivity than viral culture or other similar traditional tests, these kits and others like them will increase rapid testing capacity and are likely to result in a more accurate, rapid picture of the impact of this disease.

Similarly, also in early 2009, the U.S. Food and Drug Administration (FDA) cleared a new, more rapid test for the detection of type A/H5N1.* "This test is an important tool for helping quickly identify emerging influenza A/H5N1 infections and reducing exposure to large populations," said Daniel G. Schultz, director of the FDA's Center for Devices and Radiological Health. "The clearance of this test represents a major step toward protecting the public from the threat of pandemic flu."

Communicating during a Pandemic

As with any emergency, effective, transparent, and complete communication is of utmost importance. This section describes emerging tools and techniques relevant for pandemic events.

Social networking tools are emerging as a powerful means of communication in a broad set of scenarios. Not only are they used routinely for interpersonal or social interactions, but they also are becoming dominant for these purposes in many arenas. In fact, "...social networks and blogs...dominate Americans' time online... [and] nearly 4 in 5 active internet users visit social networks and blogs." Further, "social networks and blogs reach nearly 80 percent of active U.S. internet users, and represent the majority of Americans' time online."[58] Beyond the personal applications, business social networks also are growing. One of the fast-growing business networks, LinkedIn, is enjoying a surge of activity; "LinkedIn membership is up to 85 million [as of December 2010]...every second that ticks by, LinkedIn gets a new user."[59]

It is in this context of broad community participation that these networks are becoming effective tools for emergency communication. This is certainly true as it applies to pandemics. As examples, many such tools have set up specific sections for pandemic discussion and communication, including LinkedIn, Facebook, and Twitter. These sites are in addition to the numerous, more traditional websites devoted to these topics; excellent government sites also exist (e.g., www.flu.gov/ and www.cdc.gov/flu/pandemic-resources/), as do many outstanding private or civic sites (e.g., http://www.cidrap.umn.edu/cidrap/content/influenza/panflu/resources/personpanprep.html/). Add to these a wealth of other communication venues, such as PSAs, special public audio/video conferences and symposia,† and the paths to

* "The test, called AVantage A/H5N1 Flu Test, detects influenza A/H5N1 in throat or nose swabs collected from patients who have flu-like symptoms. The test identifies in less than 40 minutes a specific protein (NS1) that indicates the presence of the influenza A/H5N1 virus subtype. Previous tests that the FDA cleared to detect this influenza A virus subtype can take three to four hours to produce results." (U.S. Food and Drug Administration. FDA News release, April 7, 2009).

† A powerful example of this media communication is the Public Broadcast System (PBS) video, *Predicting Pandemics—How Do We Fight Both the Swine Flu Pandemic and Our Fear of It?*, of May 8, 2009.

communicate the pandemic message (in the preparedness phase and in the response and recovery phases) are many and diverse.

One particular concern regarding pandemic communications is the notion of alarming the public unnecessarily. Certainly, public awareness is important, but avoiding information fatigue (or even being perceived as "crying 'swine'" as one report quoted[60]) is equally key. Communication must be continually measured in terms of its intent, its process, and its effectiveness.

Recovering after a Pandemic

Pandemics have a powerful effect on nearly every measure of a community or nation. For example, the average life span of U.S. citizens dropped nearly 12 years in the year following the Spanish Flu pandemic of 1918–1919. The way citizens behave, and the way they see themselves and each other, fundamentally changes when a pandemic devastates their way of life. Recovery, therefore, is neither an easy nor a quick process.

Specific guidance addresses recovery from pandemics. For example, DHS has prepared a document, "Pandemic Influenza Preparedness, Response, and Recovery Guide for Critical Infrastructure and Key Resources."[61] This document focuses on businesses (stating as its purpose to "stimulate the U.S. private sector to act now"), but it specifically includes consideration of recovery activities following a pandemic. "Continuity of Operations–Essential (COP-E) is the central concept in the guidance, wherein COP-E is an extension and refinement of current business contingency and continuity of operations planning that fully exploits existing efforts and integrates. . .the suite of business disaster plans. The COP-E process assumes severe pandemic-specific impacts to enhance and complement existing business continuity plans."[62] Relevant to recovery, COP-E incorporates an approach for "survival" under distinct COP-E scenarios, and it enhances business continuity planning to address other catastrophic disasters. This holistic approach to continuity of operations is at the heart of recovery for a nation, region, or community.

Summary

Pandemic preparedness is an imperative for all communities, as rampant infectious diseases can have broad and sometimes devastating consequences. Although the public's awareness of potential pandemic events may be raised through media coverage, the same may not be true for overall preparedness. Pandemics are unpredictable, but this much is known: they will happen again, and they will more than likely have great impact on society. The impact goes beyond the health implications; the entire socioeconomic system of a population will be affected.

Rapid transmission of the disease generates enormous fear in the population. As discussed, certainly those who are ill stress the community's ability to respond;

the "worried well," driven by that fear, overwhelms the already besieged healthcare infrastructure.

When developing a tailored pandemic preparedness plan for a community or large population group, several aspects must be considered: culture; policies, strategies, and plans; economics, management, and budgeting; governance and operations; and technology.

Planning for a pandemic must be consistent with existing culture and policy. The U.S. Government has several goals for preparing for pandemics, and all lower-level plans should match these objectives: stopping, slowing, or otherwise limiting the spread of a pandemic; mitigating disease suffering and death; sustaining the infrastructure; and mitigating impact to the economy and functioning of society.

Training for and exercising of pandemic preparedness plans is critical to the ultimate success of implementing those plans. In the balance is the protection of citizens and ultimately society.

We have seen that pandemic preparedness is not only a government construct but also a process that must involve nongovernmental entities (nonprofits, faith-based organizations, and businesses) to reach a level of preparedness that enables resilience to be woven into mitigating effects and accelerating recovery.

Preparedness is neither an easy nor a quick process. Pandemics may pose one of the more difficult scenarios for planners, but also may be one of the most important—another pandemic is most certainly on its way.

References

1. Fagel, M.J. (Ed). *Principles of Emergency Management: Hazard Specific Issues and Mitigation Strategies.* CRC Press, Boca Raton, FL, December 2011.
2. Definition of Pandemic, MedicineNet.com. http://www.medterms.com/script/main/art.asp?articlekey=4751 (accessed June 1, 2012).
3. Nevondo, T. S. and T. E. Cloete. 2001. The Global Cholera Pandemic. http://scienceinafrica.co.za/2001/september/cholera.htm.
4. Hoad, P. 2003. Pandemics Timeline. http://www.guardian.co.uk/society/2003/apr/02/health.lifeandhealth?INTCMP=SRCH.
5. World Health Organization (WHO). Communicable Disease Surveillance and Response. http://www.wpro.who.int/southpacific/sites/ccd/csr/.
6. World Health Organization (WHO). Cumulative Number of Confirmed Human Cases of Avian Influenza A/(H5N1) Reported to WHO, 2003–2012. http://www.who.int/influenza/human_animal_interface/EN_GIP_20120607CumulativeNumber H5N1cases.pdf.
7. Congressional Budget Office. A Potential Influenza Pandemic: Possible Macroeconomic Effects and Policy Issues, December 8, 2005, revised July 27, 2006. http://www.cbo.gov/ftpdocs/69xx/doc6946/12-08-BirdFlu.pdf, http://www.cbo.gov/publication/17544.
8. World Health Organization (WHO), Avian Influenza: Assessing the Pandemic Threat (Geneva: WHO, January 2005); Howell, Pugh, Pandemic, The Cost of Avian Influenza,

Contingencies (September/October 2005), pp. 22–27; and Garrett, Laurie, The Next Pandemic, Foreign Affairs (July/August 2005), pp. 3–23.

9. World Health Organization (WHO). Communicable Disease Surveillance and Response. http://www.wpro.who.int/southpacific/sites/ccd/csr/.
10. Congressional Budget Office. A Potential Influenza Pandemic: Possible Macroeconomic Effects and Policy Issues, December 8, 2005, revised July 27, 2006. http://www.cbo.gov/
11. World Health Organization (WHO), Avian Influenza: Assessing the Pandemic Threat (Geneva: WHO, January 2005); Howell, Pugh, Pandemic, The Cost of Avian Influenza, Contingencies (September/October 2005), pp. 22–27; and Garrett, Laurie, The Next Pandemic, Foreign Affairs (July/August 2005), pp. 3–23.
12. U.S. Department of Health and Human Services (HHS). HHS Pandemic Influenza Implementation Plan. http://www.hhs.gov/pandemicflu/implementationplan/intro.htm.
13. CIDRAP News. WHO: H1N1 Flu More Contagious than Seasonal Virus. http://www.cidrap.umn.edu/cidrap/content/influenza/swineflu/news/may1109severity.html.
14. World Health Organization (WHO). *The Elusive Definition of Pandemic Influenza.* March 31, 2011. http://www.who.int/bulletin/volumes/89/7/11-086173/en/.
15. U.S. Department of Health and Human Services (HHS). Interim Pre-pandemic Planning Guidance: Community Strategy for Pandemic Influenza Mitigation in the United States, February 2007. http://www.flu.gov/planning-preparedness/community/community_mitigation.pdf.
16. Ibid.
17. *New York Times*, A Spotty Response to the Flu Threat. http://www.nytimes.com/2009/05/02/opinion/02sat1.html?scp=1&sq=A%20Spotty%20Response%20to%20the%20Flu%20Threat&st=cse.
18. U.S. Department of Health and Human Services (HHS). The Pandemic: The Great Pandemic—The United States in 1918–1919. http://www.flu.gov/pandemic/history/1918/the_pandemic/.
19. Osterholm, M. T. 2005. Preparing for the Next Pandemic, *New England Journal of Medicine* 352, 1839–1842.
20. Maitre, M. 2006. Heat Linked to More Than 130 Deaths. *Oakland (CA) Tribune*, July 29, 2006.
21. Sulek, D., R. Cowell, and M. Delurey. 2008. Mission Integration—A Whole of Government Strategy for a New Century. Booz Allen Hamilton white paper, December 2008.
22. The World Bank. Influenza A (H1N1): Questions and Answers. http://web.worldbank.org/WBSITE/EXTERNAL/NEWS/0,,contentMDK:22160388~pagePK:64257043~piPK:437376~theSitePK:4607,00.html.
23. Congressional Budget Office. A Potential Influenza Pandemic: Possible Macroeconomic Effects and Policy Issues, December 8, 2005, revised July 27, 2006. http://www.cbo.gov/ftpdocs/69xx/doc6946/12-08-BirdFlu.pdf, http://www.cbo.gov/publication/17544.
24. Ibid.
25. Lister, S. 2005. Pandemic Influenza: Domestic Preparedness Efforts, CRS Report for Congress RL33145 (Congressional Research Service, November 10, 2005), p. 10.
26. Congressional Budget Office. A Potential Influenza Pandemic: Possible Macroeconomic Effects and Policy Issues, December 8, 2005, revised July 27, 2006. http://www.cbo.gov/ftpdocs/69xx/doc6946/12-08-BirdFlu.pdf, http://www.cbo.gov/publication/17544.

27. Brahmbhatt, M. 2005. Avian and Human Pandemic Influenza—Economic and Social Impacts. http://web.worldbank.org/WBSITE/EXTERNAL/NEWS/0,,contentMDK:20715087~menuPK:3567553~pagePK:34370~piPK:42770~theSitePK:4607,00.html.
28. Worsnip, P. 2008. Bird Flu Pushed Back, Pandemic Threat Remains—UN. http://uk.reuters.com/article/healthNews/idUKTRE49K8BY20081021.
29. Congressional Budget Office. A Potential Influenza Pandemic: Possible Macroeconomic Effects and Policy Issues, December 8, 2005, revised July 27, 2006. http://www.cbo.gov/ftpdocs/69xx/doc6946/12-08-irdFlu.pdf, http://www.cbo.gov/publication/17544.
30. Ibid.
31. U.S. Department of Health and Human Services (HHS). Flu Terms Defined. http://www.flu.gov/definitions.html.
32. Congressional Budget Office. A Potential Influenza Pandemic: Possible Macroeconomic Effects and Policy Issues, December 8, 2005, revised July 27, 2006. http://www.cbo.gov/ftpdocs/69xx/doc6946/12-08-BirdFlu.pdf, http://www.cbo.gov/publication/17544.
33. *Scientific American*. Deadly Pandemic Bird Flu Details Finally Are Made Public. June 21, 2012. http://www.scientificamerican.com/ article.cfm?id=pandemic-bird-flu-studies-public&WT. mc_id=SA_CAT_BS_20120622.
34. CARE. WHO Raises Pandemic Threat Level to Phase 5. http://avianflunetwork.blogspot.com/search?q=phase+5.
35. U.S. Homeland Security Council. The National Strategy for Pandemic Influenza, November 1, 2005.
36. World Health Organization (WHO), Influenza (Seasonal)—Fact Sheet No. 211. http://www.who.int/mediacentre/factsheets/fs211/en/index.html.
37. Brown University. *A SMART(er) Way to Track Influenza.* June 11, 2012. http://news.brown.edu/pressreleases/2012/06/smart.
38. The White House. The National Security Strategy, May 2010. http://www.whitehouse.gov/sites/default/files/rss_viewer/national_security_strategy.pdf.
39. U.S. Department of Homeland Security. National Response Plan, December 2004.
40. U.S. Department of Homeland Security. National Response Framework, January 2008. http://www.fema.gov/pdf/emergency/nrf/nrf-core.pdf.
41. U.S. Homeland Security Council. The National Strategy for Pandemic Influenza, November 1, 2005.
42. Ibid.
43. U.S. Homeland Security Council. The National Strategy for Pandemic Influenza Implementation Plan, May 2006. http://www.flu.gov/planning-preparedness/federal/pandemic-influenza-implementation.pdf.
44. U.S. Homeland Security Council. National Strategy for Pandemic Influenza Implementation Plan One Year Summary, July 2007. http://www.flu.gov/professional/federal/pandemic-influenza-oneyear.pdf.
45. Worsnip, P. 2008. Bird Flu Pushed Back, Pandemic Threat Remains—UN. http://uk.reuters.com/article/healthNews/idUKTRE49K8BY20081021.
46. CIDRAP. *Studies: Antiviral Prescribing Reflects CDC Guidance, Flu Activity.* June 11, 2012. http://www.cidrap.umn.edu/cidrap/content/ influenza/swineflu/news/jun1112antivirals.html.
47. U.S. Homeland Security Council. National Strategy for Pandemic Influenza Implementation Plan One Year Summary, July 2007. http://www.flu.gov/professional/federal/pandemic-influenza-oneyear.pdf.

48. U.S. Office of Personnel Management (OPM). Memorandum for Heads of Executive Departments and Agencies—Advice to Federal Employees and Agencies on Preventing the Spread of the Current Flu and Maintaining Readiness to Use HR Flexibilities If Necessary, April 26, 2009.
49. Ibid.
50. World Health Organization (WHO). WHO Checklist for Influenza Pandemic Preparedness Planning. http://whqlibdoc.who.int/hq/2005/WHO_CDS_CSR_GIP_2005.4.pdf.
51. World Health Organization (WHO). Pandemic Influenza Preparedness and Response. http://www.who.int/influenza/resources/documents/pandemic_guidance_04_2009/en/.
52. Ibid.
53. Ibid.
54. U.S. Department Of Homeland Security (DHS). Top Officials (Topoff) Exercise Series: Topoff 2—After Action Summary Report, December 19, 2003.
55. FBIIC FSSCC SIFMA and U.S. Department of the Treasury. The FBIIC/FSSCC Pandemic Flu Exercise of 2007 After Action Report. https://www.fbiic.gov/public/2008/jan/Pandemic_flu_Jan08.pdf.
56. Himberger, D., J. Bishop, and M. Magoon. 2009. Business performance during a disease outbreak. *Business Today,* April 2009.
57. The United Nations. Simulation Exercises on Influenza Pandemic Responses in the Asia-Pacific Region, 2008. http://un-influenza.org/regions/asia/simex.
58. The Nielson Company. Social Media Report: Q3 2011. http://blog.nielsen.com/nielsenwire/social/.
59. Slutsky, I. Why LinkedIn Is the Social Network That Will Never Die. http://adage.com/article/digital/linkedin-social-network-die/147475/.
60. Associated Press. 2009. Swine Flu Warnings Totally Overblown, Some Say, May 7, 2009. http://www.msnbc.msn.com/id/30627377/.
61. U.S. Department of Homeland Security (DHS). Pandemic Influenza—Preparedness, Response, and Recovery: Guide for Critical Infrastructure and Key Resources." http://www.flu.gov/planning-preparedness/business/cikrpandemicinfluenzaguide.pdf.
62. Ibid.

IV

WHOLE COMMUNITY PREPAREDNESS

Chapter 16

Presidential Policy Directive 8

An Overview

Elizabeth Dawson and Jacob Dickman

PPD-8: An Introduction

Presidential Policy Directive 8 (PPD-8) was signed by President Barack Obama on March 30, 2011. The purpose of PPD-8 is to "strengthen the Nation's security and resilience against a variety of hazards, including terrorism, pandemics, and catastrophic natural disasters" (FEMA, 2011). PPD-8 directs the development of a National Preparedness Goal and a National Preparedness System.

The National Preparedness Goal will define the core capabilities necessary to prepare for the types of incidents that pose the greatest risk to national security. The goal will focus on achieving an integrated, layered, and all-of-nation preparedness approach.

The National Preparedness System will provide an integrated set of guidance, programs, and processes to enable the nation to meet the National Preparedness Goal. The National Preparedness System includes an updated series of integrated national planning frameworks and interagency operational plans. The system will also coordinate a campaign to build and sustain preparedness nationwide. A National Preparedness Report will be prepared and used as a tool to inform the President's budget annually (FEMA, 2011).

PPD-8 is the evolution of Homeland Security Policy Directive-8 (HSPD-8) and Annex 1. Both HSPD-8 and Annex 1 heralded significant achievements in the field of emergency management, specifically the nationwide acceptance of the Incident Command System and the all-hazards approach to planning. Unfortunately, the top-down approach of HSPD-8 and Annex 1 created innumerable obstacles in gathering together, organizing, and distributing the unique resources of business, faith-based organizations, communities, government, families, and individuals.

The structure of HSPD-8 and Annex 1 relied on a tactical and operational focus. Recovery and mitigation were not strongly emphasized either in research or in policy implementation. HSPD-8 and Annex 1 also derived much of their planning methodology from scenarios. Now, while scenarios are a useful and necessary tool for preparing for disaster, a completely scenario-based planning approach can fail to take into account the disaster no one imagines or thought was possible.

Over the last few years, a growing body of research has demonstrated the importance of putting a higher priority on recovery and mitigation planning. Research shows that while the number of people who plan for disasters has improved over the years, their preparation focuses only on response and not enough on recovery (Webb et al., 2000). The lack of recovery planning in business is of great concern because when too many businesses fail to remain economically viable following a disaster, the community may not be able to survive.

PPD-8 developed its philosophical core around the idea that all disasters begin and end locally. This philosophy subtly shifts the focus of emergency management from a government-based method to a coordinated community response.

This shift in focus allows for greater local and regional risk analysis. Rather than having every community, regardless of its size, prepare for all 15 national planning scenarios, PPD-8 allows a more tailored approach to disaster planning. It emphasizes flexibility, robustness, and adaptiveness of capability. In other words, it recognizes the difference between Mayberry and Metropolis in terms of both their hazards potential and their response capabilities.

PPD-8 alters the role of government to one that coordinates the involvement and collaboration between government and nongovernmental sectors to achieve "a secure and resilient nation with the capabilities required across the whole community to prevent, protect against, mitigate, respond to, and recover from the threats and hazards that pose the greatest risk" (Department of Homeland Security, 2011).

PPD-8: An Overview

PPD-8 is aimed at "strengthening the security and resilience" of the United States through "systematic preparation of threats that pose the greatest risk to the security of the nation" (Lumpkin, 2011).

PPD-8 uses four overarching elements to achieve national preparedness. These four elements are

- National Preparedness Goal—characteristics of the goal include the description of 31 core capabilities that span across five mission areas
- National Preparedness System—the methodology the whole community will employ to build, sustain, and deliver core capabilities
- Campaign for Building and Sustaining Preparedness—integrated structure for preparedness programs, research and development activities, and preparedness assistance
- National Preparedness Report—an annual report summarizing progress made toward building, sustaining, and delivering the core capabilities

The President gave the Assistant for Homeland Security and Counterterrorism 60 days (until the end of May 2011) to develop an implementation plan for completing the National Preparedness Goal and the National Preparedness System. The completion of the goal was due within 180 days and was developed by the Secretary of Homeland Security in coordination with other executive departments and agencies in consultation with state, local, tribal, and territorial governments, the private and nonprofit sectors, and the public. The goal included the use of *concrete and measurable objectives* that took into account the regional variations in security strategy while defining the core preparedness capabilities of the nation as a whole.

Once this goal was in place, President Obama directed the Secretary of Homeland Security to once again coordinate and consult with all executive departments and agencies as well as state and local governments to develop an integrated set of guidance, programs, and processes that would enable the nation to meet the National Preparedness Goal. The preparedness system included a series of integrated national planning frameworks, which, through both government-wide interaction and interagency specialization, support an "all-hazards" approach to preparedness and an "all-of-nation" approach for building and sustaining preparedness activities over time. This plan not only ensures enduring operation of the government's critical functions but also includes preparedness recommendations and guidance for businesses, communities, families, and individuals.

PPD-8 stated that on March 30, 2012, one year from the issuance of PPD-8: National Preparedness, the Secretary of Homeland Security will submit the first National Preparedness Report to the President through the Assistant to the President for Homeland Security and Counterterrorism. A primary purpose of the report is to help the President evaluate and adjust the nation's preparedness planning.

In sum, this Presidential Policy Directive represents a substantial government-wide effort to improve the sustainability of the U.S. Constitutional Government in the event of an attack or national catastrophe. The National Preparedness Goal was to be established by the end of May 2011. However, this marks only the beginning of the department's continuity efforts. Over the following year, the focus on U.S. security and resilience became much sharper and more highly defined.

National Preparedness Goal: An Overview

As stated before, the National Preparedness Goal is to achieve "a secure and resilient nation with the capabilities required across the whole community to prevent, protect against, mitigate, respond to, and recover from the threats and hazards that pose the greatest risk" (Department of Homeland Security, 2011).

The supporting components of PPD-8 are the core capabilities, capability targets, and the Strategic National Risk Assessment. The ultimate goal of these supporting components is to enable the community to contribute to as well as benefit from national preparedness.

For the emergency manager, any action that falls within the core capabilities must adhere to the guidelines set by PPD-8.

Five Mission Areas

The five overarching mission areas as described in the National Preparedness Goal are

- Prevention
  - The specific capability to avoid, prevent, or stop an act of terrorism or other imminent threats.
- Protection
  - The capabilities necessary to secure the nation against all disasters whether natural, man-made, or occurring through acts of terrorism.
- Mitigation
  - Capabilities necessary to reduce loss of life and property by lessening the impact of disasters. Examples:
    - Community-wide risk-reduction projects.
    - Risk-reduction projects for targeted vulnerable populations (e.g., children, senior citizens, or mobility-impaired individuals).
    - Efforts to improve the resilience of critical infrastructure and key resources (CI/KR).
    - Initiatives to reduce future risks after a disaster has occurred.
- Response
  - This mission area relates specifically to the capability of saving lives and protecting property and the environment. It also relates to the ability to meet basic human needs following a disaster or terrorist event.
- Recovery
  - Defined as successfully overcoming the physical, emotional, and environmental impacts of the disaster, reestablishment of an economic and social base that instills confidence in the communities' ability to remain viable, the rebuilding of functional needs of residents, the reduction to

future vulnerability, and the demonstration of the community to be prepared, responsive, and resilient in dealing with the consequences of disasters.

The integration of the five mission areas as described earlier is crucial to the success of the National Preparedness Goal. Integration between missions occurs when the purpose of one mission helps to complete and strengthen another mission. For example, *prevention*, *protection*, and *mitigation* are all interrelated missions in that they help reduce risk and increase resiliency. When they have been successfully implemented, the *response* and *recovery* following a disaster are often significantly decreased.

The 31 Core Capabilities

The National Preparedness Goal defines core capabilities as the "distinct critical elements necessary to achieve" goals (Figure 16.1).

Table 16.1 shows the 31 core capabilities as listed in the 2013 National Preparedness Report (Department of Homeland Security, 2013). The capabilities are subdivided according to their mission area. The first three capabilities span all the missions. Four of the capabilities span multiple mission areas.

Figure 16.1 Demonstration of the way in which all the mission areas contribute to the National Preparedness Goal.

Table 16.1 The 31 Core Capabilities Subdivided by Mission Area or Mission Areas

Row	*Prevention*	*Protection*	*Mitigation*	*Response*	*Recovery*
1	Planning				
2	Public information and warning				
3	Operational coordination				
4	Intelligence and information sharing	Intelligence and information sharing	Community resilience	Critical transportation	Economic recovery
5	Interdiction and disruption	Interdiction and disruption	Long-term vulnerability reduction	Environmental response/ health and safety	Health and social resources
6	Screening, search, and detection	Screening, search, and detection	Risk and disaster resilience assessment	Fatality management services	Natural and cultural resources
7	Forensics and attribution	Access control and identity verification	Threats and hazard identification	Infrastructure systems	Infrastructure systems
8		Cybersecurity		Mass care services	Housing

9		Physical protective measures		Mass search and rescue operations	
10		Risk management for protection programs and activities		On-scene security and protection	
11		Supply chain integrity and security		Operational communications	
12				Public and private services and resources	
13				Public health and medical services	
14				Situational assessment	

Note: The shading represents core capabilities that are listed under multiple mission areas and was added by the authors. This chart is taken from the first edition of the National Preparedness Goal (Department of Homeland Security, 2011).

*Common Core Capabilities: Defined**

Planning—Planning is defined as conducting a systematic process that engages the whole community in the development of executable strategic, operational, and/or community-based approaches to meet defined objectives (Figure 16.2).

- Catastrophic planning—FEMA is the lead agency for planning initiatives in different geographic areas that consider catastrophic events such as earthquakes, hurricanes, dam failures, improvised nuclear device detonation, evacuation and sheltering, and other major events. FEMA also conducts a planning framework known as the "Maximum of Maximums," which centers on collaborative, whole community planning for worst-case scenarios that exceed government capabilities.
- Emergency planning—Emergency planning efforts increasingly adhere to standardized development and maintenance processes, reflecting FEMA's Comprehensive Preparedness Guide 101 (CPG 101) and new requirements from PPD-8.
- Mitigation planning—Mitigation planning requires natural hazard analysis and encourages consideration of other threats. FEMA issues multihazard mitigation planning guidance to address the inclusion of populations with disabilities and other access and functional needs.

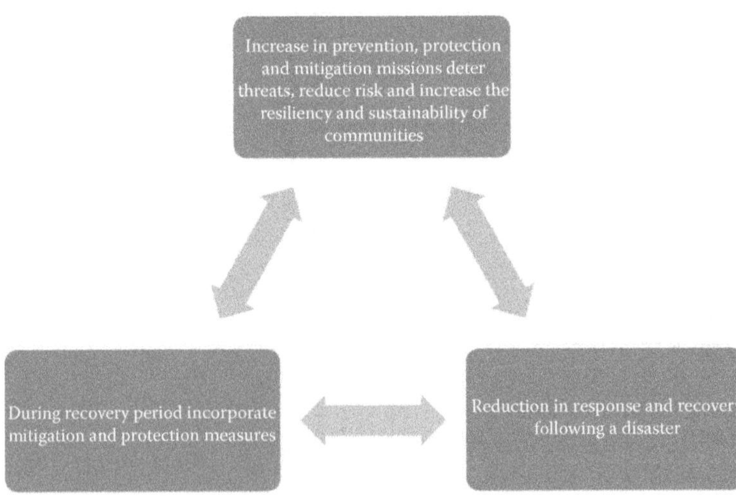

Figure 16.2 Demonstration of how one mission area can influence or the other mission areas. (From FEMA. 2012, May. *National Mitigation Framework*. U.S. Government.)

* Definitions taken from the National Preparedness Report (Department of Homeland Security, 2013).

- Private sector critical infrastructure planning—Self-organized planning and policy bodies that include broad representation from within the 18 critical infrastructure sectors. These planning bodies are tasked with establishing sector-specific plans that describe how they will identify sector-specific hazards and implement risk management techniques to enhance the protection of their critical infrastructure.
- Continuity and contingency planning—Continuity planning is the development of a plan or process that helps organizations prepare for disruptive events. Contingency planning is a course of action to be followed if a preferred plan fails or the situation changes making the primary plan obsolete.
- Recovery planning—Recovery planning is that which articulates the roles and responsibilities for long-term recovery and outlines how whole community partners can engage in implementing the six recovery support functions.

Public Information and Warning—Public information and warning is the method by which coordinated, prompt, reliable, and actionable information can be delivered to the whole community through the use of clear, consistent, accessible, and culturally and linguistically appropriate methods to effectively relay information regarding any threat or hazard, as well as the actions being taken and the assistance being made available, as appropriate.

- Integrated Public Alert and Warning System (IPAWS)—an integrated set of services and capabilities that enable local, state, and federal authorities to alert and warn their communities of a hazard.

Operational Coordination—Operational coordination means the establishment and maintenance of a unified and coordinated operational structure and process that appropriately integrates all critical stakeholders and supports the execution of core capabilities.

Forensics and Attribution—Forensics and attribution determine the means and methods of terrorist acts. It traces the act to its source in an effort to prevent initial or follow-on acts and/or swiftly develop counteroptions. Example programs are

- Counterterrorism and Forensic Science Research Unit
- Chemical, Biological, Radiological, and Nuclear Sciences Unit
- Evidence Response Teams
- Cyber Action Teams
- Criminal Justice Information Services Division's Global Initiatives Unit

Intelligence and Information Sharing—

- Intelligence provides timely, accurate, and actionable information resulting from the planning, direction, collection, exploitation, processing, analysis, production, dissemination, evaluation, and feedback of available information concerning threats to the United States, its people, property, or interests; the development, proliferation, or use of Weapons of Mass Destruction (WMD); or any other matter bearing on U.S. national or homeland security by federal, state, local, and other stakeholders.

- Information sharing is the ability to exchange intelligence, information, data, or knowledge among federal, state, local, or private sector entities, as appropriate.

Interdiction and Disruption—Interdiction and disruption provide layered defenses in the air, land, and maritime domains that enhance protection against terrorist plots for the purpose of delaying, diverting, intercepting, halting, apprehending, or securing threats and/or hazards.

Screening, Search, and Detection—Screening, search, and detection are concerned with identifying, discovering, or locating threats and/or hazards through active and passive surveillance and search procedures and includes the use of systematic examinations and assessments, sensor technologies, or physical investigation and intelligence.

Access Control and Identity Verification—Access control and identity verification applies a broad range of physical, technological, and cyber measures to control admittance to critical locations and systems, limiting access to authorized individuals to carry out legitimate activities.

Cybersecurity—Cybersecurity protects against damage to, the unauthorized use of, and/or the exploitation of (and, if needed, the restoration of) electronic communications systems and services (and the information contained therein).

Physical Protective Measures—Physical protective measures reduce or mitigate risks, including actions targeted at threats, vulnerabilities, and/or consequences, by controlling movement and protecting borders, critical infrastructure, and the homeland.

Risk Management for Protection Programs and Activities—Risk management for protection programs and activities identify, assess, and prioritize risks to inform protection activities and investments.

Supply Chain Integrity and Security—Supply chain integrity and security is concerned with strengthening the security and resilience of the supply chain.

Community Resilience—Community resilience is the integrated effort to recognize, understand, communicate, plan, and address risks so that the community can develop a set of actions to accomplish mitigation and improve resilience.

Long-term Vulnerability Reduction—Long-term vulnerability reduction is about building and sustaining resilient systems, communities, and critical infrastructure and key resources lifelines so as to reduce their vulnerability to natural, technological, and human-caused incidents by lessening the likelihood, severity, and duration of the adverse consequences related to these incidents.

Risk and Disaster Resilience Assessment—Risk and disaster resilience assessment is about assessing hazard risks and disaster resilience so that decision makers, responders, and community members can take informed action to reduce their entity's risk and increase their resilience.

Threats and Hazard Identification—Threats and hazard identification identify the threats and hazards that occur in the geographic area; determine the frequency and magnitude; and incorporate this into analysis and planning processes so as to clearly understand the needs of a community or entity.

Critical Transportation—Critical transportation provides transportation (including infrastructure access and accessible transportation services) for response priority objectives, including the evacuation of people and animals, and the delivery of vital response personnel, equipment, and services into the affected areas.

Environmental Response/Health and Safety—Environmental response along with health and safety ensures the availability of guidance and resources to address all hazards, including hazardous materials, acts of terrorism, and natural disasters in support of the responder operations and the affected communities. This includes a hazardous materials response with resources that can supplement state, local, tribal, and territorial assets in addressing large-scale disasters, including Chemical, Biological, Radiological, Nuclear, and Explosives (CBRNE) incidents.

Fatality Management Services—Fatality management services include body recovery and victim identification, working with state and local authorities to provide temporary mortuary solutions, sharing information with mass care services for the purpose of reunifying family members and caregivers with missing persons/remains, and providing counseling to the bereaved.

Mass Care Services—Mass care services provide life-sustaining services to the affected population with a focus on hydration, feeding, and sheltering to those who have the most need as well as support for reunifying families.

Mass Search and Rescue Operations—Mass search and rescue operations deliver traditional and atypical search and rescue capabilities including personnel, services, animals, and assets to survivors in need with the goal of saving the greatest number of endangered lives in the shortest time possible.

On-Scene Security and Protection—On-scene security and protection ensures a safe and secure environment through law enforcement and related security and protection operations for people and communities located within the affected areas and also for all traditional and atypical response personnel engaged in lifesaving and life-sustaining operations.

Operational Communications—Operational communications ensure the capacity for timely communications in support of security, situational awareness, and operations by any and all means available, among and between affected communities in the impact area and all response forces.

Public and Private Services and Resources—Public and private services and resources provide essential public and private services and resources to the affected population and surrounding communities, to include emergency power to critical facilities, fuel support for emergency responders, and access to community staples (e.g., grocery stores, pharmacies, and banks) and fire and other first-response services.

Public Health and Medical Services—Public health and medical services provide lifesaving medical treatment via emergency medical services and related operations and avoid additional disease and injury by providing targeted public health and medical support and products to all people in need within the affected area.

Situational Assessment—Situational assessment provides decision makers with decision-relevant information regarding the nature and extent of the hazard, any cascading effects, and the status of the response.

Infrastructure Systems—Infrastructure systems stabilize critical infrastructure functions, minimize health and safety threats, and efficiently restore and revitalize systems and services to support a viable, resilient community.

Economic Recovery—Economic recovery is about returning economic and business activities (including food and agriculture) to a healthy state and developing new business and employment opportunities that result in a sustainable and economically viable community.

Health and Social Services—Health and social services are tasked with restoring and improving health and social services networks to promote the resilience, independence, health (including behavioral health), and well-being of the whole community.

Housing—Housing implements housing solutions that effectively support the needs of the whole community and contribute to its sustainability and resilience.

Natural and Cultural Resources—Natural and cultural resources protect natural and cultural resources and historic properties through appropriate planning, mitigation, response, and recovery actions to preserve, conserve, rehabilitate, and restore them consistent with postdisaster community priorities and best practices and in compliance with appropriate environmental and historical preservation laws and executive orders.

Capability Targets

Capability targets are performance thresholds for each core capability that will guide our allocation of resources to support national preparedness. The National Preparedness Report used standardized self-assessment surveys gathered from the 50 U.S. states and 6 U.S. territories. FEMA also conducted research to identify recent, independent evaluations, surveys, and other supporting data related to core capabilities. The purpose of this research was to offer insight on critical issues, determine where the United States has improved and what areas of improvement remain (Department of Homeland Security, 2013).

Strategic National Risk Assessment

In accordance with PPD-8, a Strategic National Risk Assessment was performed and identified a wide range of threats and hazards that pose a significant risk to the nation, affirming the need for an all-hazards, capability-based approach to preparedness planning (Lumpkin, 2011).

National Preparedness System

The new National Preparedness System (NPS) components address the requirements that are outlined in PPD-8 (Lumpkin, 2011).

The components also describe the guidance for planning, organization, equipment, training, and exercises to build and maintain capabilities. The NPS addresses a national planning system that is composed of national-level frameworks focused on preparing capabilities and federal interagency operational plans to deliver capabilities. Resource guidance is included as a component, which includes arrangements enabling the ability to share personnel (Lumpkin, 2011). There is equipment guidance aimed at nationwide interoperability, guidance for national training and exercise programs, recommendations and guidance for businesses, communities, families, and individuals. The approach to the assessment has a consistent methodology (Lumpkin, 2011) (Figure 16.3).

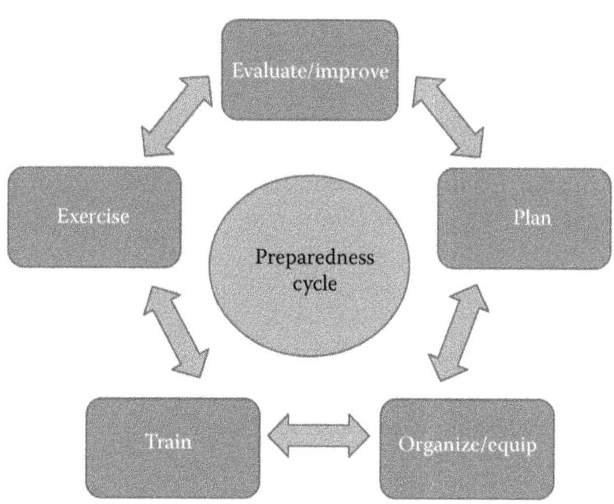

Figure 16.3 The Preparedness Cycle is an ongoing sequence of actions designed to improve communication and coordination among responding agencies. (From Department of Homeland Security. 2013. *National Mitigation Review* (Review of National Mitigation Framework). Washington, DC: FEMA.)

National Frameworks

The basic elements of the national frameworks include the following:

- Describe the roles and responsibilities of all stakeholders.
- Define the coordinating structures—either new or existing—that enable the effective delivery of the core capabilities.
- Describe how the actions taken are coordinated with relevant actions described in other mission area frameworks.
- Identify relevant planning assumptions required to inform the development of interagency operational plans and department-level plans.
- Provide information to the state, local, and tribal governments, as well as to the private sector that they can use to develop or revise their plans (Lumpkin, 2011).

Building and Sustaining Preparedness

One of the goals within PPD-8 is to sustain the preparedness that is built. HSPD-8 provided the capabilities to plan and provided the templates and frameworks for building preparedness. Sustaining the preparedness was found lacking. To sustain preparedness, four elements are identified:

- A comprehensive campaign to build and sustain national preparedness, to include public outreach and community-based and private-sector programs to enhance national resilience
- Federal preparedness
- Federal preparedness assistance (grants, technical assistance, etc.)
- National research and development efforts, which include the creation of a National Research Agenda within 5 years

Limitations

Nothing in this directive is intended to alter or impede the ability of the authorities of executive departments and agencies to perform their responsibilities under law and be consistent with applicable legal authorities and other Presidential guidance. This directive shall be implemented consistent with relevant authorities, including the Post-Katrina Emergency Management Reform Act of 2006 and its assignment of responsibilities with respect to the Administrator of the Federal Emergency Management Agency.

Nothing in this directive is intended to interfere with the authority of the Attorney General or Director of the Federal Bureau of Investigation with regard to the direction, conduct, control, planning, organization, equipment, training, exercises, or other activities concerning domestic counterterrorism, intelligence, and law enforcement activities.

Nothing in this directive shall limit the authority of the Secretary of Defense with regard to the command and control, planning, organization, equipment, training, exercises, employment, or other activities of Department of Defense forces, or the allocation of Department of Defense resources (FEMA, 2011).

Summary

The National Preparedness Goal sets the overall strategic vision for national preparedness, and establishes the core capabilities that will be used to drive preparedness activities nationwide. The National Preparedness System takes into account all the programs, processes, and tools available to build, sustain, and deliver capabilities across the nation. The national planning frameworks specify the roles and responsibilities in preparing to deliver the applicable core capabilities to include linkages across the whole community. Building and sustaining preparedness lays out how we integrate a number of efforts across the whole community to improve preparedness.

References

Department of Homeland Security. 2011, September. *National Preparedness Goal. First Edition: National Preparedness Goal.* U.S. Government.

Department of Homeland Security. 2012, March 20. *Presidential Policy Directive/PPD-8: National Preparedness.* Retrieved 2012, from Department of Homeland Security website: http://www.dhs.gov/xabout/laws/gc_1215444247124.shtm.

Department of Homeland Security. 2013. *National Mitigation Review* (Review of National Mitigation Framework). Washington, DC: FEMA.

Department of Homeland Security. 2013, March 30. *National Preparedness Report.* U.S. Government.

FEMA. 2011, March 30. *Presidential Policy Directive 8.* Washington, DC, USA: White House.

FEMA. 2011, July 27. *Presidential Policy Directive 8 National Preparedness: Frequently Asked Questions.* Retrieved June 8, 2012, from FEMA: http://www.fema.gov/prepared/ppd8_faqs.shtm.

FEMA. 2012, March 2. *National Disaster Recovery Framework Interagency Operational Plan-DRAFT. Presidential Policy Directive/PPD-8: Initial Draft-National Disaster Recovery Framework.* U.S. Government.

FEMA. 2013, May. *National Mitigation Framework.* U.S. Government.

Department of Homeland Security. 2012, March 2. *National Disaster Recovery Framework Interagency Operational Plan* (Guidance for implementation of the National Disaster Recovery Framework). Retrieved from American Veterinary Medical Association:

http://www.avma.org/Advocacy/National/Federal/Documents/NDRF_IOP_rsf_annexes_draft5B15D.pdf.

Lumpkin, M. 2011. *Presidential Policy Directive/PPD8 IAEM Session*. IAEM Website. Las Vegas, Nevada.

Webb, G. R., Tierney, K. J., Dahlhamer, J. M. 2000. Business and disasters: Empirical patterns and unanswered questions. *Natural Hazard Review* 1(2), 83–90.

Chapter 17

Emergent Group Theory and Whole Community Capability-Building

Joseph Lombardo

Introduction

The previous chapters of this book describe the prevailing policy landscape for emergency management as well as some of the significant preparedness challenges our nation faces. For example, public health threats such as pandemics or incidents affecting the safety and security of food and agriculture necessitate extensive coordination among a diverse set of stakeholders. Legal issues abound in all areas of emergency management, as crisis events often involve matters of authority or liability. The availability of resources with which to build emergency management capabilities is on an expected decline. And each of the aforementioned dynamics is taking place in an era where the impacts of disasters are increasing significantly. The increasing scope and magnitude of disasters can be explained by a number of variables, which include activity in the planet's natural environment; human and social systems that create and redistribute hazards; and hazards resulting from the human-made constructed environment (Mileti, 1999, p. 105).

Together, these problems lead emergency managers to an important consideration. The increasing complexity and magnitude of disasters, compounded with a growing scarcity of government resources needed to address preparedness capability gaps, increase the likelihood that critical response needs for a disaster will not only be met by trained volunteer organizations, but also by untrained and unaffiliated groups motivated by the crisis situation. Following a disaster, communities can expect that, as a matter of expediency, individuals will converge and commence ad hoc operations that may or may not be sustained for multiple operational periods; become more proficient and organized; or, become formal, enduring constructs with an active membership that continues to serve a critical community function postdisaster. These ad hoc associations of people during disaster events are known as emergent groups, that is, disaster-focused entities that did not previously exist in a community prior to the disaster. Without question, emergent groups are a significant aspect of the response to any major disaster or catastrophic event and undertake tasks or provide services, which cannot be done by existing groups even if the existing groups expand their functions or extend their structures (Quarantelli, 1994, p. 12).

Finding ways to tap the latent capacity of emergent groups through engaging the community before the onset of disasters provides critical force multipliers for professional emergency managers. Forging community partnerships should be a priority for all emergency managers, yet there are legitimate questions regarding how this should be done. Who are the ideal groups to engage and how do we prompt those potentially emergent groups and individuals to participate in phases of emergency management other than response and recovery? What roles are best suited for emergent, converging groups and how do they eventually receive a suitable degree of training or become qualified to provide needed contributions to preparedness capabilities?

Emergent Group Theory

The process by which emergent groups come to fruition, also known as "convergence," has basic characteristics observable in most disaster-stricken populations. Traditionally speaking, the process involves either spontaneous or unaffiliated volunteers (individuals or assemblies of people) who arrive unsolicited at the scene of a disaster (Points of Light Foundation & Volunteer Center National Network, 2002, p. 3). Typically, these groups are informal and are composed of residents of the disaster-stricken area (Tierney et al., 2001, p. 115). However, given the increasing geographical footprint of recent disasters and the ubiquity of global communications in the information age, proximity to the disaster scene and/or the affected community is not necessarily a definitive characteristic for an emergent group; convergence can happen at any physical or virtual location where motivated groups and individuals believe they have the opportunity to contribute to a disaster response. The emergent group may or may not include residents of the affected community and members of the group may or may not possess the specific skills necessary to respond to the current disaster. Most importantly, the group is not associated with any part of the existing emergency management response system (Points of Light Foundation & Volunteer Center National Network, 2002, p. 3).

A number of community factors are usually prevalent for emergent groups to originate. According to T.E. Drabek, emergent structures are likely to form in a disaster-stricken community with the following prevailing factors: a lack of overall community organization during the emergency period; ambiguity regarding authority; people isolated from organization and information; and, minimal prior disaster experience (Drabek as cited in Tierney et al., 2001, p. 117). However, predisaster cultural and societal factors, as evidenced by field studies conducted by E.L. Quarantelli, can also be catalysts for the formation of emergent groups. These factors include a legitimizing social setting, a perceived threat, a supportive social climate, preexisting social ties, and the availability of resources (Quarantelli as cited in Tierney et al., 2001, p. 117). For example, Quarantelli observed the response to flood disasters in Salt Lake City and Fort Wayne. He found the higher degree of emergence in Salt Lake attributable to specific community predisaster elements, that is, the existence of religion-focused social networks that could be easily mobilized (Quarantelli, 1994, p. 4).

Emergent group theory and the concept of convergence counter one of the more pervasive myths regarding the behavior of people and groups in disaster situations and often perpetuated by the public, the media, and even the professional emergency management community. Many contend that disaster situations result in widespread panic, looting, and/or civil unrest. However, these are rare reactions from the public during the onset and aftermath of a disaster. More specifically, panic, which can be described as "an acute fear reaction marked by a loss of self-control which is followed by nonsocial and non-rational flight behavior" (Quarantelli as cited in Lindell et al., 2007, p. 227), should be carefully differentiated from the reaction of fear. Fear, on the contrary, is a normal human reaction to disasters that does not necessarily have a negative impact on the public's ability to apply reason for problem solving. When fear is a notable reaction during an emergency situation, instances of poor decision making by members of a community may be more attributable to a lack of information about specific threats and their consequences rather than panic-fueled behavior (Perry and Lindell, Emergency Planning, 2007, p. 75).

Suffice to say, disasters do not necessarily compel people and groups to suddenly engage in antisocial behavior. In the immediate postimpact period of a disaster, communities are more likely to demonstrate a willingness to engage in productive response and recovery actions (Lindell et al., 2007, p. 230) and exhibit altruism, an enhanced sense of community solidarity, and morale (Mileti, 1999, p. 226). Research and disaster literature indicate that prosocial, altruistic, and adaptive responses are far more prevalent following a disaster (Tierney et al., 2001, p. 107). Perhaps no response is more prosocial, altruistic, and adaptive following a disaster than convergence, where spontaneous flows of people, materials, and other resources arrive at the impacted area (Perry and Lindell, 2007, p. 71). Evidence from multiple disasters shows that individuals assume responsibility over their own rescue and that of their neighbors to the fullest extent possible. These individuals are limited only by a lack of information or specific training (Poteyeva et al., 2007, pp. 206–207).

Another prevalent myth when it comes to spontaneous or emergent groups is the misconception that all those who are, at the time of the disaster, unaffiliated with an emergency-focused organization and looking to spontaneously contribute to the response efforts do not possess useful skills. The various communities that compose the nation are replete with professional groups, highly skilled private sector entities, research centers, information transfer organizations, and associations that can readily contribute people, skills, and resources to address capability gaps (Mileti, 1999, p. 280). Finding ways to employ the Whole Community approach before the onset of a disaster can harness the latent skills at a community's disposal for preparedness activities as well as help establish mechanisms for incorporating skills into structured and organized response as seamlessly as possible once the disaster occurs.

Disaster Research Center Typology

Much of the thinking around emergent groups stems from the work done by the Disaster Research Center (DRC) at the University of Delaware, which specializes in the study of social scientific issues related to disasters and emergency management.* The DRC Typology of Organizational Adaptation in Crises provides classifications of organizational responses to disasters. Observing how organizations adapt to their postdisaster environment, in terms of both their structure and the tasks for which they become responsible, is an important tool for understanding emergent group theory. Table 17.1 displays the DRC's four basic classifications of organizational adaptation.

---

* Information on the DRC is available at http://www.udel.edu/DRC/index.html.

Table 17.1 Typology of Organizational Adaptation in Crises

Organizational Structure	*Tasks*	
	Routine	*Nonroutine*
Same as predisaster	Type I Established	Type III Extending
New	Type II Expanding	Type IV Emergent

Source: Typology of Organizational Adaptation in Crises (Based on illustration Dynes as cited in Tierney et al., 2001, University of Delaware, Delaware. http://www.udel.edu/DRC/index.html, Retrieved June 2012. p. 114).

In the DRC typology, type I organizations are classified as *Established* groups, meaning they perform the same tasks during disasters that they usually do in non-disaster situations. Type II organizations are *Expanding* groups, meaning they are largely inactive during predisaster periods but increase in size during a disaster to perform essentially the same tasks. The Red Cross fits this example since they expand operations to perform regular tasks (Campbell, 2010, p. 9). Type III organizations are *Extending* organizations, which do not change structure in a disaster yet assume new disaster-focused tasks. Lastly, type IV represents informal, community-formed emergent groups designed to address unmet critical response needs (Tierney et al., 2001, pp. 114–115). As the research-defined concepts surrounding emergent groups continued to mature, the term "Emergent Citizen Group" (ECG) began to take root. Working off the type IV grouping in the DRC model, multiple studies generalized that the ECG is typically composed of three tiers: a small active core who participate for the duration of the group's existence; a larger supporting circle; and a third tier of more passive supporters (Tierney et al., 2001, p. 116).

Emergent Groups in the Context of National Preparedness Policy

New policy and doctrine emanating from the Federal Emergency Management Agency (FEMA) describe a national vision for preparedness that entails a capabilities-based, Whole Community approach in each of the preparedness mission areas, that is, Prevention, Protection, Mitigation, Response, and Recovery. The Whole Community and capabilities-based preparedness paradigms are the dominant philosophies for contemporary emergency management and describe an environment where emergent groups can be an increasingly significant element in all preparedness activities.

Whole Community Approach: Opportunities and Challenges

The complicating factors for the emergency management mission space previously described, that is, increasing magnitude of disasters combined with significant government resource constraints, have compelled FEMA and emergency managers in general to accept the limitations of government-centric solutions to emergency management. This is especially true when it comes to catastrophic events that may overwhelm traditional first responders much more rapidly (Fugate as cited in Federal Emergency Management Agency, 2011a, p. 2). The Whole Community approach, upon which current FEMA policy and doctrine are based, necessitates contributions from individuals, families, communities, private and nonprofit sectors, faith-based organizations, and all levels of government to build the various capabilities needed to achieve national preparedness (Federal Emergency Management Agency, 2011b, p. 1). FEMA describes Whole Community as "a means by which residents, emergency management practitioners, organizational and community leaders, and government officials can collectively understand and asses the needs of their respective communities and determine the best ways to organize and strengthen their assets, capacities, and interests" (Federal Emergency Management Agency, 2011a, p. 3). One can imagine the obvious overlap between the Whole Community approach and the inevitable convergence of untrained and unaffiliated individuals during the response phase of a disaster. Although a broader base of participants contributing to all phases of emergency management is an anticipated outcome of the Whole Community approach, the intent is not to have trained and qualified first responders supplanted by well-intentioned emergent individuals who lack training. Rather, the optimal process is one of preidentifying specific capability gaps and the skills and resources within a community that have the expertise and willingness to help close those gaps. Those institutions, assets, and networks that work well on a daily basis to deliver critical services before a disaster strikes can be utilized to key response and recovery factors during and after the disaster as well (Federal Emergency Management Agency, 2011a, p. 5).

Thus, the involvement of citizens in emergency management activities rather than an exclusive structure for centralized emergency management authority is critical for the purposes of community resilience and emergency preparedness (Bach et al., 2010, p. 7). However, there are inherent challenges regarding how to identify volunteers—both trained and spontaneous, unaffiliated volunteers—can best complement traditional, government-sponsored, emergency management agencies, offices, and departments. Where does the emergent group fit into this Whole Community concept once the disaster strikes? How can the phenomenon be encouraged predisaster?

The first and most obvious challenge is the fact that traditional first responders may be reluctant to widen the base of citizen participation for disaster response or engage with emergent groups and converging individuals. This is an understandable concern. The term "disaster within the disaster" is used to describe multiple unintended consequences of spontaneous response efforts, ranging from the logistical headache of sorting and warehousing unsolicited disaster donations to untrained individuals interfering with response operations or becoming causalities themselves. It is certainly not unprecedented for the convergence of spontaneous volunteers to present a host of health, safety, and security concerns and inhibit fulfillment of primary duties by first responders, particularly when response-focused organizations have not preestablished a system or structure for integrating unaffiliated volunteers (Fernandez et al., 2006, pp. 1–2).

Additionally, academics in the field of emergency management often describe the concept of "normalcy bias"—a situation where segments of a population may receive warnings regarding pending danger, but reinterpret such cues to mean that no threats or hazards are imminent (Perry and Lindell, Emergency Planning, 2007, p. 306). This concept can be extrapolated to apply to all the phases of emergency management, not only response and recovery efforts during or immediately following a disaster. Many areas of the United States are prone to disasters caused by specific hazards (e.g., hurricanes, earthquakes, and flooding) yet many communities lack an empirical basis for understanding the degree of citizen and nongovernmental engagement that is truly required for resiliency and preparedness. Thus, normalcy bias and complacency are inhibitors to preevent mobilization of a community for emergency management purposes. The latent power of emergent groups is often not realized until a focusing event, which is an actual incident such as Hurricane Katrina or the 9/11 terrorist attacks that calls attention to a specific aspect of emergency management policy (Birkland as cited in Perry and Lindell, Emergency Planning, 2007, p. 402). Motivating nongovernment actors to take part in preparedness activities prior to a disaster is imperative for the Whole Community approach to succeed.

Capabilities-Based Preparedness Policy

The Whole Community perspective is a FEMA-endorsed philosophy that encourages inclusion and creativity in emergency management efforts with the understanding that all communities are historically, culturally, demographically, and politically distinct. The preparedness policy products developed and promulgated by FEMA set the parameters for the application of the Whole Community approach as local communities follow prescribed preparedness activities and processes and develop specified products. Previous chapters in this book explain the current national preparedness policy, Presidential Preparedness Directive-8 (PPD- 8). Consistent with the implementation of PPD-8, emergency management and capability-building are guided at all levels of government by

1. A national preparedness goal that defines core capabilities necessary to prepare for the specific types of incidents that pose the greatest risk to the security of the nation (The White House, 2011, p. 2)
2. A national preparedness system that enables the nation to build, sustain, and deliver those capabilities (Federal Emergency Management Agency, 2011b, p. 1)

These efforts are anchored in Presidential Policy Directive/PPD-8, which focuses on national preparedness. Consistent with the previous national efforts to build preparedness capabilities, FEMA's implementation approach for the directive was intended to define national preparedness using a capabilities-based approach.

The approach for building capability entails using any combination of properly planned, organized, equipped, trained, and exercised personnel to achieve intended outcomes (109th Congress, 2nd session, 2006, § 641).

Building Capabilities in a Whole Community Context

Before considering how to broaden community participation in building preparedness capabilities, it is worthwhile to clarify what is meant by the term "capability." Capabilities are defined as the means to accomplish a mission, function, or objective based on the performance of related tasks, under specified conditions, to target levels of performance (Federal Emergency Management Agency, 2011b, p. 1).

For emergency managers, the process for building capabilities is fairly straightforward, although by no means a simple process. In the emergency management community, practitioners engage in a deliberate and recurring process of assessing threats, risks, hazards, and vulnerabilities; determining specific capability needs; and prioritizing efforts intended to address capability gaps. As described by FEMA, this continuous preparedness cycle for capability-building consists of planning, organizing, training, equipping, exercising, evaluating, and taking corrective action. This ongoing cycle includes all elements of the emergency management community and ensures effective coordination during times of crisis (Federal Emergency Management Agency, 2012).

When jurisdictions prioritize those capability-building activities intended to address capability gaps, one of the more important features of capabilities-based preparedness is highlighted; the idea that building capability is predicated on economic choice and feasibility. Essentially, capabilities-based planning entails planning, under uncertainty, to provide capabilities suitable for a wide range of modern-day challenges and circumstances, while working within an economic framework (Davis, 2002, p. 1). This economic consideration for capabilities-based

National Incident Management System (NIMS) preparedness cycle. (Courtesy of FEMA. http://www.fema.gov/emergency/nims/Preparedness.shtm.)

preparedness underscores the importance of volunteers; force multipliers from nongovernment sources become a necessity when there are community resource constraints. Ideally, these nongovernmental sources are fully integrated with the existing emergency management operation in a given community. For any community, the obvious intent is for emergency responder and management personnel to address capabilities in plans; train and exercise capabilities using credentialed personnel; and regularly assess capability performance through examining exercise and real-world after action reports. However, there is an inherent challenge to ascertaining the true extent to which a community possesses a capability suitable for extreme disasters and catastrophes when some of the emergent groups performing critical capability tasks will not be active participants until the actual disaster or catastrophe strikes.

Emergency management offices are challenged to understand and incorporate emergent groups since they generally utilize nonroutine organizational constructs to accomplish nonroutine tasks using nonroutine resources and activities (Majchrak et al., 2007, p. 150). Considering that the professional emergency management mission area relies on planning, exercising, training, and equipping the use of multiple capabilities predisaster and an integrated incident command structure once a disaster occurs, the emergent group is not easy to incorporate. Emergent response groups are by nature fluid in boundaries and membership can have dispersed leadership, and "learn-by-doing." (Majchrak et al., 2007, pp. 150–151). These attributes belie the approach to traditional emergency management where courses of action are carefully and deliberately planned.

Examples of Emergent Groups Filling Capability Needs

As a matter of expediency, finding critical resources to bridge capability gaps, especially during a catastrophe, involves the ad hoc involvement of the untrained and/or unaffiliated in the response phase. The following are some examples of how this process can transpire the following major events.

Search and Rescue

Perhaps the most recurring example of postdisaster emergent group formation and action is in the field of search and rescue (SAR). Research on the topic has proven that for SAR to be effective, it requires the involvement of volunteers and emergent group responders, who accomplish many of the initial SAR activities (Trainor et al., 2008, p. 2). While the emergency management community spends considerable time and resources to develop highly specialized SAR teams, there is significant evidence that supports the importance of emergent groups in searching damaged structures and rescuing victims. It is not unusual for professional SAR teams to arrive at a disaster scene too late to have an impact on victim survival; oftentimes, the SAR teams locate and remove deceased victims while survivors are located by relatives and neighbors (Tierney et al., 2001, p. 115). The location and extrication of victims using technically proficient teams and standardized processes and equipment remain an important capability to employ in multiple hazardous situations. However, the fact remain that in the initial hours of a disaster, there will likely be emergent groups trying to fill this role as a matter of expediency and with varying rates of success.

Logistics

The September 11 terrorist attacks in New York were a horrific no-notice event with the potential for significant disaster shock or disaster syndrome, meaning a reaction of apathy, confusion, and/or insensitivity to cues in the environment (Perry and Lindell, Emergency Planning, 2007, p. 67). On the contrary, thousands of emergent volunteers formed staging areas and comfort stations around the disaster perimeter, which were utilized by off-duty first responders. Given the considerable damage to the operational and bureaucratic functions of New York's Office of Emergency Management and Fire Department, the emergent groups served in a fluid capacity that official response capabilities could not (Poteyeva et al., 2007, p. 212). What began as a single person taking requests for items from rescue workers and soliciting items from local vendors became a repeatable process involving many individuals working in a coordinated fashion. Over the ensuing weeks, a 24/7 operation involving a thousand unaffiliated volunteers (using some estimates) provided critical services to the response and recovery effort (Voorhees, 2008, p. 1).

Situational Awareness/Communications

The ubiquity of modern information communications is an enabler of emergence over a wider geographic footprint and gives motivated individuals the opportunity to participate in emergent groups regardless of their location. For example, in the hours immediately following Hurricane Katrina, the Katrina Help Wiki was created and hosted by a student in Amsterdam, The Netherlands. An emergent group quickly formed providing critical information on shelters in operation and people that were missing or located. Over time, as the membership remained fluid, different features were added to the site (Majchrak et al., 2007, p. 149). Currently, photo-sharing websites such as Flickr give emergent groups the opportunity to form around the function of documenting and assessing disaster situations (Liu et al., 2008). The current FEMA administrator, Craig Fugate, has been a proponent of using information technology and social media to collect data for situational awareness from the public following a disaster. During major emergencies, the data received from social networking technology are often more timely, actionable, and informative for situational awareness than reporting via traditional, official channels (Spellman, 2010).

Areas for Future Study

How can emergency managers harness the public's inherent desire to contribute to resilient and prepared communities? Every community has individuals with the skills, abilities, and other desired qualities who are otherwise uninvolved in disaster or preparedness-focused organizations prior to the onset of a disaster. These individuals may be (1) looking for opportunities to participate and unaware of how to connect to local emergency management offices and organizations or (2) lacking awareness of the inherent threats, risks, and hazards within their community and the potential benefits of their contributions before the disaster strikes. To address both these community segments, finding ways to inform the public of the need for their participation in preparedness activities prior to the next major incident is one of the areas warranting future study by those in the emergency management field. Finding emergency management roles and opportunities for the potentially emergent individual and groups to learn preparedness and disaster-related tasks makes a community better positioned to benefit from the knowledge, skills, and abilities among its community base. By doing so, community resources can be more readily and seamlessly integrated with other trained disaster-focused groups without the confusion, duplication of effort, or lag time associated with waiting for self-motivated and self-directed groups to find and become proficient in the ideal activities for them to perform. The following are some actions that warrant additional attention by the emergency management community at the local level in regard to more effective interactions with potentially emergent groups.

Understand Community Perceptions of Threat and Risk

Emergency managers looking to connect to viable community groups and institutions need to understand the significant motivations and perceptions within the community that compel people to act. One way to do so is to seek to observe public responses to protective action recommendations and the means by which the public might try to reduce their hazard exposure (Perry and Lindell, Emergency Planning, 2007, p. 117). When disasters strike, why do members of a particular community choose to engage in certain behaviors? When it comes to warning their communities of pending dangers, emergency managers must pay attention to social-structural (i.e., degree of community integration), social-psychological (i.e., how warnings are assessed), and cognitive (i.e., how people reach decisions) processes at play (Tierney et al., 2001, p. 83). During a disaster, warning messages are more likely to result in timely action by the public if it creates the perception of threat as certain to occur and the consequences as severe (Tierney et al., 2001, p. 85). Conversely, communities lacking a historical basis for disasters, an appreciation of their threat and risks, and an understanding of potential disaster consequences, may disbelieve that an improbable event is about to transpire. These populations may not see protective actions as effective uses of time, costs, or effort (Tierney et al., 2001, p. 85). Such considerations hold for not only response and recovery periods when communities mobilize during or immediately following a disaster, but also may offer insight regarding what motivates a community during all phases of emergency management. Understanding these public perceptions and addressing them effectively can extend the parameters of convergence into predisaster periods and encourage broader community participation in preparedness activities.

Identify and Partner with Community Leaders and Prominent Organizations

Individuals motivated to contribute to disaster response and recovery efforts may already be involved in community-focused organizations with which emergency management offices can forge partnerships predisaster. Community organizations can be a trusted agent offering emergency managers access to an array of people, skills, and resources that may be able to address critical preparedness needs but are untrained or currently unaffiliated with any particular emergency management function. Such groups include not only those groups with members that would be expected to have the ability to work in a structured environment and follow specific directions to complete discrete tasks, such as veterans' groups or organizations that deal regularly with collecting and distributing donations from the public. Other entities involved in local problem solving and providing important community services are ideal candidates for groups that can widen their focus to involve preparedness activities.

The resulting analyses from FEMA's national dialog on Whole Community emergency management indicate that disaster resilient communities also have the tendency to solve problems well under nonemergency conditions (Federal Emergency Management Agency, 2011a, p. 10) using institutions, assets, and networks that address important community issues on a daily basis (Federal Emergency Management Agency, 2011a, p. 5). For example, in San Diego, a community center that existed to offer pro bono advice to residents affected by the housing foreclosure crisis also offered information regarding emergency preparedness. This piqued the interest of many local residents, previously uninvolved in emergency-related activities, who then volunteered for local events and were cataloged in a voluntary database used to organize meetings on specific emergency-related issues (Bach et al., 2010, pp. 24–25). As discussed previously, proactively looking in the community for potential partnering organizations that understand threats and hazards offer a supportive social climate and can furnish an array of resources such as people, skills, or facilities (Quarantelli as cited in Tierney et al., 2001, p. 117) is time well spent by an emergency management office.

Keep Existing Volunteer Groups Engaged and Informed about Any Relevant Activity

Once partnerships have been established in the community for the purposes of emergency management, these partnerships need to be cultivated and maintained through regular interaction. For example, the Red Cross has trained volunteers who can be deployed to perform mass care and other response activities during a large disaster through a broad national and international network. Red Cross volunteers can also maintain membership on local Disaster Action Teams (DAT) for smaller-scale disasters or regularly engage in other community-focused activities.

Another example is the Community Emergency Response Team (CERT) concept developed and implemented by the City of Los Angeles Fire Department (LAFD) in 1985 with the understanding that citizens would likely be on their own in the early stages of a catastrophic disaster and require basic survival and rescue skills (Federal Emergency Management Agency, 2003, p. i). Trained CERT members provide a first-response capability through extinguishing small fires, turning off natural gas inlets to damaged homes, performing light SAR efforts, and offering basic medical treatment (Federal Emergency Management Agency, 2003, p. iii). Although the opportunity for CERT teams to engage in disaster response activities may not be steady in many communities, simply maintaining an active roster of informed and motivated CERT participants can be a worthwhile endeavor. The CERT roster can be an asset for many other preparedness and outreach activities, such as simulating victims for first-responder exercises or helping set up staff outreach efforts (booths, flyer distribution, etc.) at community events such as fairs or entertainment venues. Maintaining affiliations and information channels with trained volunteer bases such as neighborhood CERT teams during all phases of emergency management should be a priority for emergency management offices at the local level.

Be Creative with Training and Exercise Opportunities

As the FEMA preparedness cycle detailed previously in this article, training and exercising are critical components of building capability. They also represent an area where members of the community can learn skills preincident in coordination with emergency managers instead of determining the mission in which they want to engage and learning critical tasks by delivering them as part of a disaster response. A proactive, inclusive, and creative training and exercise program developed by a local emergency management office can significantly flatten the learning curve for volunteers looking for a structured learning environment and focus-needed attention on community preparedness capability gaps.

While engaging trained volunteers is important, finding creative ways to keep the public's attention on disaster preparedness is also a good practice. Most jurisdictions conduct training and exercise activities that involve the public; the successes of these predisaster efforts vary based on some of the aforementioned community characteristics (e.g., disaster history and credibility of threats). In many communities, a creative approach to engage the public in preparedness activities yields significant returns. Many members of the public will not become trained volunteers or members of a disaster-focused organization even if involved in or made aware of a training or exercise event that is accessible to the public. However, introducing as many people as possible to basic preparedness practices, providing them actionable preparedness steps, and making them familiar with the emergency management processes in their respective jurisdictions will enable them to make smarter decisions in situations when they need to self-sustain until the situation is stabilized or feel compelled to spontaneously converge.

For example, both the Centers of Disease Control and the State of Kansas have reached broad audiences by promoting or sponsoring zombie preparedness activities at the federal, state, and local levels (Pittman, 2011). For some, preparing for a fictitious and unrealistic hazard may seem a frivolous enterprise. However, in the context of capabilities-based preparedness, participants in such exercises can learn all-hazards concepts, such as sheltering in place, incident command, public safety and security, and so on, applicable for almost any disaster. At the very least, getting the public engaged in personal preparedness activities such as making emergency plans and stockpiling supplies relevant for any disaster will build community resilience. Creating opportunities to learn emergency operations basics and becoming familiar with members of the emergency response community increases the likelihood of community members: participating in other preparedness activities; seeking affiliations with disaster-focused groups so that they can join an organized disaster response; or, being part of a more effective and expedited convergence process should one transpire.

Implement Solutions Using Social Media

Social media is of increasing use for emergency management entities across all levels of government. The considerable array of evolving communications

Washington, D.C., June 10, 2009—Michael Moore and Mike McCormack train on video software at the first FEMA multimedia workshop at FEMA Headquarters. FEMA brought videographers into the training so they could learn new methods of story telling and keep up with the expansion of the use of social media tools. (Courtesy of FEMA.)

tools, with continually increasing capacity and functionality, has become the cornerstone of many public outreach efforts by emergency management offices. However, in regard to emergent groups, the social media streams can be considered bifurcated. Emergency management offices send messages, warnings, and protective action recommendations, and preparedness/hazard mitigation information to the public through multiple platforms. During a disaster, unofficial communications occur among members of the public and various groups via public forums or peer-to-peer communication. These communications, while not emanating from an official emergency management entity, are vital for generating information, helping survivors, and potentially emergent groups coordinate their activities (Reuter et al., 2012, p. 2). The latter, even though it can contain unsubstantiated or unofficial information regarding a disaster scene, is often used by emergency managers, first responders, and the public to help establish situational awareness. During a disaster, social media supports convergence and the resourcefulness of emergent groups as well as broad interaction that generates useful information for a response effort that cannot otherwise be easily obtained (Sutton et al., 2008).

Ideally, the future state of disaster information sharing among emergency management offices and the public will entail a strong connection between social platforms intended for specific use by emergent groups (which are yet to be developed) and those information communications platforms commonly used by emergency management offices (Reuter et al., 2012, p. 91).

Foster and Support Evolution of Emergent Groups into Enduring Organizations

As disaster research literature establishes, emergent groups are spontaneous and largely concerned with immediate disaster impacts. However, there are instances where an ECG can function or endure as a more formal organization (Campbell, 2010, p. 10). For example, an ECG in the preparedness or mitigation phase is more likely to be focused on broad-based community concerns, such as a conservationist group concerned with environmental issues that may increase the risk of disaster. A response-focused ECG is more task-oriented, less formal, and focused on issues such as SAR (Tierney et al., 2001, p. 117). However, the ECG can change its focus depending on the most pressing needs of the community, that is, it may become more concerned with broader preparedness activities, specific response tasks, or recovery imperatives. For example, a National Science Foundation study observed ECGs in a community devastated by flooding. Initially, these groups focused on response and recovery yet over time began to engage in mitigation (buying flood-prone property) and preparedness (plan development) activities (Neal, 1997, pp. 249–250). The difference in disaster phase focus has also been described as "emergency times," or the time immediately following a disaster event, and "nonemergency times," or all other times to include recovery (Campbell, 2010, p. 10).

Build a Structure and Have a Plan for Volunteer Reception

The benefits of effective management of volunteers are many. Successfully bringing spontaneous, unaffiliated volunteers into the fold allows first responders to focus on their primary missions while encouraging citizens to maintain involvement through additional community service opportunities and preparedness activities such as mitigation or recovery. A well-managed volunteer network can also be seen as a therapeutic, empowering, and healing dynamic for the community (Points of Light Foundation & Volunteer Center National Network, 2002, p. 4). There is a common and often paraphrased saying among emergency managers in regard to managing disaster volunteers. "If you plan for volunteers, they will come—if you don't plan they will come!" (Ohio Citizen Corps, 2011, p. 2). Volunteer reception centers (VRCs) provide communities with a means facilitating a smoother convergence process and can help compose an organized community response among process preregistered members of various volunteer groups, professional communities (e.g., medical), and spontaneous, unaffiliated volunteers (Ohio Citizen Corps, 2011, p. 2). A review of the various VRC plans and guides employed by different jurisdictions will show some basic and consistent functions that need to be performed. However, some suggested issues and areas to explore in a VRC plan include

- A designation for a lead agency or organization to coordinate VRC operations
- A means by which the VRC is activated

- A process for placement interviews with spontaneous volunteers that captures (1) existing skill sets and (2) existing affiliations with disaster-focused or other relevant organizations
- A process for training (to include safety) with the appropriate degree of specification for the jobs individuals may be tasked to assume (i.e., mass-feeding operations, medical assistance, sandbagging, etc.)
- A mechanism to address liability

Conclusion

Research indicates that emergence can be preplanned, if a community anticipates its capacity for people and groups to converge and creates the conditions necessary for the phenomenon to materialize; this includes encouraging various social dynamics and establishing in the community a perceived need to address a problem (Quarantelli, 1994, p. 16). In an era of anticipated increases in disaster impacts and fewer government-centric resources with which to address them, communities need to find ways for citizens to be genuine assets and active participants for all phases of emergency management. A growing body of research indicates that the motivation and ability of the public to assume a greater role in disaster response clearly exists and becomes apparent in crisis situations. The challenge remains harnessing this latent capacity predisaster and engaging the Whole Community for the purposes of building specific capabilities that address community threats, hazards, and vulnerabilities. Regardless of their degree of training, there are many emergency management activities that are not suitable for the public; for reasons of public safety and efficacy, these missions should be reserved for professional responders or emergency managers with the requisite technical specialization. However, there are many areas where the public and nongovernmental entities are quite capable of contributing to a community's resiliency and preparedness posture. Determining and proactively engaging with those likely force multipliers in advance of the next disaster should be an activity rating high on the emergency management agenda.

References

109th Congress, 2nd session. 2006. *Post-Katrina Emergency Management Reform Act, Public Law*, 109–295.

Bach, R., Doran, R., Gibb, L., Kaufman, D., and Settle, K. 2010. Policy challenges in supporting community resilience. In: *London Workshop of the Multinational Community Resilience Policy Group* (pp. 24–25). London: Multinational Community Resilience Policy Group.

Campbell, D. 2010. *Stand by Me: Organization Founding in the Aftermath of Disaster.* New York: Baruch College: Center for Nonprofit Strategy and Management.

Davis, P. K. 2002. *Analytic Architecture for Capabilities-Based Planning, Mission-System Analysis, and Transformation.* Santa Monica, CA: RAND Corporation.

Federal Emergency Management Agency. 2003. *Community Emergency Response Team: Participant Manual.* McLean, VA: Human Technology, Inc.

Federal Emergency Management Agency. 2011a. *A Whole Community Approach to Emergency Management: Principles, Themes, and Pathways for Action.* Washington, DC: Federal Emergency Management Agency.

Federal Emergency Management Agency. 2011b. *National Preparedness System.* Washington, DC: Federal Emergency Management Agency.

Federal Emergency Management Agency. 2012. *Preparedness.* Retrieved June 6, 2012, from Federal Emergency Management Agency: http://www.fema.gov/emergency/nims/Preparedness.shtm.

Fernandez, L., Barbera, J., and van Dorp, J. 2006, October. Strategies for managing volunteers during incident response: A systems approach. *Homeland Security Affairs*, 2(3). http://www.hsaj.org/?article=2.3.9.

Lindell, M. K., Prater, C., and Perry, R. W. 2007. *Introduction to Emergency Management.* Hoboken, NJ: John Wiley & Sons.

Liu, S. B., Palen, L., Sutton, J., Hughes, A. L., and Vieweg, S. 2008. In search of the bigger picture: The emergent role of on-line photo sharing in times of disaster. In: F. Fiedrich, and B. Van de Walle (Eds.), *Proceedings of the 5th International ISCRAM Conference.* Washington, DC: connectivIT Lab & The Natural Hazards Center, University of Colorado, Boulder.

Majchrak, A., Jarvenpaa, S. L., and Hollingshead, A. B. 2007, January–February. Coordinating expertise among emergent groups responding to disasters. *Organization Science*, 18(1), 147–161.

Mileti, D. S. 1999. *Disasters by Design.* Washington, DC: Joseph Henry Press.

Neal, D. M. 1997. Reconsidering the phases of disaster. *International Journal of Mass Emergencies and Disasters*, 15(2), 239–264.

Ohio Citizen Corps. 2011. *Volunteer Reception Center: Incorporating Citizen and Medical Professional Volunteers into Disaster and Emergency Response.* Columbus, OH: Ohio Citizen Corps.

Perry, R. W. and Lindell, M. K. 2007. *Emergency Planning.* Hoboken, NJ: John Wiley & Sons.

Pittman, E. 2011, October 19. *Are Zombies and Preparedness a Perfect Match?* Retrieved May 28, 2012, from Emergency Management: http://www.emergencymgmt.com/training/Are-Zombies-Preparedness-Perfect-Match.html

Points of Light Foundation & Volunteer Center National Network. 2002. *Preventing a Disaster Within a Disaster: The Effective Use and Management of Unaffiliated Volunteers.* Washington, DC: Points of Light Foundation & Volunteer Center National Network.

Poteyeva, M., Denver, M., Barsky, L. E., and Aguirre, B. E. 2007. Search and rescue activities in disasters. In: H. Rodríguez, E. L. Quarantelli, and R. R. Dynes (Eds.), *Handbook of Diaster Research* (pp. 200–216). Newark, DE: Springer Science + Business Media.

Quarantelli, E. 1994. *Emergent Behaviors and Groups in the Crisis Time Periods of Disasters Preliminary Paper #206.* Newark, DE: University of Delaware Disaster Research Center.

Reuter, C., Heger, O., and Pipek, V. 2012. Social media for supporting emergent groups in crisis management. In: *Proceedings of the CSCW Workshop on Collaboration and Crisis Informatics, International Reports on Socio Informatics*, 9(2), 84–92 (ISSN: 1861-4280). http://www.wiwi.uni-siegen.de/wirtschaftsinformatik/paper/2012/2012_reuterhegerpipek_socialmediaemergentgroups_cscw12-ws.pdf.

Spellman, J. 2010, September 22. *Heading Off Disaster, One Tweet at a Time.* Retrieved September 23, 2010, from Cable News Network: http://www.cnn.com/2010/TECH/social.media/09/22/natural.disasters.social.media/index.html?hpt=Sbin

Sutton, J., Palen, L., and Shklovski, I. 2008. Backchannels on the front lines: Emergent uses of social media in the 2007 Southern Califronia wildfires. In: F. Fiedrich, and B. Van de Walle (Eds.), *Proceedings of the 5th International ISCRAM Conference.* Washington, DC.

The White House. 2011. *Presidential Poicy Directive PPD-8.* Washington, DC: The White House.

Tierney, K. J., Lindell, M. K., and Perry, R. W. 2001. *Facing the Unexpected: Disaster Preparedness and Response in the United States.* Washington, DC: Joseph Henry Press.

Trainor, J. E., Aguirre, B., and McNeil, S. 2008. *A Brief Summary of Search and Rescue Literature: A Report to COT Netherlands.* Newark, DE: University of Delaware Disaster Research Center.

Voorhees, W. R. 2008. New Yorkers respond to the World Trade Center attack: An anatomy of an emergent volunteer organization. *Journal of Contingencies and Crisis Management,* 16(1), 3–13.

V

EXERCISE DESIGN AND DEVELOPMENT

Chapter 18

Exercise Design and Development Challenges

Matthew Lawrence

Introduction

Designing, developing, and implementing an emergency response exercise can be one of the most time consuming and challenging nonemergency events that organizations are tasked with completing on a regular basis. Ensuring that an exercise is functional, applicable to capabilities that need to be exercised, comprehensive and thorough enough to be realistic is key to the effective implementation of a successful exercise. Functional, applicable, comprehensive, and thorough make up the basis of a FACT-based exercise.

This chapter will approach exercise design and development with a focus on the challenges that any organization may expect to experience or witness, from scoping to after-action. The U.S. Department of Homeland Security (DHS) already has well established, documented procedures for developing exercises in their Homeland Security Exercise Evaluation Program (HSEEP). The information provided in this chapter comes from best practices and experience from the viewpoint of the author, and is intended to highlight problem areas and what should be avoided, or in some instances improved, in the exercise design and development processes.

Exercise Design

Exercise design is typically defined as the top-down view of the exercise planning process. It typically begins with the identification of a need to exercise a capability, either through a pre-defined process set forth from the organization's training, exercising, and planning workshop (TEPW) that is conducted on an annual basis, or from past after-action reports (AARs) defining a need to further test or refine a specific capability. Design begins with the scoping process and moves into the planning process to include capabilities to test, goals and objective development. Included in this process is typically an overview of the scenario and evaluation criteria, although these areas are included under Exercise Development in this chapter.

Identifying the Exercise Manager

Designing an exercise takes time, and having an experienced exercise manager will aid in this process. Many times, the exercise manager is the organization's emergency manager or coordinator, based on the fact that this individual is typically responsible for maintaining the emergency response capabilities of the organization, and knows what needs to be done to ensure that the organization effectively evaluates those capabilities. Often, the exercise manager may look to subject matter experts (SMEs) who have experience in scenario and exercise design and development, and who are familiar with the capabilities that need to be evaluated. The exercise manager may delegate some of the design duties to these SMEs, such as developing injects for Master Scenario Events List (MSEL). In some cases, such as exercises that are contracted out to a private company to develop and implement, there may be two exercise managers—one for the organization and one for the private company actually developing the exercise.

Ultimately, to ensure that the exercise design and development phases maintain some consistency, the exercise manager should be the person who finalizes the MSEL, exercise documents, and exercise evaluation guides (EEGs), and keeps the exercise in focus.

Deciding Capabilities to Exercise

The decision to exercise a capability, or capabilities, is typically the first step in the design process. The emergency manager, coordinator, agency director, or whomever, may make the decision that a specific capability, or set of capabilities, needs to be tested, based on the results from previous exercises, real-world events, or through the TEPW process. This decision is normally made before a planning team is established and an exercise manager is assigned.

The move to Core Capabilities from Target Capabilities has presented challenges to many organizations as the shift from a metric-based, organizational preparedness system to a whole community preparedness system has left many groups re-learning how capabilities are tested. Some Target Capabilities have been lumped together under related, general Core Capabilities with little guidance on how those Core Capabilities relate to the organization's preparedness and response efforts. Deciding how to test these capabilities takes more time for these organizations to define metrics and activities associated with the capability. In general, most organizations have decided to merge the Target Capabilities' activities and metrics under the tasks associated with Core Capabilities to provide a basis for testing.

In the past, some organizations chose to test multiple capabilities over a short period of time, which often complicated the design process. Testing multiple capabilities in a one-day exercise made evaluating an exercise a challenge, which further complicated the development of the after-action report/improvement plan (AAR/IP). That trend has since changed, with a focus on one or two capabilities. The move to Core Capabilities has further pushed this trend along. This has been a positive trend. For example, in the past an organization may have focused on testing fatality management, structural damage assessment, medical surge, responder safety, and health and hazardous materials response in a four-hour exercise. While all of these capabilities are related, an organization would do better to focus on one or two related capabilities—more information comes out in the AAR/IP with a smaller scope.

In general, it is normally better to focus on specific aspects of your organization's response capabilities, narrowing the scope of the exercise and conducting more exercises over a longer period of time.

Establishing a Planning Team

After establishing the need to exercise a specific capability, or set of capabilities, the organization's director, or person responsible for the exercise, should begin assembling a planning team. At this point, an exercise manager should be identified and have appropriate staff. As mentioned previously, the exercise manager should be the single point of contact for ensuring consistency throughout the exercise design and development process. Often the exercise manager is delegated the responsibility of assembling a planning team. The planning team should consist of people from the organization who have direct knowledge of the organization's capabilities and understand the systems and processes that the organization uses. The planning team may be expanded to include other people from different organizations that would be participating in the exercise. It is important to note that members of the planning team should not be players in the exercise, due to the type of information they will be privy to during the design and development process.

In general, the planning team should have one to two people from each organization represented for each exercise; they should be prepared to spend considerable time attending planning meetings, exercise conferences and reviewing documents; and they should have considerable knowledge about the structure and response plans of their organization.

Establishing Trusted Agents

The use of trusted agents in an exercise should be limited. Trusted agents should not, under most circumstances, be players, due to the amount of information they will be privy to during the exercise design and development process, much like the members of the planning team. However, unlike the requirements placed on planning team members, such as being readily available for planning meetings and exercise conferences, trusted agents are typically used as SMEs within the organization. Their time requirements are less, and they are typically only used to help ensure that the flow of the exercise is realistic as far as how the organization would actually respond to a scenario. Trusted agents may be identified at any time throughout the design and development process, unlike the planning team, which should be identified before this process begins.

It is critical that the selection of trusted agents includes personnel actually familiar with the structure of the organization, and that trusted agents understand that they will not be playing in the exercise. The selection process should be discussed thoroughly with the planning team to ensure that the right people are identified as trusted agents before those people are approached, especially if the exercise is a "no notice" exercise. In some cases, a nondisclosure agreement may need to be implemented between the trusted agents and the organization, if deemed necessary.

Trusted agents, when used appropriately, can help ensure that during the exercise design and development process, events, systems and processes are consistent with the organizations that will be playing in the exercise.

Developing a Scope

As mentioned in the Deciding Capabilities to Exercise section, keeping a narrow focus on the scope typically allows for a more detailed analysis of the exercise. With a broad scope, many of the critical details may be overlooked during the evaluation process that could really provide valuable input into areas for improvement. In general, areas for improvement should never be taken as a report card for the organization; rather, they should be used as a basis for implementing best practices and working through issues that may be a bottleneck in response.

If an organization is interested in broadening the scope, this should be accomplished through an expansion of the exercise as a whole, by extending the time of play and the number of organizations involved. When developing an exercise with a broad scope, the scope should be broken into parts based on organizations participating, and each part should be closely observed in the evaluation process.

A narrow scope, with well defined evaluation criteria and the understanding that it is better to develop multiple exercises over a longer period of time, will help ensure that the organization is better prepared for real-life incidents in the future.

Developing Goals and Objectives

The keys to developing goals and objectives are to limit the number of them, keep them simple and make sure they are measurable. As repeated throughout this section, it is better to keep the exercise focused on specific aspects of the organization's capabilities and conduct more exercises over a longer period of time. An organization will get more information from these focused exercises than from one large, overly broad exercise that tries to test everything in one day. For a four-hour exercise, it is typically recommended to focus on one or two goals, and four to six objectives. For longer exercises, those lasting two or more days, the maximum number of goals should be no more than six to eight, with 12–16 objectives.

Typically, at this point, an exercise scenario has been outlined in some fashion, to ensure that the proposed scenario matches up with the objectives that are going to be tested. The scenario will further be fleshed out in the planning meetings and conferences, but the organization at this point needs to have the basic scenario down on paper before goals and objectives are considered. In addition, if the exercise development process is going to be contracted out to a different entity, that group should be included in developing goals and objectives, or at the very least, should have some input later in the development process to ensure that the exercise in general can be developed to meet the objectives.

Measurable objectives are key to a successful exercise. Developing objectives that require subjective observations, such as evaluating how well the organization cooperates, should be avoided, as it is nearly impossible to evaluate that from an independent standpoint. When developing objectives, they should match up with the organization's capabilities and should be closely linked with the Core Capabilities as a reference point, and a basis for evaluation.

Summary

The key points to take away from the Exercise Design section include the following:

1. The exercise manager should be the person who keeps the exercise organized. When developing a large exercise, injects and scripts may be built using multiple exercise staff and SMEs. The information coming in from these other groups will need to be validated and reviewed appropriately to match up with other events occurring in the exercise. The exercise manager is that single point of contact who coordinates and organizes those events.
2. Keeping a narrow focus on capabilities, scope, goals and objectives will result in an exercise that is easier to manage, appropriately tests capabilities in a structured manner and provides more specific feedback for the organization's director.

3. A good, well-constructed planning team will help the organization's director accomplish his or her objectives in an organized manner, and will help ensure that the exercise is polished and includes the appropriate amount of realism.
4. Trusted agents should be used sparingly, and should understand their role in the exercise design and development process.

Exercise Development

If exercise design is defined as the top-down view of the exercise planning process, then exercise development is everything that occurs between the design and exercise conduct. While there are some design concepts that overlap with exercise development, such as beginning to develop the exercise scenario, the majority of the work occurs in the development phase.

Establishing Planning Conferences

Depending on the type of exercise, there are several planning conferences that should be scheduled throughout the course of exercise development. This chapter will not discuss all of these planning conferences, and what goes into them; HSEEP already clearly defines these types of conferences.

The most critical issue that the exercise manager and organization director should address early in the exercise development phase is scheduling these conferences. The exercise planning team needs to be aware of these conferences so they can prepare for them, attend, and provide input. The exercise manager should set dates as early as possible for each planning conference, beginning with the Concepts and Objectives Meeting. Typically, each planning conference builds off the previous one. Setting the date for the first conference sets the tone for the exercise and establishes a timeline for the rest of the planning conferences. In short, once the Concepts and Objectives Meeting is scheduled, and the exercise planning team attends the meeting, discussion should take place as to when the next conferences should take place. The group should agree on these dates and when the exercise meeting minutes goes out to the participants, the dates for the next planning conferences should be included in the meeting minutes.

To keep the group on track, and to make the best use of everyone's time, an agenda for each conference should be sent out to the planning team in advance. It is recommended that the planning conferences be limited to two to four hours. The exercise manager should ensure that the information being conveyed in these planning conferences is organized and follows the agenda. Too much superfluous information, and straying off track from the agenda, are a poor use of time, which alienates the planning team and makes for a tough start to the development process.

If the exercise development process is to be contracted out to a different entity, as is often the case, conducting the first planning conference, such as the Concepts and Objectives Meeting, prior to the exercise being contracted out presents additional problems. Often the entity that will be responsible for developing the exercise has more experience in developing exercises in general, and may have ideas and best practices for getting the exercise development process established. Removing the entity from the first planning conference also removes these best practices from the development process if the concepts are already locked in place. This is different from the design phase, including the scoping process, as designs can be easily changed or revised. As mentioned previously, the contracted entity should have some input into the development of objectives to ensure that the final exercise product can meet the objectives established by the planning team.

Planning conferences are key to exercise development, and the exercise manager needs to keep these planning conferences organized and establish conference dates early in the development phase.

Recruiting SMEs

If the exercise development process is being conducted by the organization, it may be necessary to identify and recruit SMEs to help in the development process. If the exercise development process is contracted out, the entity developing the exercise will likely bring its own SMEs. SMEs provide the organization with expertise and experience that the organization or private entity may not have, or may not have readily available for exercise development. If the exercise is to focus on the possibility of a large-scale human health disease outbreak, something the organization may have never experienced, having the SMEs available will provide real-world experience to the development process. Likewise, if the organization's geographic region has the potential for a natural disaster, such as a hurricane, and it has been 30 years since the last hurricane hit the area, the likelihood of having someone in the organization with that type of experience is slim, and the organization may want that person to be a player in the exercise. SMEs are valuable, as they often bring that real-world experience from all over the country, and sometimes from across the world.

Recruiting SMEs to work on an exercise can be a challenging task. Often, SMEs have other time commitments that need to be taken into consideration. The exercise manager should define the role of the SME early in the development process. Will they be developing injects for the MSEL? Will they be responsible for developing the scenario? Will they be developing EEGs? Will they be working in the simulation cell (SIMCELL) during exercise conduct? The exercise manager should work individually with each SME to address any time availability issues related to the development of the exercise. SMEs need to understand the expectations of the exercise manager, and should have the option of declining any offers prior to the development process. Likewise, if an SME commits to assisting the exercise manager, the SME should stand by that commitment.

Not every SME has been involved in developing exercises, from scoping to after action, and the exercise manager should not make this assumption and place unrealistic expectations on the SMEs. The SMEs are a resource, and should be used as such. However, some SMEs have been involved in the entire exercise design and development process, and their experience may be very valuable. In fact, some organizations hire SMEs just to facilitate the exercise design and development process.

If the exercise manager brings in SMEs to develop injects for the MSEL, the exercise manager needs to understand that while some SMEs have incredible experience in real-life situations and can develop very detailed reports and supporting documentation, they may not have experience with developing actual injects to make a functioning MSEL. The exercise manager should never, under any circumstances, rely on SMEs to develop all the injects for an exercise.

Exercise development does not stop at inject development. The exercise manager is there to assemble the injects, validate the information in the injects and connect all of the dots to formulate the MSEL. While exercise conduct is not covered in this chapter, it is important to note that, when selecting SMEs to develop injects for the MSEL, these same SMEs may be brought into the exercise conduct phase as part of the SIMCELL. SMEs need to be able to think quickly on their feet, and change the thought processes throughout the exercise.

It is important that the exercise manager is aware that SMEs are there to help develop the content for injects, and not necessarily to develop everything that goes into a functioning MSEL.

Developing the Scenario

The general outline for the scenario should have been developed during the exercise design phase. This outline provides a starting point for the actual scenario development phase, which includes fleshing out the details of the overall scenario. Before inject development begins, the scenario should be fully developed, with critical events identified and expected actions noted. The scenario will ultimately act as the backdrop and outline for the MSEL. If the exercise is designed as a "no notice" event, the backdrop may be unnecessary, and in some cases the events leading up to the exercise may be unnecessary too. However, in order for the SIMCELL to function, and to ensure that events are as realistic as possible, it is still recommended that the scenario reach back as far as possible and detail events leading up the exercise "Day 1" events.

SMEs may be brought in to develop the scenario. Depending on the type of exercise, these SMEs should be involved in the exercise conduct, either in the SIMCELL or as part of the exercise staff. The SMEs should be part of the exercise conduct in the event that questions are brought up concerning intent of specific aspects of the scenario. If the SME is not available to participate in the exercise conduct, the exercise manager needs to have a thorough understanding of the scenario and should be prepared to answer any questions pertaining to specific events. If additional personnel are working in the SIMCELL, these people should have a full understanding of the events as well.

The scenario should be fully developed before injects are developed for the MSEL. The scenario acts as an outline for inject development and all injects should be validated to the scenario over the course of exercise development. In order for all exercise staff to fully understand the exercise and avoid any issues in exercise conduct, all exercise staff should be fully aware of the scenario and understand the intent of all exercise events.

Developing the MSEL

As mentioned previously, the scenario should constitute an outline for the MSEL. The events that go into the MSEL should be carefully crafted to ensure that exercise evaluation is possible—most injects should have an expected action tied to each event. The exercise team should have an understanding of how the events unfold, why they unfold the way they do, and what actions should occur following the events.

When developing injects for the MSEL, the exercise manager should work with the planning team to identify critical events that are relevant to the scope, goals, and objectives of the exercise. This can be accomplished during planning conferences or through informal emails or conference calls. Inject development should occur around these critical events. These critical events can become the basis of exercise evaluation, if the format of the exercise allows it. It is important to note that too many critical events can be overwhelming and take away from the thoroughness of the exercise evaluation.

The format for the MSEL should be established early in the exercise development phase. Typical formats for MSELs should include sections for inject number, delivery time, who the inject is going to, who the inject is coming from, inject delivery method, inject text, expected actions, additional notes, and a section for linking supporting documents (i.e., scripts, reports, etc.) to the MSEL. Operations-based exercises typically follow this format, and there may be several hundred injects for a one-day exercise. Discussion-based exercises may include a series of events instead of injects, with discussion based on break-out groups and questions pertaining to response actions needed for the specific events.

Writing injects can be the toughest part of developing an exercise. At this point, the planning team fully understands what the end product should be. SMEs may have been brought in to begin developing damage reports and resource requests. At this point, the question becomes, how does this information "start" an exercise? With a tabletop exercise, it becomes fairly easy; there is a lot more leeway on how to get information out to players. In a real-time scenario, commonly applied to operations-based exercises, that process becomes a little more difficult. Here are some issues to think about when writing injects. These questions and points of issue relate more to natural disasters, but can be applied to any type of exercise.

1. Once the exercise starts, how long does it take to get that initial call to someone who can relay information? How long will the average citizen take to size up their situation?
2. How does that first call work and who does it come from and who does it go to? If cell towers and landlines have been damaged, how does that work? What about Internet capability? It may take a while before situation reports start coming in.
3. To get an initial damage report, someone has to make that call into dispatch, local emergency management, the sheriff's office, and the like. It will take some time weeding out calls to size up damage reports.
4. Most importantly, how do you write such information requests for this exercise? It needs to be realistic and coming from a realistic source. It may take an hour or more before the state emergency operations center (EOC) is fully operational. Injects should mirror the form of how that information is being relayed in a real event and should utilize whatever system the organization currently uses for disaster management.
5. Injects need to be scripted with enough supporting information that anyone in the SIMCELL can respond to calls coming back into the SIMCELL. For example, instead of "three buildings damaged" the inject should say, "This is Bob at dispatch. I have been receiving calls over the last 10 minutes concerning initial damage reports. I have reports of three buildings downtown that have been damaged—a pharmacy, a small clothing store, and a filling station. Addresses to come shortly. There were three people in the pharmacy, two were injured and need assistance. The filling station has two large underground 2000-gallon storage tanks. It received significant damage and the tanks are ruptured. People are in the process of being evacuated. There were 12 people inside the clothing store, with four people being reported as injured. We need additional manpower to evacuate the area and a hazardous materials team to respond to the gas spill."

There has been a move to develop more consequence-based exercises, with a limited MSEL and more "on-the-fly" inject development. Consequence-based exercises typically utilize a small number of written injects and a fully functional SIMCELL to conduct the exercise. Injects are developed so that players have options on how their response would work, much like in real life, and based on the actions the players take, the SIMCELL responds in kind. If players decide to take one course of action over another, the SIMCELL develops injects related to those actions "on-the-fly" and delivers them appropriately. Deploying a large consequence-based exercise is not feasible with most organizations, due to the large number SMEs needed and agencies that would need representation in the SIMCELL; however, small drill-like exercises can easily take the form of a consequence-based exercise and can prove to be quite valuable to the organization.

Developing the MSEL takes time, and, depending on the type of exercise a large number of resources are needed to make the exercise function as it would in real life. As mentioned previously, and throughout this chapter, the exercise manager should act as the line that connects all the dots in the exercise development phase.

Validating the Events

It is critical that all events outlined in the MSEL are applicable to the organization, and there is a level of realism contained within the events so that participants can accept the events and respond appropriately. The difference between reality and realism can be apparent here. While it may be difficult to define the difference between reality and realism, those who have experience developing an exercise can certainly understand the difference. Reality is defined as those events that actually happen in real life. Realism is defined as events that are realistic and have some relationship to what could happen in normal, real-life events. The difference is what happens when events are scaled up. Sometimes reality events can be so far "out in left field" that it becomes difficult for players to accept them, even though they may have actually happened in real life. If an exercise manager chooses to model a real-life event, while there is a defensible basis for the scope of the exercise, players may become overwhelmed and not focus on exercise play but on their precept that an event of this magnitude would never happen. In this case, exercise play does not represent the organization's true capability.

Many real-life incidents are worsened by the "perfect storm" effect. The exercise manager should not get caught up in developing a "perfect storm" exercise that ignores realism and thereby alienates the players to the point where the exercise shuts down because events are not realistic.

It is also critical that injects get validated before publishing the final MSEL for review. By validating the injects, the exercise manager ensures that an inject is not physically impossible, or improbable. If the staging area for an organization is noted as being at parking lot of the court house, but the court house was wiped out in the natural disaster, then it is probably not a good idea to use the court house as a staging area. Simply, the exercise manager, and anyone that the exercise manager uses to help develop injects (i.e., SMEs) should take the time to review the organization's plans, look at maps of the area and review other published reports (such as traffic counts) to make sure that the necessary level of realism is added to the exercise. This process involves "ground truthing" the injects.

Developing an exercise is not so much about pushing people or systems to failure; it is more about testing capabilities in a controlled environment, to gain an understanding of any areas that need to be improved. Events that are outlined in the MSEL should have an acceptable level of realism that allow players to respond appropriately, and avoiding that "perfect storm" scenario that can shut down an exercise almost immediately.

Developing Evaluation Criteria

The final product of exercise development is exercise evaluation criteria. However, exercise evaluation should be considered in the exercise design phase and should be kept in mind throughout the entire development process. Exercise evaluation is the cornerstone of all exercises, and is the key reason why exercises take place. Organizations need to be able to identify areas of improvement as part of their emergency preparedness cycle, and to ensure that preparedness is linked directly to response. As mentioned previously, areas for improvement should never be taken as a report card for the organization; rather, they should be used as a basis for implementing best practices and working through issues that may be a bottleneck in response.

In general, HSEEP defines how exercise evaluation should be developed and how it should be implemented, based on the performance levels of the exercise. This section does not look at how evaluation relates to performance; rather it looks at how evaluation criteria should be developed. The following questions can be used by evaluators to further understand what they should be looking for in the evaluation process of the exercise:

1. How would response personnel perform this task?
2. What decisions would need to be made and who would make them?
3. Are personnel trained to perform this task?
4. Are other resources needed and how will they be obtained?
5. Do plans, policies, and procedures support the performance of this task? Are response personnel familiar with these documents?
6. Do response personnel from multiple agencies or jurisdictions need to work together to perform this task? If yes, are the agreements or relationships in place to support its performance?
7. What should be learned from this task?
8. What improvement actions are recommended?

HSEEP also includes standardized EEGs for organizations to use, but these EEGs contain multiple pages and have a complicated user interface. Some form of these EEGs, modified to fit to one page, and focused on critical tasks, is the format many organizations use. A checklist format for critical tasks may be used, along with areas for additional comments and a section for identifying strengths and areas for improvement. If the organization decides to develop its own customized version of EEGs, the following information provides useful instructions for the use of the EEGs.

- *Tasks:* The tasks section includes a checkbox-style listing of tasks that may be associated with the core capability. If the task is addressed during the exercise, evaluators check it off and make general observation notes. It is important to note that this list of tasks and actions is not all-inclusive. If one of these tasks is not applicable to the exercise, evaluators are instructed to write "N/A" in front of the checkbox.
- *Observations:* The observations section includes an area to insert short, complete sentences that describes the general observation as it relates to the strengths and areas of improvement noted on the EEG.
- *References:* The references section includes an area to list relevant plans, policies, procedures, laws, and/or regulations, or sections of these plans, policies, procedures, laws, and/or regulations. If no references apply to the observation, the evaluator is instructed to list "N/A" or "Not Applicable."
- *Strengths:* This section includes an area to note strengths associated with the core capability.
- *Areas for Improvement:* This section includes space to note areas for improvement associated with the core capability.

The final item that should be noted in exercise evaluation is the inclusion of a participant feedback form. This form is handed out to all players before the exercise begins and allows the players to comment on the exercise itself. Some questions should be included to address the overall structure of the exercise, with additional questions asking the players for feedback as to what they believe the strengths and areas for improvement for their organization should be following the exercise. HSEEP has a good participant feedback form available for the exercise manager to incorporate into the exercise development process.

> Exercise evaluation is the cornerstone of all exercises. The information from the evaluators should be used to develop an after-action report and identify actions that should be taken for the organization's improvement plan.

Summary

The key points to take away from the Exercise Development section include the following:

1. Planning conferences are key to exercise development, and the exercise manager needs to keep these planning conferences organized and establish conference dates early in the development phase.
2. It is important that the exercise manager is aware that SMEs are there to help develop the content for injects, and not necessarily to develop everything that goes into a functioning MSEL.

3. The scenario should be fully developed before injects are developed for the MSEL. The scenario acts as an outline for inject development and all injects should be validated to the scenario over the course of exercise development. In order for all exercise staff to fully understand the exercise and avoid any issues in exercise conduct, all exercise staff should be fully aware of the scenario and understand the intent of all exercise events.
4. Developing the MSEL takes time and, depending on the type of exercise, a large number of resources are needed to make the exercise function as it would in real life. The exercise manager should act as the line that connects all the dots in the exercise development phase.
5. Developing an exercise is not so much about pushing people or systems to failure; it is more about testing capabilities in a controlled environment to gain an understanding of areas that need to be improved.
6. Events that are outlined in the MSEL should have an acceptable level of realism that allow players to respond appropriately. Exercises should avoid that "perfect storm" scenario that can shut down an exercise almost immediately.
7. Exercise evaluation is the cornerstone of all exercises. The information from the evaluators should be used to develop an AAR and identify actions that should be taken for the organization's improvement plan.

Chapter 19

Operational Exercise Design

Derek Rowan

Introduction

Drills, functional exercises, and full-scale exercises. These are the exercise types that have your players perform skills and interact with others in real time. Writing how to design, develop, conduct, and evaluate these operations-based exercises in a single chapter is challenging. We hope to provide enough information to get you started and avoid some of the common mistakes, but we certainly cannot provide all guidance or tricks of the trade.

Good exercise design requires a huge amount of minute detail. It can be done without that level of detail, but the simulation will suffer. When the simulation suffers, the players may make up their own simulation as they go, and then your evaluation suffers. When that occurs—your exercise may not actually provide a validation for the items you wanted. This point is key. While operation-based exercises can indeed be the best learning and best way to experience items prior to a real incident, they need to be executed in an environment where players are prepared adequately.

One of the most common errors I see in exercise design is picking the exercise type, scenario, and timeframe without going through the proper process to develop them. This is like the boss saying, "I'd like an earthquake functional exercise next month." For small teams, this can be effective, but for most agencies, departments, or disciplines, this results in wasted planning effort, and a less than ideal exercise experience.

Resist the urge to pick the scenario first. Instead, walk through what your players are going to demonstrate and who is going to do it. These two elements, called "objectives" and "extent of play," will provide a much better framework for determining the details of the scenario. Otherwise, if you start with the scenario, all players may not have a role and you will end up trying to shoehorn who is playing in it and getting their play to conform to the scenario.

Before we start, we need to pick our design team. In some cases, it may just be you handling all aspects. In larger more complex exercises, it may be a large dedicated team. In either case, it's important to identify subject matter experts that can be consulted within your organization on how tasks are done in the real world. This is especially important for agencies and entities that are not playing in the exercise, but would be involved in a real-world response. These simulated elements are critical to be presented in a realistic manner. For operational exercises, we need to schedule important meeting dates. Many times, we have a date that the exercise must be conducted by, so we must backwards plan meeting dates. Most operations-based exercises have a minimum of three formal planning meetings with four or five meetings not uncommon. These meetings are where you are getting the player stakeholders together and presenting and approving exercise design work. The typical meetings are:

1. Concepts and Objectives (C&O) Meeting—This is the formal beginning of the planning process. According to FEMA, "it is held to identify the type, scope, objectives, and purpose of the exercise." When exercises are relatively small with fewer players, this conference can be held in conjunction with the Initial Planning Meeting.
2. Initial Planning Meeting (IPM)—This conference is a validation of all objectives, scope, and should provide a forum to outline in detail the extent of play for each organization. The scenario should be ratified and additional details provided along with exercise conduct details such as venue, times, duration, and so on.
3. Midterm Planning Meeting (MPM)—This conference should provide comments for the draft documentation of the exercise, agreement on final logistical concerns for the exercise, and input and agreement on the Master Scenario Events List (MSEL) events.
4. Final Planning Meeting (FPM)—This conference is the final forum for approving all documentation, plans, logistics, evaluation, and conduct procedures for the exercise. No major changes to the exercise should take place at or after the FPC.

Scope

First, determine who wants (or is directed by management) to play in the exercise. What is that team role? How do the multiple players interact in various real-world incidents? Once we determine our "major" players, then we look at the overall outcomes. They may not be the same for each playing group.

Looking at the overall outcomes, purpose, and capabilities for the exercise is a simpler and easier approach to determining exercise type and objectives at this point. Over the years of design, I have found that most teams need to consistently practice and improve the same eight items. I have put these into the 8 C's of operational exercise design:

1. Command—Who is in charge, what are the roles and responsibilities of the command personnel?
2. Coordination—What are the coordination procedures with multiple teams or agencies for an incident?
3. Collaboration—What is the procedure for getting input from different agencies, jurisdictions, levels of government, or teams to solve problems?
4. Communications—What is the process for communications to occur from the field through the command element, between each other, and how do we communicate with other agencies, levels of government, and the public?
5. Common operating picture—Does critical information flow to all entities and does each person know the appropriate amount of information for their position?
6. Commodities—Are resources managed, ordered, delivered, allocated, and tracked for our team, our patients/public, and others that may need it?
7. Cents—Do we have the right documentation, payment procedures, money distribution, and tracking for financial elements?
8. Capacity—Are we prepared for large numbers of patients, people, resource influx, and other items?

These elements span all disciplines and are the perfect starting point to determine what players would like to accomplish during an exercise. You can ask players, or you may be in a position to already know, which of these items would be the ideal starting platform. If they have just performed training, updated plans or policies, or have resolved a previous recommendation, then it is likely they need to exercise that component again. Note, you are not limited to these items. Indeed, choosing your own capabilities and outcomes is preferred. However, if you are not sure where to start—these may provide a good framework for the discussion.

For those in the public health and hospital fields, you can look at your Joint Commission requirements, or the Center for Disease Control's Public Health Capabilities list to also provide some capability guidance. This can be quite helpful in determining which items need to improve in your player's organization. For those that have developed a multi-year training and exercise plan, the starting point is to look at that plan, where you are on the cycle, and continue down the list as you make improvements.

This chapter only provides an overview of determining the capabilities that you should be addressing in your training and exercise program. It is imperative that you seek additional guidance on this step so that you are not incorrectly exercising or focusing on capabilities that are not improving your performance. Ideally, you will base these on your Training and Exercise Plan and your Threat and Hazard Identification and Risk Assessment (THIRA).

Exercise Type

Once you have identified the major players and the outcomes and capabilities each want to address, the next step is to determine which exercise type is the best choice to demonstrate or validate your outcomes. To assist with this, I have included a simple flowchart that will walk you through determining which exercise types should be considered (Figure 19.1). I have outlined all seven Homeland Security Exercise and Evaluation Program (HSEEP) exercise types. More information on these exercise types can be found in the HSEEP guidance at http://hseep.dhs.gov.

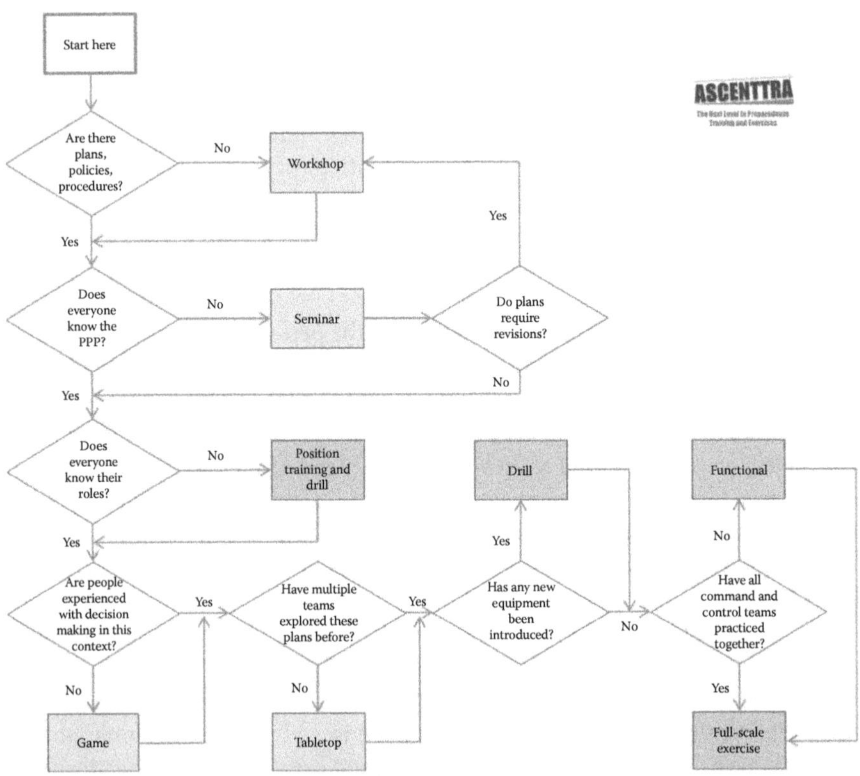

Figure 19.1 Flowchart: determine the right exercise type.

Objectives

Now that we identified the major players, main capabilities each want to address, and the appropriate exercise type, we need to determine each player's exercise objectives. We want to ensure our objectives are SMART (Specific, Measureable, Achievable, Relevant, and Time-Bound) (Table 19.1).

Good exercise objectives are written from the player's perspective. They state which player is going to perform the specific task or action to a standard that is measureable and observable. Exercise objectives that start with the word "evaluate" or "examine" should be revised to more active and measureable objectives.

Detailed exercise objectives will make your design process simpler. They are the basis of your simulation plan and your staffing plan, which I will discuss in a moment.

We all know that exercises can be fantastic training events where people learn a tremendous amount. However, when designing exercises, I think of exercises as "validation" events instead of training events. Design exercises to make sure your players, systems, and plans are adequately prepared.

The players, agency, or jurisdiction should possess the following prior to the exercise:

1. Adequate plans, policies, and procedures for the capabilities being exercised.
2. Adequate numbers of players and the players for each of the roles in the capabilities exercised.
3. Known organizational structure for various scenarios including roles and responsibilities.
4. Correct types of equipment if required to function in those positions.
5. Adequate training on all of the above.

Table 19.1 April 2013 HSEEP Revision SMART Guidelines for Exercise Objectives

Specific	Objectives should address the five Ws—who, what, when, where, and why. The objective specifies what needs to be done with a timeline for completion.
Measureable	Objectives should include numeric or descriptive measures that define quantity, quality, cost, etc. Their focus should be on observable actions and outcomes.
Achievable	Objectives should be within the control, influence, and resources of exercise play and participant actions.
Relevant	Objectives should be instrumental to the mission of the organization and link to its goals or strategic intent.
Time-bound	A specified and reasonable timeframe should be incorporated into all objectives.

This is critical to designing, developing, conducting, and evaluating a good exercise! If a player or group is missing one of the items from this list prior to exercise start, you already know one of the evaluation outcomes. This does not mean you cannot do the exercise, but it will change how you design the exercise around this already-known issue. This will also ensure that players are well prepared prior to exercise start—which is exactly what you want.

I have seen some exercises designed by others that have as their sole intent to throw people under the proverbial bus. This results in ineffective evaluation, and only creates distaste for future exercises. Exercises designed to provide players unrealistic or ridiculous "Godzilla" scenarios do not serve the intended purpose of making us better. It only causes players to not play realistically which actually destroys valid evaluation.

I have found that the more realistic the scenario is, the more realistic the simulation is, the better people play. This does not mean easy—people want to be challenged—but they want to be challenged realistically. Having difficult and hard exercises is good, as long as they are realistic. Note that there are some rooms in exercise design for "fantasy scenarios" such as zombie apocolypse, and so on. However, choosing a fantasy scenario requires the training audience having "bought in" to the idea, and it requires the exercise to be focused on specific processes versus scenario response. Use them sparingly and wisely.

Now we know who wants to play, their major outcomes they want to examine, and looked at the player's preparedness for the exercise.

Since this chapter is about operational-based exercises, we will assume that we are working on either a drill, functional, or full-scale exercise. It is now time to work on the scenario.

Scenario

For those that are closet novel writers, I have bad news—exercise scenarios should not read like a novel. They are designed to determine the framework of the simulation and to ensure players have an environment to address their exercise objectives. The scenario is the mechanism that provides players a realistic method to validate their exercise objectives using their own decision making. The novel must stay in the closet!

Scenarios do not have to be long. In fact, I have found that writing long narrative scenarios is inefficient and takes away from the important elements that must be in the scenario!

Scenario development is easy once you know who is playing and their objectives. Here is an example. These are objectives created for a hospital functional exercise. The players within the hospital are:

1. The Hospital Command Center staff (25 positions)
2. One hospital evacuation team comprised of four persons and an evacuation sled

3. Hospital pharmacy cache team comprised of four persons and one security guard
4. Emergency department triage teams (20 persons)
5. Hospital and patient record group (four persons)

These were the SMART objectives created for this exercise:

1. The Hospital Command Center will be fully staffed within 30 min of code announcement.
2. The hospital will deliver patient medical records and transfer documentation to the evacuation staging area for each evacuated patient within 10 min of request.
3. The hospital will establish and operate backup communications systems after loss of both internal and external conventional power.
4. The hospital will move the Hospital Command Center to the backup location during event response without loss of communications.
5. The hospital pharmacy will provide and deliver a 96 h medical sustainment supply to the evacuation staging area for each evacuated patient within 15 min of request.

On the basis of the extent of play, we know that our scenario and hazard must include:

1. Need, or perceived need, to activate all Hospital Incident Command Positions in the Hospital Command Center very quickly
2. Evacuate multiple patients from the hospital
3. Reason to move the command center to operate in the backup location

Looking at the THIRA for the hospital or the surrounding jurisdiction would then be your next logical step to determine which hazards would cause those items to occur and be the right size for the extent of play. This is a balancing act, as you also must make sure that the players have the correct personnel to respond to that hazard and address their objectives. If these do not align, you may need to adjust your hazard, objectives, or extent of play.

In our example, one of the major hazards for this fictional hospital is earthquake. An earthquake would cause a large activation of command system, is realistic to cause damage to the main command center forcing use of the backup, and possibly having damage to critical facilities causing certain patients to be evacuated.

Certain scenarios could be unrealistic or not allow the existing extent of play to address their objectives. For example, an active shooter scenario would not likely allow the players to achieve these objectives while the shooter was active. A fire in the hospital would not necessarily cause the full staffing as they had outlined for the command structure. Each hazard should be analyzed against the extent of play and against the objectives.

At this point you should have:

1. Major players
2. Major outcomes, capabilities, scope
3. SMART objectives
4. Scenario hazard

Exercise Development

With these data, you can begin the more detailed exercise development. I have developed three key questions to ask during the creation of operations-based exercises.

1. How will the objectives be stimulated?
2. How will the objectives be simulated?
3. How will the objectives be evaluated?

The answers to these questions will help you develop your simulation requirements, MSEL, staffing, and evaluation plan. For example, let us look at the first objective above:

The Hospital Command Center will be fully staffed within 30 min of code announcement.

How will this objective be stimulated? What will be required for the players using their own decision making to address this objective? Here is where the major attention to detail starts to play. You now have to analyze how this would occur in the real world, who would do it or initiate it, how they would do it or initiate it, what would they know at the time, and what communications processes would be used.

For example, in this scenario, if the plans call for automatic activation of the command center based on the ground shaking, then the stimulation is built into the scenario. However, if someone needs to make a decision to activate, then that person (assuming they are playing) needs to be presented with the information they need to make that decision.

The second question is "how will the objectives be simulated?" This can point to the objective itself, or the scenario that supports that objective.

If the stimulation for this objective is the earthquake, then we have to figure out how we will simulate it. We are not able to actually shake the ground, so what is our procedure? What would people actually see? What damage would be obvious? What news reports would be realistic? What communications lines would be working? Would we affect power? What people need to provide direct simulation to get the players to address those objectives? For example, if the "charge nurse" is responsible for activating the command system during the proposed play hours, and that person is not playing, then we must provide a simulator to act as that individual. Do we need to provide photographs of damage of outlying buildings? What information would the command center receive?

The final question is "how will the objective be evaluated?" This one question results in five discrete items for the evaluator:

1. Numbers—How many evaluators are required to evaluate this objective? Generally speaking you want no more than one person. However, if it requires more than one, then how do they communicate the effectiveness?
2. Locations—Where do you physically have the evaluator so they can observe the behavior?
3. Equipment—Does the evaluator have any special equipment in order to observe the objective? For example, portable radio, access to computer logs, access keys, and so on.
4. Expertise—What expertise does the evaluator need to possess in order to provide a proper, objective, and informative evaluation?
5. Time—What time does the evaluator need to be at this position in order to properly observe?

Once you answer and record the responses to all these data, your staffing plan and scenario data are starting to emerge. You can now start working on the next major component of the design—play versus simulation.

Revealing the Simulation

The "Play/Sim line" is a concept that shows the delineation between who is actually playing, and who would normally be responding to that same scenario in the real world. Anyone who would normally be responding but is not playing needs to be simulated directly or indirectly. This concept allows you to create the correct simulation information and put in place the right simulation cell (SimCell) which will contain the people that will be acting as all nonplaying entities.

The easiest way to start this is to draw an organizational chart for each playing entity that corresponds to how they would organize (based on their plans, policies, and procedures) for that specific scenario/hazard. Once you have it drawn, then determine who in that organization chart is going to actually be playing in the exercise and draw a line delineating. This is your Play/Sim line. Everything below the line needs to be simulated—all information, reports, requests, and so on. Look at the entities that are directly below the line and determine what information would they know, how would they get it, and how would they communicate it, and at what times (Figure 19.2). This information becomes part of the scenario data you need to have or create. Then look at the other player's organizations or other nonplaying

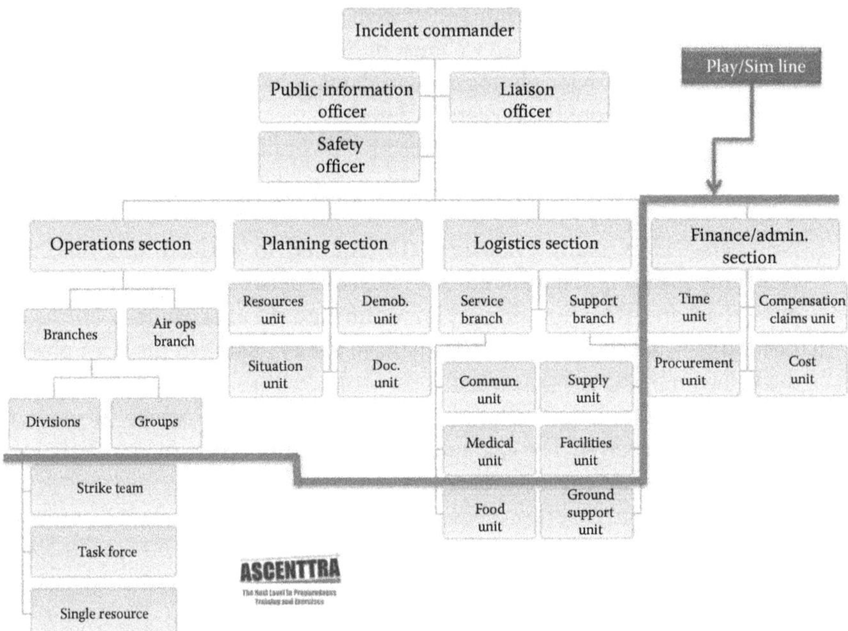

Figure 19.2 Developing a Play/Sim line chart for each playing entity will provide you a graphical representation of the needed scenario and simulation elements. (Image courtesy of Ascenttra.)

organizations that would be involved in the real-world incident and repeat the process. Now you can start building the process of how you will reveal this scenario data to the players. One method is the MSEL. The MSEL contains information about the scenario that is revealed to players over the course of the exercise.

You now have the basic information you need to start developing the MSEL for your exercise. When you are developing a drill exercise type, a MSEL may not be needed. Since drills focus on a very specific task, having a large simulation may not be required. If it is, it could be done with the exercise staff (called a controller for operations-based exercises) providing any answers to simulation questions from players.

For a functional exercise, since many of the real-world response elements may not be playing, it is critical to have the scenario outlined and a plan to reveal that scenario to the players. The MSEL allows for a procedure to do this. Building an MSEL is easy. The key is to go back to those Play/Sim diagrams. Determine who would normally respond during this incident. What would they normally do and when would they normally do it?

Start writing down the major actions that both the players and the simulated entities would be doing over time. Determine how players would be notified—what would the stimulation be to get them involved? How would that work? Is it coming from a simulated entity? If so, that stimulation needs to be entered into the MSEL.

When your exercise contains multiple disciplines or multiple agencies, it is typical that subject matter experts (SMEs) will need to be consulted to determine how that team or agency performs those tasks and how they would normally respond and interact. This provides the most realistic simulation for the players.

Expected communications from entities that are not playing in the exercise that involve communication to the players become "contextual injects." Inject is the term used to describe the event that is injected into play to reveal the scenario to the players at a particular point. Contextual injects reveal the scenario as the players would normally receive it over time. You may then add additional contextual injects based on the normal communication flow of nonplaying entities and how the players would normally receive scenario information over time. Using our earthquake example, it may be standard procedure for the city emergency operations center or the hospital's additional facilities to contact the playing hospital to check on the status. Since this is a normal expected communications involving nonplaying entities, these should be injected at the proper time and become contextual injects.

"Contingency injects" are injects designed to keep players moving in the proper direction toward the objectives. For example, if a secondary improvised explosive device is placed at an incident site and it is required to be found by the players in order for them to address their objectives, a contingency inject may be needed if the players do not actually find the device. If required, the contingency inject could be that "a citizen reported seeing someone place a backpack at the scene" prompting players to investigate. When entering contingency injects, it should be noted that it may be an evaluation point that should be recorded if the players did not perform as anticipated.

Evaluation

Designing the evaluation is a critically important aspect of any exercise. The intent of the exercise is to validate our plans, organizational structure, roles, equipment, and training. Making sure that we are able to realistically demonstrate these items, and capture observations related to them is the only way we can produce after action reports (AARs) that provide for a detailed roadmap on recommendations for improvement along with existing strengths that should be maintained.

Good evaluation starts with good objectives and a realistic exercise. By answering the questions outlined above on how will you evaluate your objectives, the key now is to provide your evaluators with the proper tools to assist them in capturing observations. One of these tools is the Exercise Evaluation Guides (EEGs) that give your evaluators a framework to capture observations, record information, and analyze the data. The HSEEP site (http://hseep.dhs.gov) provides some preexisting EEGs, related to core capabilities, that can (and should) be customized to your specific exercise and exercise objectives.

Choose your evaluators based on their knowledge of the plans, organizational structure, roles, equipment, and training of the players so they can adequately observe actions and make recommendations if required for areas for improvement. Alas, this segment is too short to provide additional guidance on writing up your improvements, but additional information can be found on the HSEEP website. Select "Sample Exercise Materials" from the home page and search for "EEG."

Immediately following the exercise, a "Hot Wash," or exercise debriefing, should be performed by the controller or evaluator at each functional area. The Hot Wash should focus conversation on the player-identified strengths and areas for improvement. It also allows evaluators to receive clarification on areas of play that was not observed during conduct. Participant feedback forms should be distributed during the Hot Wash and collected to provide additional information on the exercise for evaluation purposes.

Conduct

Exercise conduct is when you put into place everything you have worked so hard to design! You should setup any play areas with props or other items for an accurate simulation as required. Perimeters should be checked and verified if outdoors. Administrative items such as feedback forms, player briefing materials, and so on should be checked, printed as required, and ready for conduct. Communication capabilities for the exercise control staff should be readied and checked. Phone directories as needed for players to know who is playing or not, and how to communicate with the SimCell should be verified.

Just prior to the start of the exercise (StartEx), the controllers for each playing venue should provide a "Player Briefing" that outlines how the exercise will start and end, how players interact with the SimCell, who is playing or not, safety procedures, security concerns, and what will be happening after the exercise such as a Hot Wash. If live actors are to be used during the exercise, they should receive an actor briefing outlining how they are to interact with the players, what they "know" about the scenario and should report to players, real-world emergency procedures, schedule, communications, and safety.

Other briefings may be required based on your exercise for observers, very important persons (VIPs), real-world media, or agency executives.

Training

Controllers, simulators in the SimCell, and evaluators should all receive training on the exercise prior to conduct. This training should provide for their specific role within the exercise, how to utilize the tools for them such as the MSEL and the

SimCell, how and when to interact with the players, communications procedures for both control staff and players, and safety procedures.

Documentation

During the development of the exercise, information and design of the exercise should be recorded. Documentation is a critical component of effective exercise development, and there are several documentation items that will assist you in the delivery of the exercise. The more common documentation for an operations-based exercise includes:

1. Exercise Plan (ExPlan)
2. Controller and Evaluator (C/E) Handbook
3. Controller and Evaluator (C/E) Packets
4. Master Scenario Events List (MSEL)
5. Exercise Evaluation Guides (EEGs)
6. Player Briefing
7. Actor Briefing
8. Actor Waivers
9. Weapon and Safety Policy
10. Press Release and/or Media Guidance
11. Participant Feedback Form
12. After Action Report/Improvement Plan (AAR/IP)

More information for these along with templates that you can download and use is available at the HSEEP website (http://hseep.dhs.gov).

Chapter 20

Exercises

Testing Your Plans and Capabilities in a Controlled Environment

James A. McGee

Introduction

Emergency managers should conduct exercises to test plans and capabilities while promoting awareness of various roles and responsibilities during an incident. This chapter introduces students to the value of exercises for crisis planning and emergency response. The benefits and types of exercises available for emergency responders are discussed. The knowledge of material presented in this chapter is a critical prerequisite for subsequent discussions regarding personal protection and safety, strategies, and tactics for response to a critical incident.

The financial impact and potential loss of life from disasters are significant. Many industry governing bodies have mandated preparedness training and exercising requirements. In the United States, nuclear power plants must exercise their plans annually and conduct a full-scale exercise (FSE) every 2 years, which is subsequently evaluated by the Nuclear Regulatory Commission (NRC). Airports, hospitals, and healthcare facilities must conduct an FSE every 2 years to maintain a license to operate. Additionally, the Occupational Safety and Health Administration (OSHA) requires many employers to develop an emergency action plan (EAP) and exercise it at least annually (Federal Emergency Management Agency, 2008a).

Importance of Testing Plans and Capabilities

Successfully conducting an exercise involves considerable coordination among participating agencies and officials. The Homeland Security Exercise and Evaluation Program (HSEEP) methodology divides individual exercises into five overarching phases (U.S. Department of Homeland Security, 2007a):

- Foundation—The following activities must be accomplished to provide the foundation for an effective exercise: create a base of support from the appropriate entities or senior officials, develop a project management timeline and establish milestones, identify an exercise planning team, and schedule planning conferences.
- Design and development—Building on the exercise foundation, the design and development process focuses on identifying objectives, designing the scenario, creating documentation, coordinating logistics, planning exercise conduct, and selecting an evaluation and improvement methodology.
- Conduct—After the design and development steps are complete, the exercise takes place. Exercise conduct steps include setup, briefings, facilitation, control, evaluation, and wrap-up activities.
- Evaluation—The evaluation phase for exercises includes a formal exercise evaluation, an integrated analysis, and an after action report (AAR) and improvement plan (IP) that identify strengths and areas for improvement in entities preparedness, as observed during the exercise. Recommendations related to areas for improvement are identified to help develop corrective actions to be tracked throughout the improvement planning phase.
- Improvement planning—During improvement planning, the corrective actions identified in the evaluation phase are assigned with due dates to responsible parties, tracked to implementation, and then validated during subsequent exercises.

Establishing a Foundation to Exercise Plans

Before plans and capabilities can be tested through exercises, first responders must have the necessary foundation of training to perform appropriately during a crisis. There are three different levels of training that must be addressed in emergency management:

- Awareness—All emergency responders should have awareness training because the more people that are aware of potential threats and response procedures, the more efficient and effective the response will be during an incident.
- Performance—This training is for those who will need to execute response plans as well as those who will evaluate the conduct of others during these types of incidents.
- Planning and management—Management training is for officials who will have direct management responsibilities in an emergency situation.

Senior management personnel and first responders are responsible for making sure they receive the appropriate levels of training. Designated senior management personnel are responsible for developing management plans and ensuring coordinated response. Senior management must make rapid decisions when implementing an emergency response plan (ERP). They should know the basics of all hazard planning, the Incident Command System (ICS), the National Incident Management System (NIMS), the Unified Command System (UCS), and the relevant ERP. Senior management must understand how they will perform under the IC and UC systems with other jurisdictions and response agencies when reacting to and managing a terrorist event or natural disaster.

During an emergency, first responders will need to perform specific duties. Properly trained emergency response personnel will achieve the most effective response. In regard to training, exercises, and equipment, it is necessary for executive leaders to answer the following questions:

- What do we need to prepare?
- What do we need to prepare for?
- How prepared are we?
- How do we pay for these needs?

First responders who are working under a UCS structure will normally take control of the incident site and play a lead role in safeguarding lives, isolating the site, and notifying the appropriate response agencies. Training in all hazard planning and response is essential for the emergency response team and responding personnel. This information helps to identify what type of event has occurred and the procedures used when responding to specific types of human-made events or natural disasters. These emergency response team members and responding personnel should know what types of personal protective equipment (PPE) are necessary for different situations and understand how to use time, distance, and shielding to minimize exposure. Responding personnel require training in critical thinking, self-protection, proper notification procedures, chain of command, and event documentation.

Exercises are not a one-time event. They must be conducted on a continuous basis to address evolving threats, new plans and procedures, new equipment, training new personnel, and other contingencies (see Figure 20.1).

Design and Development of Exercises

The HSEEP is a capabilities- and performance-based exercise program. The program provides a standardized methodology for exercise design, development, conduct, evaluation, and assessment. Adherence to the policy and guidance presented in the HSEEP ensures that exercise programs conform to the USDHS established best practices. Furthermore, it helps provide consistent and complementary effort for exercises at all levels of government. There are seven types of exercises defined within HSEEP, each of which is either discussion-based or operation-based (see Table 20.1).

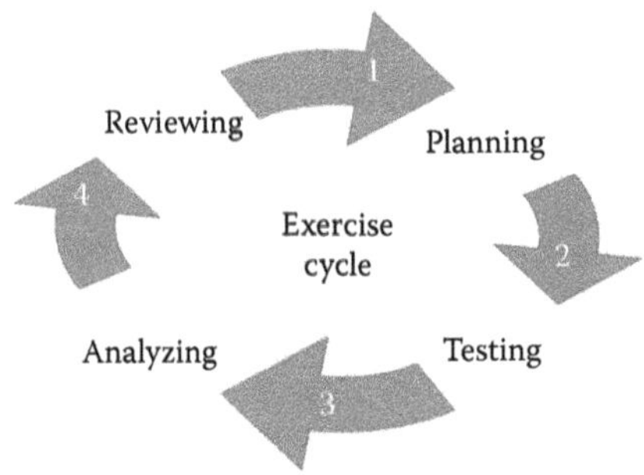

Figure 20.1 The exercise planning process. The model is cyclical, showing the importance of conducting exercises on a continuous basis and evaluating the results of each exercise. (Adapted from Hall, S. A., Cooper, W. E., Marciani, L., and McGee, J. A. 2012. *Security Management for Sport and Special Events—An Interagency Approach to Creating Safe Facilities*. United States: Human Kinetics.)

Table 20.1 Exercise Types and Examples

Exercise	*Example*
Discussion-Based Exercise	
Seminar	New personnel attend a lecture and PowerPoint presentation on general security procedures
Workshop	The command group attends a multiagency management and leadership workshop to develop an all-hazards emergency response plan
Tabletop exercise (TTX)	The command group tests the emergency response plan by working through a critical incident scenario
Game simulation	The command group attends a virtual learning laboratory to participate in a decision-making exercise
Operation-Based Exercise	
Drill	Facility managers test the alert and notification system before the day of an event
Functional exercise (FE)	The command group and local first responders work through a critical incident scenario to test available resources and communication capabilities (no deployment of assets)
Full-scale exercise (FSE)	The command group and local first responders work through a critical incident by deploying real-world assets and testing available resources

Discussion-based exercises familiarize participants with current plans, policies, agreements, and procedures, or may be used to develop new plans, policies, agreements, and procedures. The various types of discussion-based exercises include

- *Seminars* use a number of instructional strategies such as lecture, panel discussions, case studies, and multimedia presentations. Seminars are informal and productive for small and large groups and are used for orientation to organizational policies and procedures, protocols, response resources, or concepts and ideas. Seminars normally last a maximum of 1–2 h.
- *Workshops* increase participant interaction and are effective for solving complex problems, team building, information sharing, and brainstorming. Workshops differ from seminars in that they emphasize producing a product or goal such as a new policy or plan, mutual aid agreement, and standard operating procedures (SOP). Workshops also involve greater participant discussions and often use breakout sessions to explore parts of an issue with smaller groups.
- *Tabletop exercises* (TTXs) consist of informal facilitated discussions of simulated emergencies among key personnel. Basic TTXs involve a constant, unchanging simulation, whereas advanced TTXs present the group with injects (message updates) that progress the initial scenario. TTXs are a useful tool for first responders who want to assess current plans and identify gaps in security operations. The purpose of a TTX is to test existing plans without incurring costs associated with actually deploying resources. The TTX can involve many people and many organizations who can contribute to the planned discussion topics, typically those entities with a planning, policy, or response role. A TTX usually lasts 1–4 h. A sample TTX situation manual is provided at the end of this chapter.
- *Games* are provided in a computer simulation of operations that often involve two or more teams, usually in a competitive environment, using rules, data, and procedures designed to depict an actual or assumed real-life situation. Game simulations conduct "what if" analysis of existing plans and potential strategies without actually deploying resources to explore the processes and consequences of decision making.

Operation-based exercises validate plans, policies, agreements, and procedures; clarify roles and responsibilities; and identify resource gaps in an operational environment. The various types of operational-based exercises include

- *Drills* are coordinated, supervised activities usually employed to test a specific operation or function within the organization. Participants may gain training on new equipment, practice, and maintain skills. The time required to conduct a drill is usually one half to two hours.

- *Functional exercises* (FEs) were previously referred to as a command post exercise (CPX). An FE examines and validates the coordination, command, and control between various agencies responding to an incident. An FE involves a notional deployment of resources and personnel in a highly stressful environment requiring rapid problem solving. FEs can be used to evaluate management of emergency operation centers (EOCs), multiagency coordination centers, and command posts. An FE normally requires 3–8 h to complete. This type exercise does not involve any "boots on the ground" or real-world deployment of assets.
- FSE were previously referred to as field training exercises (FTXs). An FSE is a multiagency, multijurisdictional exercise involving a functional response of assets that replicates a real-world response. Real-world deployment of assets occur in support of the exercise scenario. Participants are able to assess plans and evaluate coordinated responses under crisis conditions. An FSE may be designed to last several hours or several days.

The characteristics of discussion-based and operational-based exercises are further defined in Table 20.2.

Identify Key Personnel to Be Involved in the Exercise Process

Exercise planning team members should be determined based upon the scope and type of exercise as well as the scenario or subject to be tested. Table 20.3 provides recommendations in terms of what agencies and areas of expertise might be involved in exercise planning. The lists should be modified to meet the needs of the jurisdiction or organization.

Exercise Conduct

The type of exercise selected by the entity should be consistent with the entity's multiyear training and exercise plan, which includes

- The entities' training and exercise priorities.
- The capabilities from the target capabilities list (TCL) that the entity will train for and exercise against.
- A multiyear training and exercise schedule that reflects the training activities that will take place prior to an exercise, allowing exercises to serve as a true validation of previous training.
- Reflects all exercises in which the entity participates.
- Employs a "building block approach" in which training and exercise activities gradually escalate in complexity.
- The multiyear training and exercise plan must be updated on an annual basis (or as necessary) to reflect schedule changes.

Table 20.2 Characteristics of Exercise Types

Type of Exercise	*Utility or Purpose*	*Type of Player Action*	*Duration*	*Real-Time Play*	*Scope*
Discussion based	To familiarize players with current plans, policies, agreements, and procedures; to develop new plans, policies, agreements, and procedures	Notional player actions are imaginary and hypothetical	Rarely exceeds 8 h	No	Varies
Seminar	To provide an overview of new or current plans, resources, strategies, concepts, or ideas	Not applicable	2–5 h	No	Multi- or single agency
Workshop	To achieve a specific goal or build a product (e.g., exercise objectives, SOPs, policies, or plans)	Not applicable	3–8 h	No	Multiagency or multiple functions
Tabletop exercise	To assist senior officials in the ability to understand and assess plans, policies, procedures, and concepts	Notional	4–8 h	No	Multiagency or multiple functions
Game	To explore decision-making processes and examine the consequences of those decisions	Notional	2–5 h	No (some simulations provide real-time or near-real-time play)	Multiagency or multiple functions

(Continued)

Table 20.2 (continued) Characteristics of Exercise Types

Type of Exercise	*Utility or Purpose*	*Type of Player Action*	*Duration*	*Real-Time Play*	*Scope*
Operation based	Test and validate plans, policies, agreements, and procedures; clarify roles and responsibilities; identify resource gaps	Actual player action mimics reaction, response, mobilization, and commitment of personnel and resources	May be hours, days, or weeks depending on purpose, type, and scope	Yes	Varies
Drill	Test a single operation or function	Actual	2–4 h	Yes	Single agency or function
Functional exercise	Test and evaluate capabilities, functions, plans, and staffs of incident command, unified command and intel centers, or other command or operation centers	Command staff actions are actual; movement of other personnel, equipment, or adversaries is simulated	4–8 h or several days or weeks	Yes	Multiple functional areas or multiple functions
Full-scale exercise	Implement and analyze plans, policies, procedures, and cooperative agreements developed in previous exercises	Actual	One full day or longer	Yes	Multiple agencies or multiple functions

Source: Reprinted from the U.S. Department of Homeland Security, 2007a, Homeland Security exercise and evaluation program (HSEEP), Volume I.

Table 20.3 HSEEP Recommended Planning Team Members for Exercises

Discussion-Based Exercises	*Operation-Based Exercises*
Emergency Management	***Emergency Management***
Emergency manager	Emergency manager*
Homeland security	Homeland security*
Public Safety	Public health*
Fire	Public works
Hazardous materials (HAZMAT)	Transportation or transit authority
Law enforcement	Public affairs
Emergency medical services (EMS)	Exercise venue management
Special operations/bomb squad	***Fire***
Federal Bureau of Investigation (FBI)	Fire department*
Public Health	Communications or dispatch*
Public health department	Special operations (e.g., HAZMAT, Metropolitan Medical Response System (MMRS))*
Communicable/infectious disease	Mutual aid fire*
Epidemiologists	***Law Enforcement***
Pathology	Police*
Poison control	Special operations (e.g., bomb squad, SWAT)*
Medical	Sheriff's department*
Hospital administrators	Federal Bureau of Investigation (FBI)*
Coroner or medical examiner	Mutual aid law enforcement*
Hospital infection control	***Medical***
Hospital lab managers	Hospital representatives (primary trauma center or hospital association)*
Hospital emergency room	Emergency medical services (EMS)*
Private practitioners	Mutual aid

(*Continued*)

Table 20.3 (continued) HSEEP Recommended Planning Team Members for Exercises

Discussion-Based Exercises	*Operation-Based Exercises*
Veterinary	Medical examiner or coroner
Other	***Other***
Public works	Volunteer organizations
Public information officer (PIO)	Subject matter experts
Volunteer organizations	Private security
Communications or dispatch	Government officials
Government officials	Meteorologist

Note: The agencies marked with asterisks are most critical to be present during all planning conferences.

Exercise objectives should be based on capabilities and their associated critical tasks, which are contained within the exercise evaluation guides (EEGs). For example, if an entity, based on its risk/vulnerability assessment, determines that it is prone to hurricanes, it may want to validate its evacuation capabilities. The scenarios used in exercises must be tailored toward validating the capabilities and should be based on the entity's risk/vulnerability assessment. Exercise planners should develop the following documents, in accordance with HSEEP, to support exercise planning, conduct, evaluation, and improvement planning:

- *For discussion-based exercises:*
  - Situation manual (SitMan) for facilitators
  - Situation manual for participants
- *For operation-based exercises:*
  - Exercise plan (ExPlan)
  - Player handout
  - Master scenario event list (MSEL)
  - Controller/evaluator handbook (C/E handbook)

The exercise planning team determines the timelines for completion of the exercise plan, acquisition of necessary resources, and development of sufficient support. The exercise scope and statement of purpose must be clearly defined. A statement of purpose identifies the issue(s) to be addressed in detail. These issues may have been identified based on a past crisis, observation, or determined during a risk assessment.

Table 20.4 describes the important document types associated with most exercises.

A detailed *exercise scenario* must be developed based on the statement of purpose. Either in narrative format or depicted as an event timeline, the scenario

Table 20.4 Exercise Document Types

Discussion-Based Exercises	*Operation-Based Exercises*
Situation manuals (SitMan) are individual facilitator and participant handbooks for discussion-based exercises, particularly TTXs. It provides background information on exercise scope, schedule, and objectives. It also presents the scenario narrative that will drive participant discussions during the exercise.	The exercise plan (ExPlan), typically used for operation-based exercises, provides a synopsis of the exercise and is published and distributed to players and observers prior to the start of the exercise. The ExPlan includes the exercise objectives and scope, safety procedures, and logistical considerations such as exercise schedule. The ExPlan does not contain detailed scenario information.
Multimedia presentation supports the SitMan, concisely summarizing written information. Enhances exercise realism with audio or visual depiction of the scenario. Focuses and drives the exercise.	The controller and evaluator (C/E) handbook supplements the ExPlan for operation-based exercises, containing more detailed information about the exercise scenario and describing exercise controllers' and evaluators' roles and responsibilities. Because the C/E handbook contains information on the scenario and exercise administration, it is distributed only to those individuals specifically designated as controllers or evaluators.
Exercise evaluation guides (EEGs) are necessary for all evaluated exercises. An EEG helps evaluators assess performance of capabilities, tasks, and objectives during an exercise.	The master scenario event list (MSEL) is a chronological timeline of expected actions and scripted events (i.e., injects) to be inserted into the operation-based exercise play by controllers to generate or prompt player activity. It ensures necessary events happen so that all exercise objectives are met.
	A player handout is a 1–2 page document, usually handed out the morning of an exercise, which provides a quick reference for exercise players on safety procedures, logistical considerations, exercise schedule, and other key factors and information.

provides a storyline that drives the exercise. For a discussion-based exercise, the scenario provides the backstop that is the basis for participant discussion. For an operation-based exercise, the scenario provides background information regarding the incident and is fueled by periodic injects as the exercise unfolds.

A number of factors should be taken into consideration when developing a scenario, including level of realism, type of threat or hazard, site selection, optimal date and time for conducting the exercise, and cost. The scenario selected for the exercise should be a realistic representation of potential threats faced by the exercising entities.

Exercises should adhere to the planning timelines laid forth in HSEEP and reflect the principles of NIMS and ICS. A consistent terminology and methodology for exercises is critical to avoid confusion and to ensure that entities can exercise together seamlessly.

The below list describes the important key personnel involved with conducting the exercise:

- Planners—Designs, develops, conducts, and evaluates exercise; determines exercise objectives, creates scenarios, and develops documentation; develops and distributes preexercise materials; and conducts exercise briefings and training sessions.
- Players—Participating agency personnel who discuss their roles and responses to the scenario during the exercise; drawn from participating agencies to accomplish exercise objectives; and familiar with the agency SOP and emergency operation plans (EOPs) being tested.
- Facilitators—Facilitate discussion and coordinate issues between groups; focuses the group's discussions on specific areas and questions; elicit resolutions to issues; monitor the recorder and prepare notes on the group's discussions; comfortable talking in front of large groups of people; and knowledgeable on plans and policies.
- Evaluators—Observe and record player discussions; do not interfere with exercise play; chosen for their knowledge of a particular functional area; familiar with the jurisdictions SOPs and EOPs; and provide input to the AAR.
- Recorders—Record discussions during break-out sessions; should have working knowledge of commonly used terms and acronyms in the area they are recording; have no input into the exercise process; and require no formal training.
- Subject matter experts (SMEs)—Add functional knowledge and expertise to the exercise planning team; help make the scenario realistic and plausible; and ensure jurisdictions have appropriate capabilities to respond.
- Observers/VIPs—Do not have an active role in exercise play; should be allowed to see or hear appropriate aspects of exercise play; may be senior or elected officials, neighboring jurisdictions or media representatives; observe the exercise from a designated area; require no formal training; and may attend an observer briefing.

Design and Develop an Exercise to Include an After Action Report

Before a discussion-based exercise begins, the planning team must deliver the necessary exercise materials and equipment to the exercise location (U.S. Department of Homeland Security, 2007b):

- Exercise manuals (SitMans) for facilitators and participants
- Multimedia presentation (PowerPoint)
- AV equipment (including televisions, projectors, projection screens, microphones, and speakers)
- Table tents (for agency names)
- Name tents (for participants)
- Badges identifying the role of each exercise participant (e.g., player, observer, VIP, facilitator, and evaluator)
- Sign-in sheets and registration information
- Participant feedback forms

The following checklist contains all of the tasks that need to be completed for each discussion-based exercise (seminar, workshop, TTX, and game simulation). This checklist should be customized to each exercise, identifying any missing activities, and removing any redundant activities.

Discussion-Based Exercise Master Task List

Exercise Planning Tasks	*Responsible Party*	*Contact Information*	*Date Due*	*Date Completed*	*Remarks*
I. Foundation					
Develop exercise budget					
Develop exercise planning timeline					
Identify exercise planning team					
Schedule first planning conference					

(Continued)

Exercise Planning Tasks	*Responsible Party*	*Contact Information*	*Date Due*	*Date Completed*	*Remarks*
II. Design and Development					
Planning Conferences					
Initial Planning Conference (IPC)					
Prepare and send invitations and read-ahead packets					
Develop agenda, presentation, and sign-in sheets					
Determine exercise scope (see *Scope* section)					
Determine exercise scenario (see *Scenario* section)					
Determine date for next planning conference					
Assign responsibilities and due dates for tasks					
Develop IPC minutes					
Begin development of exercise documentation (see *Documentation* section)					
Final Planning Conference (FPC)					
Prepare and send invitations					
Develop agenda, briefing, and sign-in sheets					

Exercise Planning Tasks	*Responsible Party*	*Contact Information*	*Date Due*	*Date Completed*	*Remarks*
Review all exercise materials, documents, and tasks					
Assign responsibilities and due dates for tasks					
Develop FPC minutes					
Scope					
Identify exercise design objectives					
Identify exercise participants					
Scenario					
Identify threat/hazard and/or specific agent					
Identify exercise venue					
Documentation					
Develop situation manual (SitMan)					
Develop multimedia exercise presentation					
Develop participant feedback forms					
Develop exercise evaluation packets (including exercise evaluation guides (EEGs))					

(*Continued*)

Exercise Planning Tasks	*Responsible Party*	*Contact Information*	*Date Due*	*Date Completed*	*Remarks*
Exercise Site Areas					
Designate media/ observer area					
Designate registration area					
Designate parking area					
Media/Public Information					
Develop media policy					
Develop media release/public information handout					
Logistics					
Arrange for use of exercise venue (reserve room/use of facility)					
Arrange for participant parking at venue					
Arrange for audio/ visual equipment (e.g., microphones, screens, and projectors)					
Arrange for exercise supplies (e.g., pens, markers, and flipcharts)					
Develop mailing lists (players, facilitators, exercise planning team)					

Exercise Planning Tasks	*Responsible Party*	*Contact Information*	*Date Due*	*Date Completed*	*Remarks*
Develop ID badges, name/table tents, and sign-in sheets					
Arrange for restrooms					
Provide food and refreshments					
Develop signage					
Arrange for videotaping of exercise					
Exercise Staffing					
Determine exercise staff requirements					
Select and train exercise staff					
III. Conduct					
Briefings					
Present multimedia exercise briefing					
Documentation					
Distribute SitMan					
Distribute exercise evaluation packets					
Distribute participant feedback forms					
Exercise Control					
Set up exercise site					
Conduct/facilitate exercise					
Conduct postexercise hot wash					

(*Continued*)

Exercise Planning Tasks	*Responsible Party*	*Contact Information*	*Date Due*	*Date Completed*	*Remarks*
IV. Evaluation					
After Action Review					
Develop hot wash minutes					
Conduct controller and evaluator (C/E) debriefing					
Develop C/E debriefing minutes					
Develop draft AAR					
Send draft AAR to exercise planning team for review					
V. Improvement Planning					
After Action Conference					
Schedule conference					
Prepare and send invitations					
Conduct after action conference					
Finalize AAR					
Develop improvement plan (IP)					
Improvement Planning					
Share lessons learned, best practices, and successes identified in AAR/IP					
Implement AAR/IP					
Track AAR/IP implementation					

Source: Reprinted from the U.S. Department of Homeland Security. 12/18/2009. *Homeland Security Exercise and Evaluation Program (HSEEP), Master Task List for Discussion-Based Exercises*, Volume IV. https://hseep.dhs.gov/hseep_vols/allDocs.aspx?a=M. (Accessed August 9, 2012).

The setup for an operation-based exercise begins as many days before the event as necessary, depending on the scope of the scenario. The setup entails arranging briefing rooms, and testing AV and communications equipment, placing props and effects to add realism to the incident, marking the appropriate areas and their perimeters, and checking for potential safety issues. Safety is the most important consideration in planning an operation-based exercise. The following actions must take place to ensure a safe environment (U.S. Department of Homeland Security, 2007b, p. 24):

- Identify safety controllers (not to be confused with a safety officer designated by the incident commander as part of the response to the exercise scenario).
- Dedicate advanced life support or basic life support ambulance units for real-world emergencies only.
- Identify real-world emergency procedures with a code word or phrase.
- Identify safety requirements and policies.
- Consider other safety issues outside the scope of exercise control (e.g., weather, heat stress, hypothermia, fire or pyrotechnics, weapons, and traffic accidents).

The following checklist contains all of the tasks that need to be completed for each operation-based exercise (drill, FE, FSE). This checklist should be customized to each exercise, identifying any missing activities and removing any redundant activities.

Operation-Based Exercise Master Task List

Exercise Planning Tasks	*Responsible Party*	*Contact Information*	*Date Due*	*Date Completed*	*Remarks*
I. Foundation					
Develop exercise budget					
Develop exercise planning timeline					
Identify exercise planning team					
Schedule first planning conference					

(Continued)

Exercise Planning Tasks	*Responsible Party*	*Contact Information*	*Date Due*	*Date Completed*	*Remarks*
II. Design and Development					
A. Planning Conferences					
1. Concepts and Objectives (C&O) Meeting					
Prepare and send invitations					
Develop agenda, presentation, and sign-in sheets					
Determine exercise scope (see *Section B: scope*)					
Develop C&O meeting minutes					
2. Initial Planning Conference					
Schedule IPC					
Prepare and send invitations					
Develop and distribute read-ahead packet					
Develop agenda, presentation, and sign-in sheets					
Determine exercise scope (if no C&O meeting is held prior. See *Section B: Scope*)					
Determine scenario (see *Section C: Scenario*)					
Assign responsibilities and due dates for each task					

Exercise Planning Tasks	*Responsible Party*	*Contact Information*	*Date Due*	*Date Completed*	*Remarks*
Schedule next planning conference					
Develop IPC minutes					
Begin development of exercise documentation (see *Section D: Documentation*)					
3. Midterm Planning Conference					
Prepare and send invitations					
Develop agenda and sign-in sheets					
Schedule next planning conference					
Assign responsibilities and due dates for tasks					
Conduct visit of exercise site(s)					
Develop MPC minutes					
4. Master Scenario Events List (MSEL) Conference					
Prepare and send invitations					
Review MSEL					
Schedule next planning conference					
Assign responsibilities and due dates for tasks					
5. Final Planning Conference					
Prepare and send invitations					
Develop agenda and sign-in sheets					

(*Continued*)

Exercise Planning Tasks	*Responsible Party*	*Contact Information*	*Date Due*	*Date Completed*	*Remarks*
Review all exercise materials, documents, and tasks					
Assign responsibilities and due dates for tasks					
Develop FPC minutes					
B. Scope					
Identify exercise design objectives					
Identify exercise participants					
C. Scenario					
Identify exercise venue					
Determine exercise weather conditions (predetermined or real-world)					
Determine date and time for scenario to take place					
Identify the threat/hazard and/or specific agent					
D. Documentation					
Develop exercise plan (ExPlan)					
Develop controller and evaluator (C/E) handbook					
Develop evaluation plan (EVALPLAN)					

Exercise Planning Tasks	*Responsible Party*	*Contact Information*	*Date Due*	*Date Completed*	*Remarks*
Develop control staff instructions (COSIN)					
Develop master scenario events list (MSEL)					
Develop simulation cell (SIMCELL) messages					
Develop exercise evaluation packets (EEPs)					
Develop controller and evaluator packets					
Develop multimedia presentation					
Develop exercise schedule					
Develop deployment timetable for assembly area					
Develop list of controller and evaluator assignments					
E. Exercise Site Areas					
Determine exercise site					
Determine dispatch requirements (e.g., whether dispatch will be off-site or on-site; provide dispatchers)					
Define response routes					
Designate parking area					

(Continued)

Exercise Planning Tasks	*Responsible Party*	*Contact Information*	*Date Due*	*Date Completed*	*Remarks*
Designate registration area					
Designate Assembly Area					
Designate observer/ media area					
Designate SIMCELL and master control cell (MCC) locations					
F. Actors					
Determine number of actors required					
Identify source(s) of actors					
Confirm recruited actors					
Develop actor waiver forms					
Develop actor instructions					
Arrange for moulage (actual materials, staff, and location for actor moulage)					
Arrange water and food for actors					
Arrange necessary transportation for actors (determine mode, schedule, pick-up and drop-off locations, actor tracking system)					
Develop casualty matrix					

Exercise Planning Tasks	*Responsible Party*	*Contact Information*	*Date Due*	*Date Completed*	*Remarks*
Develop symptomology cards/victim tags					
Identify number and type of victim actors that will be at each location/exercise site					
Provide necessary protection after decontamination process (i.e., blankets)					
G. Media/Public Information					
Develop media release/public information handout					
Identify media/ public liaison (communicates with media and public prior to exercise; escorts and briefs media and observers/VIPs before and during exercise)					
Develop public announcement					
Disseminate information to public and media (via print, television, radio, etc.)					
Schedule and conduct press conference					
Develop media policy					

(*Continued*)

Exercise Planning Tasks	*Responsible Party*	*Contact Information*	*Date Due*	*Date Completed*	*Remarks*
H. Logistics					
Develop correspondence letters (invitations, thank you letters)					
Develop mailing lists (players, controllers, VIPs, evaluators, exercise planning team)					
Develop ID badges					
Procure necessary color hats, vests, armbands, and so on					
Provide food and refreshments					
Determine location for food and water stations					
Arrange for restrooms					
Develop communications plan					
Provide radios for controllers					
Designate radio channels for the exercise					
Provide videotaping for the exercise					
Provide necessary props (e.g., debris, mannequins)					

Exercise Planning Tasks	*Responsible Party*	*Contact Information*	*Date Due*	*Date Completed*	*Remarks*
Provide necessary devices (e.g., flash bangs, smoke machines, pyrotechnics)					
Provide secondary device for render-safe procedures					
Provide site security					
Arrange for perimeter barricading and signage					
Develop weapons policy					
I. Safety					
Identify safety controller					
Develop exercise play rules					
Arrange for dedicated ALS/BLS ambulance unit for real emergencies only					
Determine real-world emergency procedures					
Develop safety policy (to include section on weapon safety protocols)					
J. Exercise Staffing					
Determine exercise staff requirements					
Select and train exercise staff					

(Continued)

Exercise Planning Tasks	*Responsible Party*	*Contact Information*	*Date Due*	*Date Completed*	*Remarks*
III. Conduct					
A. Briefings					
Controller and evaluator (C/E) brief					
Hospital brief					
Actor brief					
Observer brief					
C/E debrief					
Hospital debrief					
B. Documentation					
Distribute ExPlan					
Distribute actor waiver and information sheet					
Distribute participant feedback forms					
Distribute C/E packet (C/E handbook, COSIN, EVALPLAN, MSEL, EEPs)					
C. Exercise Control					
Conduct communications check					
Conduct pyrotechnic and device check					
Conduct safety check					
Conduct weapons check					
Announce start of exercise					
Conduct player hot wash					

Exercise Planning Tasks	*Responsible Party*	*Contact Information*	*Date Due*	*Date Completed*	*Remarks*
IV. Evaluation					
Develop hot wash minutes					
Conduct C/E debrief					
Develop C/E debrief minutes					
Develop draft AAR					
Distribute draft AAR to exercise planning team for review					
V. Improvement Planning					
A. After Action Conference					
Schedule conference					
Prepare and send invitations					
Conduct after action conference					
Finalize AAR					
Develop improvement plan (IP)					
B. Improvement Planning					
Share lessons learned, best practices, and successes identified in AAR					
Implement AAR/IP					
Track implementation of AAR/IP					

Source: Reprinted from U.S. Department of Homeland Security. 10/07/2009. *Homeland Security Exercise and Evaluation Program (HSEEP), Master Task List for Operations-Based Exercises*, Volume IV. https://hseep.dhs.gov/hseep_vols/default1.aspx?url...?q... (Accessed August 9, 2012).

Evaluation and Improvement Planning

EEGs help evaluators collect and interpret relevant exercise observations. EEGs provide evaluators with information on what tasks they should expect to see accomplished during an exercise, space to record observations, and questions to address after the exercise as a first step in the analysis process. To assist entities in exercise evaluation, standardized EEGs have been created that reflect capabilities-based planning tools, such as the TCL and the universal task list (UTL). The EEGs are not meant as report cards. Rather, they are intended to guide an evaluator's observations so that the evaluator focuses on capabilities and tasks relevant to exercise objectives to support the development of the AAR/IP.

AAR/IP is the final product of an exercise. The AAR/IP has two components: an AAR, which captures observations and recommendations based on the exercise objectives as associated with the capabilities and tasks, and an IP, which identifies specific corrective actions, assigns them to responsible parties, and establishes targets for their completion. The lead evaluator and the exercise planning team draft the AAR and submit it to conference participants prior to an after action conference (AAC). The draft AAR is distributed to conference participants for review no more than 30 days after the exercise is conducted. The final AAR/IP is an outcome of the AAC and should be disseminated to participants no more than 60 days after the end of the exercise (ENDEX).

The HSEEP methodology defines a variety of planning and AACs. The need for each of these conferences varies depending on the type and scope of the exercise. They include

- Concepts and objectives meeting
- IPC
- Midterm planning conference (MPC)
- MSEL conference
- Final planning conference (FPC)
- AAC

Following every exercise, an AAC must be conducted in which key personnel and the exercise planning teams are presented with findings and recommendations from the draft AAR/IP. Following each exercise, a draft AAR/IP must be developed based on information gathered through the use of EEGs. AAR/IPs created from the exercise must conform to the templates provided by HSEEP. Corrective actions addressing a draft AAR/IP's recommendations are developed and assigned to responsible parties with due dates for completion. A final AAR/IP with recommendations and corrective actions derived from the discussion at the AAC must be completed within 60 days after ENDEX.

IP include broad recommendations from the AAR/IP organized by target capability as defined in the TCL. Corrective actions derived from an AAC are associated with the recommendations and must be linked to a capability element as defined in the TCL. Corrective actions included in the IP must be measurable and designate a projected start date and completion date. Corrective actions included in the IP must be assigned to an organization and a point of contact (POC) within that organization. Corrective actions must be continually monitored and reviewed as part of an organizational corrective action program. An individual should be assigned the responsibility of managing a corrective action program to ensure corrective actions resulting from exercises, policy discussions, and real-world events are resolved and support the scheduling and development of subsequent training and exercises.

Lessons learned:

- There are three different levels of training that must be addressed in emergency management.
- Types of discussion-based exercises.
- Types of operational-based exercises.
- The important key personnel involved with an exercise process.
- A consistent terminology and methodology for exercises are critical to avoid confusion and to ensure that entities can exercise together seamlessly.

Active Shooter Threat

An active shooter is defined by the U.S. Department of Homeland Security as an individual actively engaged in killing or attempting to kill people in a confined and populated area. Historically, active shooters use firearms and there is no pattern or method to their selection of victims. The shootings at Sandy Hook elementary school in Newton, Connecticut can be added to the increasing list of locations attacked by an active shooter. Within the last 5 years, there have been at least 15 prominent, high-casualty-producing active shooter incidents. Most of these cases have occurred in locations where the shooter has been undeterred and unobstructed from carrying out their attack. The crime scenes have been described as soft targets with limited active security measures or armed personnel available to provide protection for members of the public. In most instances, shooters have taken their own lives, been shot by police, or surrendered when forced into a confrontation with law enforcement.

Provided below is a TTX facilitators guide that includes a generic active shooter scenario. The scenario is a college campus but can be adjusted to address any location. Each TTX module concludes with questions directed to the campus and emergency responder community.

Situation Manual
(SitMan)
Table Top Exercise
Facilitator's Guide
SCENARIO
(Active Shooter Threat)
Place Date Here

Preface

What follows is a capabilities- and performance-based exercise experience. The TTX is based upon a standardized methodology for exercise design, development, conduct, evaluation, and assessment. The format is in adherence to the policy and guidance presented in the U.S. Department of Homeland Security Exercise and Evaluation Program (DHS-HSEEP). HSEEP ensures that exercise programs provide consistent and complementary effort for exercises at all levels of government. This TTX is designed in accordance with the DHS-HSEEP exercise methodology.

To mitigate gaps and deficiencies in preparedness for incidents, this TTX is developed to assist in training and exercising emergency preparedness for critical incidents. The TTX experience is intended to continuously develop and enhance preparedness for incident avoidance, management, response, and recovery.

This is an unclassified exercise. The control of the information is based more on public sensitivity regarding the nature of the exercise than on the actual exercise content. Some exercise materials are intended for the exclusive use of exercise planners, facilitators, and evaluators. Players may view other materials deemed necessary to their performance. All exercise participants may view the SitMan.

All exercise participants should use appropriate guidelines to ensure the proper control of information within their areas of expertise and to protect this material in accordance with current jurisdictional directives.

Handling Instructions

1. The title of this document is Situation Manual (SitMan) Table Top Exercise SCENARIO (Active Shooter Threat).
2. The information gathered in the SitMan is *for official use only (FOUO)* and should be handled as sensitive information not to be disclosed. This document should be safeguarded, handled, transmitted, and stored in accordance with appropriate security directives. *Reproduction of this document, in whole or in part, without prior approval is prohibited.*
3. At a minimum, the attached materials will be disseminated only on a need-to-know basis and when unattended, will be stored in a secure container

or area offering sufficient protection against theft, compromise, inadvertent access, and unauthorized disclosure.

4. For more information, please contact the following:

Sample Exercise Schedule

Facilitator Instruction: Show "Exercise Schedule" slide. Go over the exercise schedule with participants.

8:30 a.m.	Participant sign-in
9:00 a.m.	Introductions
	Discuss general instructions and ground rules of the exercise
9:15 a.m.	Exercise overview
	Discuss exercise objectives, and schedule of exercise
9:30 a.m.	Read Module 1
9:45 a.m.	Module 1 Discussion
10:00 a.m.	Read Module 2
10:15 a.m.	Module 2 Discussion
10:30 a.m.	Read Module 3
10:45 a.m.	Module 3 Discussions
11:00 a.m.	After action hot wash/final comments
11:30 a.m.	End exercise

Introduction

This facilitator manual is designed as a guidance for designated exercise controllers to facilitate this TTX. It contains general instructions to the facilitator on the overall exercise process, necessary materials, and discussion questions. Detailed notes for the facilitator's consideration are shown in bold and italicized font.

General Instructions

This TTX begins with a PowerPoint presentation as it outlines the content of the participant manual. The presentation will detail, in the following sequence, the rules, objectives, and scenario included in this TTX. Please note that although the scenario presented is fictitious, it realistically represents a probable event affecting a campus environment.

Players are strongly encouraged to participate in in-depth discussions as the primary purpose of the exercise is to evaluate and improve skills, knowledge, and ERPs. It is important for players to keep the exercise objectives in mind as all issues raised by the scenario will be thoroughly discussed.

Exercise Structure

Players will participate in the following three distinct modules:

- *Module 1:* Warning (Credible Threat)
- *Module 2:* Notification and Initial Response
- *Module 3:* Continued Response/Evacuation and Recovery

Each module begins with an update summarizing the key events occurring within the time period. Following the updates, participants review the situation and engage in functional group discussions of appropriate response issues. Participants then enter into a facilitated caucus discussion in which they present their group's actions based on the scenario.

Although they are encouraged to move among tables to ensure thorough and thought-provoking discussion, participants will be divided into functional groups to discuss aspects of the situation as presented. Exchanges among functional groups will be necessary to coordinate actions and decisions. The functional groups will be determined based upon the participating agencies and their areas of expertise.

Following the functional group discussions, participants then hold a functional caucus discussion, in which a spokesperson from each group presents a synopsis of the group's actions based on the scenario.

Each exercise participant will receive a copy of the SitMan, which provides a written scenario and situation updates. Following each module is a series of questions highlighting pertinent issues for consideration. These questions are supplied as a catalyst for the group discussions. Participants are not required to answer every question, nor are they limited to those topics. Participants are encouraged to use the SitMan as a reference throughout the exercise.

Following each module, players will have a set time period to review the module and discuss the suggested issues. During this exercise, the following rules apply:

- This TTX is conducted in an artificial environment where time compression is necessary to examine and resolve issues. Some aspects in terms of resources and response will be notionalized.
- The scenario represents a plausible critical incident.
- There are no trick questions or "hidden agendas" associated with this TTX.
- Players have no previous knowledge of the scenario and will receive all information at the same time.
- Players will respond using existing plans, procedures, and other response resources.
- Decisions are not precedent setting and may not reflect your organization's final position on a given issue.

Note to facilitator: Before showing the slide on "Exercise Rules," brief the group on emergency exits, bathroom locations, and other relevant housekeeping items.

Facilitator instructions: Show "Exercise Objectives" slide. Read the narrative as written below.

Exercise Objectives

Exercise design objectives are focused on improving the understanding of a response concept, identifying opportunities or problems, and/or achieving a change in attitude. This exercise will focus on using the four phases of emergency management (prevention–mitigation, preparedness, response, and recovery) as a foundation to the following design objectives selected by the exercise planning team.

- *Intelligence/information gathering and dissemination*: Discuss plans, policies, and procedures for ensuring the proper gathering, analyzing, sharing, and dissemination of incident-related information during all stages of a critical incident.
- *ICS/unified command*: Responders will demonstrate the ability to implement a functional ICS, transition to unified command and effectively direct, coordinate, and manage a response to a critical incident. Responders will activate their respective ERP and all relevant annexes (evacuation plan, HAZMAT/WMD response plan, etc.). Responders will establish an incident command post (ICP) and EOC in a timely matter after the initial call for services. Responders will designate/recognize a lead agency on scene commander (OSC) for crisis response/mitigation.
- *Communications*: Understand communication channels and procedures to conduct incident management activities. Determine strengths and weaknesses in the communication of response activities. Identify critical issues and potential solutions. Identify and activate a primary and alternate communication system.
- *Threat assessment*: Assess existing hazard prevention measures by addressing threats.
- *Preparedness*: Assess preparedness, such as maintaining sufficient supplies and providing training to staff in prevention–mitigation, preparedness, response, and recovery, in anticipation of a critical incident.
- *Recovery*: Assess the ability to recover from a critical incident and restore a safe environment.

Note to facilitator:

1. These objectives should be displayed on the screen throughout the duration of the exercise if multimedia presentation capability allows.
2. Provide participants a few minutes and/or review with them the appendices and inform them of the tools they may want to reference during the scenario.

Purpose

The purpose of this exercise is to provide participants with an opportunity to evaluate current response concepts, plans, and capabilities for response to a critical incident. The exercise will focus on key emergency responder coordination, critical decision making, and the integration of assets necessary to save lives and investigate the incident.

Scope

The exercise is a 3-h interactive exercise consisting of three modules, each portraying a milestone responding to a critical incident involving an active shooter.

This TTX will focus on the role of various agencies in response to the consequences of the critical incident. Emphasis is on decision-making emergency response processes, coordination, integration of capabilities, problem identification, and resolution.

Participants

- *Players*—Players respond to the situation presented based on expert knowledge of response procedures, current plans, and procedures in place in their agency, and insights derived from training.
- *Observers*—Observers support the group in developing responses to the situation during the discussion. However, they are not participants during the moderated discussion period.
- *Facilitators*—Facilitators provide situation updates and moderate discussions. They also provide additional information or resolve questions as required. Key planning committee members may also assist with facilitation as SMEs during the TTX.
- *Evaluators*—Evaluators observe and record the discussions during the exercise, participate in the data analysis, and assist with drafting the AAR.
- *SMEs*—The SMEs are similar to observers but may be asked specific questions about their agencies, certain policies, or areas of expertise.

Exercise Guidelines

This is an open, low-stress, no-fault environment. Varying viewpoints, even disagreements, are expected.

- Respond based on your knowledge of current plans and capabilities (i.e., you may use only existing assets) and insights derived from training.
- Decisions are not precedent setting and may not reflect your organization's final position on a given issue. This is an opportunity to discuss and present multiple options and possible solutions.
- Assume cooperation and support from other responders and agencies.

- Issue identification is not as valuable as suggestions and recommended actions that could improve response and preparedness efforts. Problem solving should be the focus.
- The situation updates, written material, and resources are the basis for discussion. There are no situational injects.
- Do not read ahead.
- List options if no plan exists.

Module 1: Warning (Credible Threat)

The scenario will present an overview of actions that may be necessary to handle an all-hazards campus emergency. It is critical for participants to understand why it is essential to plan, train, coordinate, exercise, and integrate personnel and operations for an effective all-hazards response. This multifaceted, integrated, all-hazards response will enhance a community's ability to effectively prevent, prepare for, respond to, and recover from natural and man-made events.

One important component of a campus all-hazards response is the ability of the jurisdiction and the campus to work together. Planning, training, exercising, building relationships, and integrating operations will promote a successful management of an all-hazards campus event. These five factors, established in advance, will assist the campus community in overcoming difficult challenges. The process is continuous and constantly evolving. It will need refinement as the campus and community grow and change. Campus jurisdictions must understand that when a crisis occurs it is the established process that will enable them to successfully manage the incident. University officials and community leaders may change; however, an established process ingrained and exercised will continually assist campus communities in preventing, preparing for, responding to, and recovering from all-hazards incidents.

Friday: Game Day preparations are underway to host a Friday evening home football game. Kick-off is scheduled for 7:00 p.m.

07:00 a.m.

Neighbors hear what appears to be another loud argument coming from off-campus apartment 102 in the university apartment complex. The residents of the apartment are Jane and Tom Smith. Both are PhD candidates and employees of the local university. Neighbors call the local police to report the argument. The dispatcher advises it will be approximately 25 min before a unit can respond due to a high number of concurrent calls requiring police assistance/response.

07:29 a.m.

A local police patrol unit arrives at the Smith apartment. Jane Smith answers the door. She is visibly upset and when questioned states that her husband and she had been arguing. She states that her husband left the premises and went to see their marriage counselor. When asked if there are any weapons in the residence, Jane Smith tells the police that her husband is a hunter and keeps a shotgun in the bedroom. Jane is provided with contact information regarding domestic violence and the police officers depart. After the police depart, Jane notices that the shotgun, normally stored in the corner of the closet, is not there.

08:10 a.m.
Tom Smith is successful in getting an emergency appointment with his counselor. Tom vents to the counselor about his frustration with his marriage. He becomes increasingly angry as he recounts the morning argument. He talks about divorce and ultimately states, "I should end this whole mess." The counselor, who is aware that divorce has been discussed during past sessions, assumes Tom intends to file for a divorce. Tom eventually settles down and the session ends.

09:30 a.m.
Jane Smith departs her apartment and goes to the university to attend an employee meeting at 10:00 a.m. Activities on the campus are preoccupied with preparations for the home football game scheduled for that evening. The weather is clear and hot. Numerous tailgaters are congregated in various designated parking areas. Other Game Day preparations are in the works to include those associated with the pregame festivities that occur adjacent to the stadium.

10:30 a.m.
Tom Smith departs his counselor's office and travels to the university. He intends on meeting his wife to have some dialogue. When he arrives at her office, she is not there. One of her coworkers advises she left to attend a meeting with HR regarding possible changes in her employee benefits package. The coworker states that after the meeting she was going to the student union for lunch. Tom becomes visibly upset and leaves the office. He returns to his vehicle and retrieves a jacket and his 12-gauge shotgun. He places a box of ammunition in his jacket pocket. The weapon is already loaded with five rounds of buckshot. Tom then makes his way across campus to the student union on foot. As he walks, his anger intensifies. He had no idea she was planning on changing her benefits.

11:35 a.m.
Jane has begun her lunch break with two friends from her department. They are all sitting at a centrally located table in the student union. As they begin their lunch, one of the friends notices Tom approaching their table from behind where Jane is sitting. Jane turns to greet her husband. Tom calmly continues to walk toward the table. As Tom nears the table, he reaches beneath his jacket and draws the shotgun. He then fires one round into his wife's chest. The two friends immediately dive under the table. Pandemonium erupts in the student union and Tom turns and runs out of the area.

11:37 a.m.

Across the commons area, campus police officer Adams hears what sounds like gunfire and then notices a man running from the student union carrying a long weapon. Officer Adams calls for the man to stop. Tom stops and turns toward the officer and fires a round in the officer's direction.

Key Issues

- Game Day preparations are underway on campus for a football game that evening.
- Police respond to a domestic disturbance from people who work/attend the university.
- The disgruntled husband (Tom Smith) gets an emergency appointment with his counselor.
- Smith arrives on campus at his wife's office. He has a firearm in his vehicle.
- After retrieving the firearm from his vehicle, Smith locates his wife at the student union. He shoots her and then flees the scene.
- Smith is confronted by campus police and fires a round in their direction.

Questions

Based on the information provided, participate in the discussion concerning the issues raised in Module 1. Identify any additional requirements, critical issues, decisions, and/or questions, which should be addressed at this time.

The following questions are provided as suggested general subjects you may wish to address as the discussion progresses. These questions are not meant to constitute a definitive list of concerns to be addressed, nor is there a requirement to address every question.

University Critical Incident Response Team

1. Would coordination with surrounding agencies and jurisdictions be done at this time?
2. Would the campus police have been made aware of the earlier domestic dispute?
3. How would the university president be notified of the threat? How would the university critical incident response team be notified? Would other universities in the state and surrounding area be notified?
4. Would an evacuation of the campus be ordered at this time?
5. Would this impact Game Day preparations? In what way?
6. What interagency coordination is necessary at this point?
7. Does your university ERPs address an active shooter threat? What actions would you take at your level? What other plans do you currently have that would need to be reviewed for potential implementation?

8. What factors would support a decision to alert and/or preposition selected emergency assets based on available information? How many and what assets would be included? Would you request external assistance? What type of support would you request and from whom?
9. Is your current level of emergency response training adequate? What public affairs guidance will be provided to your personnel?
10. Would an activation of the campus EOC be directed?
11. What potential response issues should be addressed at this point (e.g., communications interoperability)?
12. Are university officials aware if the local community, city, and county have resources to respond appropriately?
13. Would the medical community be alerted to take any preliminary action? Are medical facilities adequate to handle mass casualties? Is there a local trauma center? What provisions must be made to accomplish the task?
14. Will you need to ensure that mutual-aid support can be obtained if and when necessary? What must be done to ensure this response?
15. Would you request assistance? What would you ask for and from whom?
16. What security issues will arise?

Module 2: Notification and Initial Response

Friday, 11:37 a.m.–11:39 a.m.

Officer Adams immediately responds in the direction of the gunfire. While responding, Officer Adams calls, via radio, that shots have been fired at the student union.

Tom Smith, having shot his wife, begins to panic. He sees the campus police officer and fires another round in the officer's direction. He begins to run and a student attempts to obstruct him. Tom fires a round at the student and the student immediately falls to the ground.

Tom makes his way toward a multistory on-campus residence hall. He enters the residence hall lobby.

Friday, 11:41 a.m.–11:50 a.m.

Tom enters the residence hall and confronts several students. He immediately orders them to get out of the area and states that he is going to "blow himself up."

Officer Adams sees the subject enter the residence hall. The students who were confronted by Tom inform Officer Adams of the threat made by Tom to "blow himself up." Officer Adams passes this information over the radio as well as to other officers who have arrived on scene to assist. Other Officers stop to render assistance to the down student.

Friday, 11:50 a.m.–12:15 p.m.

One of the individuals leaving the residence hall informs a campus police officer that a man, armed with a rifle, was seen running into the dormitory administration area, adjacent to the lobby. The student was unsure if other people were still in that area.

Tom finds the residence hall administration office empty. He turns the shotgun on himself and fires. He dies of a self-inflicted gunshot wound to the head.

Key Issues

Officer Adams notifies dispatch, via the radio, that shots have been fired on campus.

- Officer Adams attempts to confront the armed subject/shooter.
- The subject, Tom, fires his weapon toward Officer Adams and then fires another round and shoots a student.
- Tom enters an on campus residence hall and states that he is going to "blow himself up."
- Campus police respond to the residence hall and are advised of a possible location of the subject.
- Tom turns the shotgun on himself and fires. He dies of a self-inflicted gunshot wound to the head.

Questions

Based on the information provided, participate in the discussion concerning the issues raised in Module 2. Identify any additional requirements, critical issues, decisions, and/or questions, which should be addressed at this time.

The following questions are provided as suggested general subjects you may wish to address as the discussion progresses. These questions are not meant to constitute a definitive list of concerns to be addressed, nor is there a requirement to address every question.

University Critical Incident Response Team

1. Who is in charge based on the current circumstances?
2. What resources do you anticipate will be needed from various local agencies, including Public Works, Parks and Recreation, Transportation Authority, Public Water, community service groups, and other mutual-aid or private resources? How will these resources be integrated into the response? Will there be difficulties getting resources from other municipalities?
3. Would local elected officials be notified? Who would notify these officials and how much information are they provided?

4. Would an evacuation be ordered at this time?
5. What are potential limitations in your emergency response capabilities? Will any limitations be placed on the actions of campus law enforcement personnel? What are your alternatives?
6. What capabilities do you have to conduct bomb threat and rescue operations?
7. How will responding organizations coordinate?
8. What level of response would you activate? Where would additional personnel be staged? How would transportation issues be handled?
9. What security concerns would you have as a result of this incident regarding Game Day preparations?
10. What dispatch protocols are in place for this type of incident? What mutual-aid resources could be activated?
11. Do you have a callback plan to meet needs such as this? Do you have communications interoperability capabilities with other agencies possibly involved in this incident?
12. What actions would be taken for traffic and access control in the affected area? Who would perform the functions? How would you notify the public/students?
13. How will information be exchanged to support decision making?
14. What specialized federal, state, mutual-aid, or military support do you anticipate possibly needing?
15. Are communication systems adequate if commercial and cellular systems experience overload? Does a backup communications plan exist?
16. Would coordination with surrounding communities be done at this time? Would any consideration be made for EOC activation?
17. How are you communicating with medical treatment centers or hospitals?
18. What actions would be taken to protect the public at this point? Who would perform the functions? How would you notify the public?
19. What is the closest medical facility to the incident site? Is the facility adequate?
20. What is the protocol for a shooting on campus? Would officers immediately enter the residence hall in pursuit of the shooter?

Module 3: Continued Response/Evacuation and Recovery

Recovery/Remediation

The following priorities for the postincident phase of the crisis are identified:

- Mitigating further incidents
- Treatment and care of casualties
- Public information
- Individual and family assistance
- Site restoration

- Volunteer management
- Critical incident stress management (CISM)
- Business resumption and recovery

Key Issues

- *Prevention/deterrence/protection*: The planning and execution of this event would require significant interagency coordination.
- *Emergency assessment/diagnosis*: Actions require the delivery and distribution of vital supplies and resources.
- *Emergency management response*: Actions required include search and rescue, alerts, activation and notification, traffic and access control, protection of special populations, resource support, requests for assistance, and public information/media. Establishment of an IPC and EOC is necessary.
- *Incident/hazard mitigation*: Primary hazards include possible additional subjects, secondary threats/devices, crowd control, and traffic control.
- *Public protection*: Evacuation is required as well as additional threat assessments. The areas must be secured and cordoned.
- *Victim care*: Injuries range from gunshot victims to mental/physical trauma.
- *Recovery/remediation*: Must be coordinated with search and recovery efforts. Restoration of public confidence could take months.

Questions

Based on the information provided, participate in the discussion concerning the issues raised in Module 3. Identify any additional requirements, critical issues, decisions, and/or questions, which should be addressed at this time.

The following questions are provided as suggested general subjects you may wish to address as the discussion progresses. These questions are not meant to constitute a definitive list of concerns to be addressed, nor is there a requirement to address every question.

University Critical Incident Response Team

1. Who is in charge of the incident at this point?
2. Would the university president be kept informed about the situation? Where would this occur? How will personal property of victims be treated?
3. What protocols will be used for contact with victims' families?
4. What local resources are available for recovery?
5. What actions are necessary to restore preincident capabilities? Who will fund these measures? What process will be followed?
6. What critical incident stress support would you consider? How long would you need to use these assets? Are sufficient assets locally available?

7. Are current record-keeping requirements adequate for an event of this magnitude during the short term and the long term?
8. Is there a mechanism for updating plans, policies, and procedures as a result of this incident? Who is responsible for coordinating these changes?
9. What contingencies are in place for adapting to a response plagued by communication overload? What is your media strategy at this time? What plans are in place to establish a joint information center (JIC)? How would they be implemented?
10. What advisories would be issued to preclude widespread panic? Who would coordinate and issue these advisories?
11. What actions are necessary to restore preincident capabilities? How would you expect supplies to be replenished? Who will fund these measures?
12. Is there a plan for demobilization and release of assets? How would public inquiries be managed during the long term?
13. Is there a plan or policy in place to reconstitute supplies used in the incident?
14. Is postincident counseling available for victims and emergency responders?
15. What are 10 immediate recovery issues that need to be addressed?

- Housing
- Transportation
- Parking
- Classroom space
- Law enforcement
- Administrative space
- Media relations
- Financial
- Registration
- Educational

Acronyms

AAC	After action conference
AAR	After action report
AAR/IP	After action report/improvement plan
CPX	Command post exercise
C/E handbook	Controller/evaluator handbook
EEG	Exercise evaluation guide
ENDEX	End exercise
EOP	Emergency operation plan
ExPlan	Exercise plan
FE	Functional exercise
FPC	Final planning conference

FSE	Full-scale exercise
FTX	Field training exercise
HSEEP	Homeland Security Exercise and Evaluation Program
ICS	Incident Command System
IPC	Initial planning conference
MPC	Midterm planning conference
MSEL	Master scenario event list
NIMS	National Incident Management System
POC	Point of contact
PPE	Personal protective equipment
SitMan	Situation manual
SME	Subject matter expert
SOP	Standard operating procedure
TCL	Target capabilities list
TTX	Tabletop exercise
UCS	Unified Command System
UTL	Universal task list
VIP	Very important person

References

Active Shooter Booklet, U.S. Department of Homeland Security Active Shooter Response (October 2008).

Federal Emergency Management Agency, Emergency Management Institute. 2008a. Introduction to exercise design. In *Exercise Design: IS-139.*

Hall, S. A., Cooper, W. E., Marciani, L., and McGee, J. A. 2012. *Security Management for Sport and Special Events—An Interagency Approach to Creating Safe Facilities.* United States: Human Kinetics.

U.S. Department of Homeland Security. 2007a. *Homeland Security Exercise and Evaluation Program (HSEEP),* Volume I.

U.S. Department of Homeland Security. 2007b. *Homeland Security Exercise and Evaluation Program (HSEEP),* Volume II.

U.S. Department of Homeland Security. 12/18/2009. *Homeland Security Exercise and Evaluation Program (HSEEP), Master Task List for Discussion-Based Exercises,* Volume IV. https://hseep.dhs.gov/hseep_vols/allDocs.aspx?a=M. (Accessed August 9, 2012).

U.S. Department of Homeland Security. 10/07/2009. *Homeland Security Exercise and Evaluation Program (HSEEP), Master Task List for Operations-Based Exercises,* Volume IV. https://hseep.dhs.gov/hseep_vols/default1.aspx?url...?q... (Accessed August 9, 2012).

U.S. Department of Homeland Security. *Homeland Security Exercise and Evaluation Program Terminology, Methodology, and Compliance Guidelines.* https://hseep.dhs.gov/support/HSEEP_101.pdf. (Accessed August 9, 2012).

U.S. Department of Homeland Security. *Homeland Security Planning For Campus Executives Participant Guide.* www.hsp.wvu.edu/r/download/27354. (Accessed August 7, 2012).

U.S. Department of Homeland Security. *National Incident Management System* (March 1, 2004).

VI

VULNERABILITY ASSESSMENTS AND CRITICAL INFRASTRUCTURE

Chapter 21

Determining Your Impacts

Impact Assessment Teams

S. Shane Stovall

Introduction

Following a disaster, it is critical to be able to figure out how severely a jurisdiction or organization has been affected. This is true both in the public sector, as well as in the private sector. Information that is needed includes: (1) What areas have been affected (this can be on a community basis in the public sector or a departmental basis in the private sector)? (2) Where do resources need to be prioritized response and recovery? Answers to these questions are important not only for emergency managers and business continuity professionals to be able to handle immediate needs, but also the information can be used to request assistance to supplement local forces. Such tasks can be accomplished effectively by performing a multidisciplinary impact assessment. In this chapter, topics will be covered such as what an impact assessment is, how to develop an impact assessment team, why impact assessment teams are needed, and how to staff, equip, and train an impact assessment team.

Impact Assessment Defined

Oftentimes, there is confusion as to what an impact assessment is. Many confuse the term "impact assessment" with "damage assessment," when in fact they are two different processes. See Table 21.1 for a description of the differences.

Damage to homes and property in Lower 9th Ward due to Hurricane Katrina. (Photo courtesy of Andrea Booher/Federal Emergency Management Agency, September 18, 2005.)

For the purposes of this chapter, we are going to focus on "impact assessment," or the process of examining the overall effects of a disaster on a jurisdiction during the first 72 h following a disaster. During the impact assessment, there are many questions to answer in order to build a true picture of how a disaster has affected a jurisdiction or a business. Some of the questions that may be asked include:

- Do we have power/electricity? If not, where is it out?
- Do we have water? If not, where is water service interrupted? Is available water potable?
- Do we have natural gas? If so, is it leaking anywhere?
- Are all areas accessible? Are there any isolated areas?
- Is there going to be a need for human services (shelter, mass feeding, comfort stations)?
- Have critical processes been interrupted or destroyed?

Table 21.1 Assessment Types

Assessment Type	*Duration of Assessment*	*Responsible Party*
Impact assessment	First 72 h following a disaster.	Affected jurisdiction
Damage assessment	Depends on severity of incident. Usually 1–2 weeks. More detailed than impact assessment.	Affected jurisdiction

Table 21.2 Examples of Public Sector Critical Facilities

Public Sector Critical Facilities	
Communication facilities (911 dispatch, radio towers, cellular towers, etc.)	Community centers (possibly used for disaster recovery centers or mass feeding sites)
Police stations	Telephone switching stations
Fire stations	Power infrastructure
Emergency operation centers	Main road arteries and bridges
Healthcare facilities	Animal control facilities and locations housing large number of animals (zoos, shelters, etc.)
Shelter facilities	Schools (public and private)

Table 21.3 Examples of Private Sector Critical Facilities and Services

Private Sector Critical Facilities and Services	
Data processing centers	Customer relations functions
Payroll	Manufacturing processes
Accounts receivable and payable	Security/proprietary information or product control
Telecommunications	Reporting processes
Quality control functions	Network/database functions
Asset/inventory management	Operational processes

Additionally, it is important that jurisdictions get an idea of how their critical facilities are affected. Critical facilities include those locations (both public and private sector) that provide a critical service to the community. Some examples of critical facilities and process can be found in Table 21.2 for the public sector and Table 21.3 (although it can be argued that there are some cross-over functions between the two sectors).

Need for Impact Assessment Teams

The information collected from impact assessments can provide useful information as to where resources may need to be prioritized to focus the disaster response, or to begin the disaster recovery process. In the public sector, information derived from impact assessments can also assist officials in determining what sort of assistance to request through mutual aid processes. Such information can also aid in obtaining a local state of emergency (LSE), a state emergency declaration, and/or a presidential emergency or disaster declaration. Impact assessment teams often serve as the initial "eyes and ears" of officials and staff who are trying to obtain a comprehensive idea of how a disaster has affected an area or an organization.

It is also important to consider developing a jurisdictional-based impact assessment team in the public sector, or an organizational-based impact assessment team in the private sector. There are many benefits in using this approach for impact assessments. One advantage is to understand that the people who work or live in the affected areas typically know the demographics and the geographic layout of the affected jurisdiction or organization, as well as the location of critical facilities and functions. Relying on someone from an outside agency or organization to conduct and impact assessment can potentially lead to delays and misunderstandings due to learning curves and a lack of knowledge on locations and functions of what is being assessed. Therefore, it is highly recommended that a local impact assessment team be developed.

Missions and Functions of Impact Assessment Teams

Before an impact assessment team is developed, it must be determined what missions, or functions, the teams will be assigned. This must be considered very carefully, for mistakes can be made by assigning functions that take away from the primary task of the teams, which is to provide an impact assessment. As the name mentions, impact assessment should remain the primary function of the team. In the public sector, another consideration for a function of the team may be debris clearance (not debris hauling or removal). Oftentimes, following a disaster, debris may block roads that emergency traffic will need to ingress and egress of an area. If debris clearance is considered as a mission, the organization will need to make sure that they are aware and knowledgeable of public assistance rules relating to debris under the federal emergency management agency's policies and guidelines (i.e., only dealing with debris in the public right of way).

While there are some functions that could be added to the impact assessment team's primary missions, there are others that should be strongly discouraged. Due to many jurisdictions and organizations having limited staff to participate in impact assessment, adding additional assignments for the teams to perform can cause significant delays in performing the initial impact assessment. The following are some missions that may want to be avoided:

- *Search and Rescue*—This is a time-consuming task that also takes specialized training. Teams should be able to call for someone to perform this function. Liability must also be considered.

Eureka, MO, March 22, 2008—Members of the Missouri Emergency Response Service Team, a nonprofit organization that does large animal rescues, launch a boat to take part in a large animal rescue along with the Humane Society to rescue 13 cattle that were stuck in flood waters. (Photo courtesy of Jocelyn Augustino/FEMA).

- *Human Needs Counseling*—More often than not, human needs are going to be identified during an impact assessment. Stopping to talk to survivors of the disaster can be time consuming, and should be minimized. Impact assessment teams should have the ability to refer those with needs to the appropriate person or location.
- *Damage Assessment*—As mentioned above, this is a much more detailed and time-consuming process than impact assessment, and should be avoided during the very initial phases following a disaster when impact assessment information is necessary.
- *Animal Control*—Impact assessment teams will more than likely find animals that have been displaced roaming freely. Trying to capture animals can be dangerous and time consuming. Animal control issues should be referred to another agency.

There are several different post-disaster needs that surface in both the public and private sectors following a disaster. The functions of the impact assessment teams can vary based on jurisdictional and organizational wants and needs. However, the mission of performing an initial impact assessment should never be lost sight of when determining what functions the teams shall perform.

Staffing Impact Assessment Teams

Whether in the public sector or private sector, the staffing of the impact assessment team is just as critical as the missions and functions that the teams perform. It is strongly recommended that jurisdictions and organizations take a multidisciplinary approach to staffing their teams. The multidisciplinary approach involves staffing the impact assessment teams with people from different departments and agencies. This allows for multiple skill sets and areas of expertise to be represented on the teams. There are several reasons for this approach. First, as seen above, there are many different areas to be examined and assessed in a truly comprehensive impact assessment. To depend on one agency or department with one set of skill sets to provide a comprehensive impact assessment is unrealistic and unreasonable. Additionally, there are often situations that are encountered during impact assessments that take more than one organization to assess. For example, A bridge has been washed out by a recent flood. Barricades probably will be needed to block routes leading to the bridge. There also may have been utilities that ran along the underside of the bridge that are now severed. This type of situation would require more than one skill set to assess. In the private sector, an example could include a building fire: How has it affected hardware, software, product, overall operations? All of these will take a separate skill set to assess. As can be seen, multidisciplinary impact assessment teams certainly have the best potential for being most comprehensive and effective.

Certain jurisdictions or organizations may decide upon multiple teams. This could be due to geographic boundaries, politics, or any number of different variables. However, it is important to ensure that there is consistency among multiple teams. It is not operationally sound to have one team to have a full complement of resources and capabilities, while another has very limited capability for performing assessments. This type of scenario would lead to inconsistent and incomplete assessments, which do not allow for a true comprehensive impact assessment in an organization or in a jurisdiction. Consistency across multiple teams is necessary.

Once it is determined that an impact assessment team is to be established, it must be staffed with the appropriate personnel. Keeping with the multidisciplinary approach, there are a multitude of agencies that could potentially be involved. Table 21.4 gives some examples of agencies to involve on a public sector impact assessment team, while Table 21.5 provides some examples of who private sector entities may want to include on their teams.

The types and numbers of personnel on each impact assessment team is dependent on the availability of personnel and the needs of the organization or jurisdiction. As can be seen, multidisciplinary teams can provide the most comprehensive look at the effects of a disaster.

Managing Impact Assessment Teams

As impact assessment teams are being developed, it is important to ensure that they are coordinated and managed by the appropriate personnel. The impact assessment teams should be structured to where they have an overall coordinator (especially in the instance of having multiple teams), and then managers or chiefs for the teams in the field. When possible, the coordinator of the impact assessment teams should

Table 21.4 Examples of Potential Impact Assessment Team Members for the Public Sector

Potential Impact Assessment Team Members: Public Sector	
Law enforcement	Insurance claims adjustors
Fire/emergency medical services	Power company
Public works—engineering/road and bridge	Phone company
Utilities (public and private)	Cable company
Building inspectors	Gas company
Fleet management	Communications (public and private)
Environmental health	Facilities management
Animal control	Parks and recreation
Salvation Army	American Red Cross

be someone of a neutral party. Oftentimes this can be an emergency manager who can coordinate the day-to-day activities, such as training, but who will not be in the field in the teams following a disaster (assuming that they are working in an emergency operations center following a disaster).

Whether it is the overall impact assessment team coordinator(s), or the team managers, the success of the team revolves around having proper leadership. Personality types run from one extreme to the other. Most individuals fall somewhere in between. The next few sections will describe different types of leaders and how they can affect the overall missions of the team.

Table 21.5 Examples of Potential Impact Assessment Team Members for the Private Sector

Potential Impact Assessment Team Members: Private Sector	
Security	Facilities management
Information technologies/data processing	Risk management
Business process team leaders	Production recovery team
Business continuity coordinator/planner	Telecommunications
Company executives or managers	Department/division head
Business process team leaders	Logistics managers

Management by Intimidation

These types of managers often feel it necessary to scare people into accomplishing tasks. These types of leaders also do not allow for much input from others—assuming that their way of accomplishing tasks is the only way that they can be completed. Very rarely do these managers gain the respect of their peers or the people who they are leading. This type of manager will typically get very poor performance, and thus is not the type of manager who should be coordinating or managing an impact assessment team, or any team for that matter.

Absentee Managers

These types of managers often disappear when things need to be accomplished or the going gets tough. They may give direction, but very rarely follow through. These types of leaders also do not seem to have a vested interest in what is occurring around them. These managers also do not earn the respect of others, and should not be managing an impact assessment team.

Management by Example

This type of manager is one who will get in the trenches and help others accomplish tasks, a true leader if you will. They are typically open to input, and are able to give direction and follow-up where needed. This manager will also allow others to accomplish tasks and get credit and recognition for activities. Respect is usually easily earned by this type of manager, and is the type of person that would be ideal for running an impact assessment team.

Training Impact Assessment Teams

Just as important as the staffing of the impact assessment team, is the training provided to impact assessment team members. Administrative training, safety training, impact assessment training, and cross-training can all be considered. However, before the types of training can be considered (to be discussed below), there must be a determination on how often these teams will train. Impact assessment teams should train often enough to keep knowledge fresh and comprehensive. Waiting too long between training, or not covering certain topics frequently enough, can lead to a decrease in memory of what has been taught. While a monthly training may be more ideal, such frequency may not be feasible. At a minimum impact assessment teams should train on a bi-monthly basis. With the amount of training that an impact assessment team should have, this is not considered to be too frequent.

Additionally, it is recommended that a set date and time be allotted for the team trainings (i.e., first Thursday of every month at 10 a.m.). This makes it easier for team members to schedule and remember the training dates and times. This rule helps to retain attendance and gives people little reason to "forget" meetings.

Administrative Training

All impact assessment teams should have some sort of administrative training for its members. This type of training can include rules of conduct, attendance policies, impact assessment forms, standard operating procedures, and other training such as National Incident Management System (NIMS) training. These are necessary in order to build a foundational knowledge on the expectations and conduct of the teams.

Functional Training

There are various training subjects that can prove useful for impact assessment teams and their missions and functions. There is no "one size fits all" training program for these teams due to the varying needs of each organization or jurisdiction sponsoring the teams. Training classes that can be considered include:

- First aid/CPR/AED
- Critical incident stress
- Hazardous materials awareness
- Impact assessment
- Global positioning system (GPS)
- Communications
- Power line safety (downed power lines are common following disasters)
- Cross training in functions of other departments of agencies on the team
- Equipment usage and safety
- Critical facility familiarization

Firefighters learn how to use GPS devices, City of Portland, Oregon.

SAMPLE
IMPACT ASSESSMENT FORM

INCIDENT ________________ TEAM# _____ REPORT COMPILED BY ________________

LOCATION	DESCRIPTION OF DAMAGE	MIN. (0–39%)	MAJ (39–50%)	DEST. (over 50%)	TYPE OF DAMAGE	Picture #

Sample impact assessment form.

Drills and Exercises

As part of the training curriculum for impact assessment teams, drills and exercises should be included. Annual drills are important to ensure that team members get hands on experience in a mock scenario to test certain skills and methodologies. Training programs can be built toward the execution of an exercise to allow team members to apply skills that they have recently learned.

Equipping an Impact Assessment Team

For an impact assessment team to properly complete its mission, it must be properly equipped with forms, safety equipment, and other tools and equipment to complete the tasks for which it is assigned. At the same time, there are pieces of personal equipment that individual team members should consider when assembling their "go kit." This section will provide lists of those items, and some discussion around why some items are on the list for consideration.

Personal Equipment

Some teams may be required to be self-sustaining for a period of time depending on the severity of the disaster, and the availability of resources. Most teams should consider personal items for a period of 72 h to ensure that they are personally equipped to take care of themselves. Such equipment may include:

- Clothes
- Insect repellant

Personal hygiene items, Federal Prison Industries.

- Food (nonperishable)
- Sunscreen
- Water
- Personal hygiene
- First aid kit
- Medications
- Cash

Each of these items should be considered a personal responsibility for each member of the impact assessment team.

Administrative Team Equipment

There are some basic items that an impact assessment team should be equipped with to perform their primary function of impact assessment. These include:

- Maps
  - Street maps
  - Aerial maps
  - Custom maps
  - Building plans
- GPS devices (landmarks and street signs may be damaged or destroyed, making location identification difficult)
- Laptop
- Pencils
- Forms
- Camera (video and/or still)
- Paper (blank or notebook)

(a)

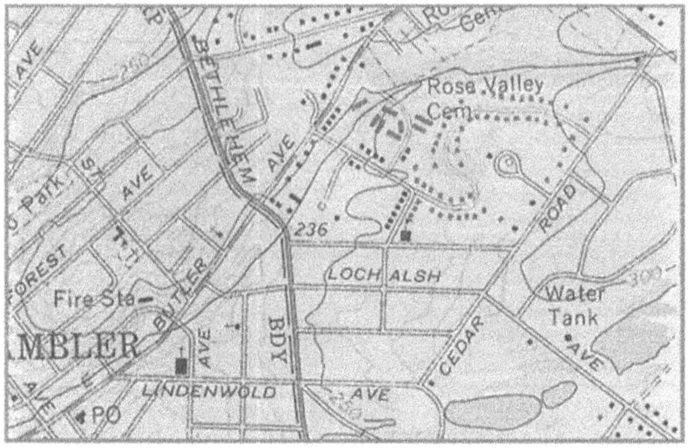

(b)

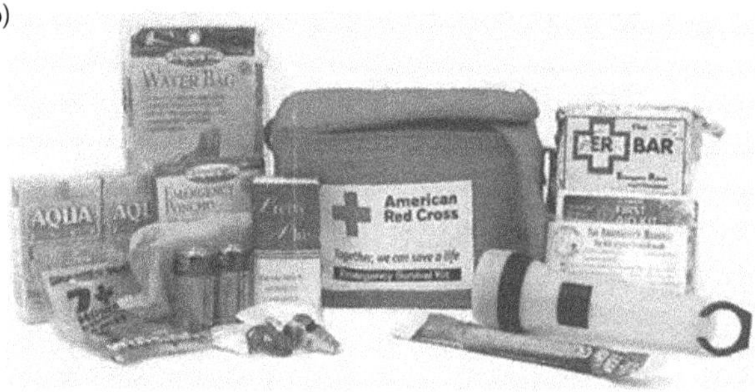

(a) Sample topographic map, the United States Geological Survey. (b) First aid kit, the United States Geological Survey.

- First aid kit
- Critical facilities list

Functional Team Equipment

The remaining functional equipment is really dependent on what additional missions that the impact assessment teams are assigned. One anticipated the need of impact assessment teams, particularly those sponsored by jurisdictions, is the ability to clear debris to maneuver through impacted areas. At a minimum, some equipment that may want to be considered (depending on available resources) may include:

- Heavy equipment
  - Front-end loader
  - Bulldozer
  - Dump truck
  - Crane

Articulating front-end loader, Center for Disease Control.

  - Backhoe
  - Boat (Jon boat, airboat, other)
- Light equipment
  - Chainsaws with pre-mix
  - Air compressor
  - Stone-cutting saws

As mentioned previously, impact assessment team equipment will vary in every jurisdiction and in every organization. The key is to ensure that the proper tools are available to allow the impact assessment teams to effectively perform their assignments.

Financing Impact Assessment Teams

Financing impact assessment teams is certainly a question that will need to be addressed prior to the development of the teams. Most organizations and jurisdictions have financial restraints that disallow them to procure equipment specifically for impact assessment efforts. Such an approach, in all actuality, is unreasonable except for possibly specialized equipment needed by the impact assessment teams. With some creative thinking, one may be able to come up with low- or no-cost methodologies for equipping their impact assessment teams. Some possibilities for equipping or finding impact assessment teams may include:

- Using existing resources within the organization or jurisdictions;
- Developing public–private partnerships (PPPs) that could assist in impact assessment team efforts (through funding or participating with personnel);
- Grant funding;
- General funding from jurisdictional or organizational budgets.

In some cases, it may be necessary to take a phased approach to funding by developing a prioritized list of items/projects that may need to be funded, and seeking funding accordingly. This can prove to be challenging, but yet something needs to be addressed when proposing the development of impact assessment teams to jurisdictional or organizational leadership.

Other Considerations for Impact Assessment Teams

In many jurisdictions and organization, there may be a variety of challenges that are encountered either when developing an impact assessment team or once the team has been developed. For the purposes of this section, we will be primarily focusing on inner-team challenges that may surface within the impact assessment teams during their development, or after they have been established.

Lack of Commitment/Lack of Interest

Once an impact assessment team is developed, it will quickly become clear which members are going to be "dead weight" to the team efforts. These people may join the impact assessment teams just to get out of some other duty, they may have been forced to volunteer for the teams, or they simply just do not understand the mission/tasks of the teams. These personnel need to be replaced, for they typically will not carry their weight, nor be an active member of the team. These people will become the weakest members of the team and will more or less become a liability to the impact assessment team efforts in a post-disaster situation.

Lack of Planning/Training

Impact assessment teams do not establish themselves, nor do they maintain themselves. To have effective impact assessment teams, it is critical that time, planning, and training have been invested into the teams. Lack of planning for the teams, or a lack of training of the teams, will lead to an ineffective impact assessment that could ultimately be a waste of money, man-hours, and effort. As mentioned earlier in this chapter, impact assessment teams must be trained regularly, and must drill regularly, in order to be effective.

Lack of Post-Disaster Critical Incident Stress Debriefing

During a disaster, and immediately following a disaster, there are sights, sounds, smells, and activities that easily act together to overwhelm the senses and the mental well-being of responders. The impact assessment team members are included in this group. Many first responders, following a traumatic incident, are required to go through a critical incident stress debriefing. These debriefings are designed to allow first responders to talk about what they experienced and how they personally feel about what they saw, heard, and so on. The intent is to counsel responders through their feelings in an effort to limit posttraumatic stress. This is very critical to the well-being of the individual team members, as well as to the team as a whole. Ensure that steps have been taken so that such debriefings are available for team members following a deployment during a disaster. Contact your local fire or law enforcement agencies to see who they may have to do their critical incident stress debriefings.

Summary

In this chapter, the need for impact assessment teams has been explained, along with the elements of what's needed to plan for, equip, fund, and train an impact assessment team. As mentioned, there is no "one size fits all" approach to developing or training an impact assessment team. However, it is believed that the information in this chapter will provide food for thought for organizations in the private and volunteer sectors, and for jurisdictional entities in the public sector. Having an impact assessment team will provide an effective tool and source for the collection of information following a large-scale emergency or disaster. Additionally, having such a team will take another step toward the resilience and self-reliance that communities and organizations should ultimately strive for during disaster response and recovery operations.

Chapter 22

Vulnerability Assessments

James Peerenboom, Ronald E. Fisher, and Wade Townsend

Introduction

The Homeland Security Act of 2002 provides the primary authority for the overall homeland security mission in the United States. This act charged the Department of Homeland Security (DHS) with the responsibility for developing a comprehensive national plan to secure critical infrastructure (CI) and key resources and recommended "the measures necessary to protect the key resources and critical infrastructure of the United States." This comprehensive plan is the National Infrastructure Protection Plan (NIPP), first published by the DHS in June 2006 and updated in 2009 (DHS, 2006). As defined in the 2009 NIPP, CI are the systems and assets, whether physical or virtual, so vital that the incapacity or destruction of such may have a debilitating impact on the security, economy, public health or safety, environment, or any combination of these matters, across any federal, state, regional, territorial, or local jurisdiction (DHS, 2009). Key resources are publicly or privately controlled resources essential to the minimal operations of the economy and government. The NIPP provides the unifying structure for integrating a wide range of efforts for the protection of CI into a single national program.

Homeland Security Presidential Directive/HSPD-7, *Critical Infrastructure Identification, Prioritization, and Protection,* was established as a national policy for federal departments and agencies to identify and prioritize United States CI and to protect them from terrorist attacks (White House, 2003). The NIPP provided the follow-up plan to implement HSPD-7. The NIPP called out the need to conduct risk assessments to deter threats, mitigate vulnerabilities, and minimize consequences.

The White House released an update on this guidance with *Presidential Policy Directive 21—Critical Infrastructure Security and Resilience.* This policy directive calls out the need to conduct comprehensive assessments of the vulnerabilities of the Nation's critical infrastructure in coordination with stakeholders and infrastructure owners and operators.

Vulnerability Assessment

Vulnerability assessment methodologies are generally intended to identify any weakness that can be exploited by an adversary to gain unauthorized access to or to disrupt an asset, facility, or system. Terrorism is often the primary focus; however, vulnerabilities can take an all hazards approach. Vulnerabilities can result from, but are not limited to, the following:

- Asset, building, site, or system characteristics
- Equipment properties
- Personal behavior
- Operational and personnel practices
- Security weaknesses (physical and cyber)

Vulnerability assessment methodologies can be characterized in terms of four assessment elements—physical, cyber, operations security (OPSEC), and interdependencies. Each of the above is briefly described below.

Physical. A physical security assessment typically evaluates the physical security systems in place or planned at a site, including access controls, barriers, locks and keys, badges and passes, intrusion detection devices and associated alarm reporting and display, closed circuit television (CCTV) (assessment and surveillance), communications equipment (telephone, two-way radio, intercom, cellular), lighting (interior and exterior), power sources (line, battery, generator), inventory control, postings (signs), security system wiring, and protective force. These systems are generally reviewed for design, installation, operation, maintenance, and testing. It may also include an evaluation of sites housing critical equipment or information assets or networks dedicated to the operation of the physical systems.

Cyber. A cyber security assessment evaluates the security features of the information network(s) associated with an organization's critical information systems. This could include an examination of network topology and connectivity, principal information assets, interface and communications protocols, function and linkage of major software and hardware components (especially those associated with information security such as intrusion detectors), and policies and procedures that govern security features of the network. It may also include internal and external scanning for vulnerabilities (penetration testing).

Operations security. OPSEC is the systematic process of denying potential adversaries information about capabilities and intentions of the host organization. This is accomplished by identifying, controlling, and protecting generally nonsensitive activities concerning planning and execution of sensitive activities. An OPSEC assessment typically reviews the processes and practices employed for denying adversary access to sensitive and nonsensitive information that might inappropriately aid or abet any individual's or organization's disproportionate influence over system operation. This should include a review of security training and awareness programs, a review of personnel policies and procedures, discussions with key staff, and tours of appropriate principal facilities. It should also include a review of information that may be available through public access (e.g., the Internet).

Interdependencies. Infrastructure interdependencies refer to the physical and electronic (cyber) linkages within and among our nation's CI (i.e., within and among the 16 CI defined in the NIPP). An interdependency assessment typically identifies the direct infrastructure linkages between and among both the internal infrastructures at a site as well as the linkages to external infrastructures outside the site. The process of identifying and analyzing these linkages requires a detailed understanding of how the components of each infrastructure and their associated functions or activities depend on, or are supported by, each of the other infrastructures. For example, a SCADA (supervisory control and data acquisition) system that operates a natural gas pipeline depends on the local electric power and telecommunications infrastructures to function. The failure of a separate, external infrastructure could prevent the SCADA system from operating, thus impacting natural gas deliveries to or within a system. Interdependencies can create subtle interactions and feedback mechanisms that often lead to unintended behaviors and consequences, including the potential disruption of CIs.

Methodological Approaches to Vulnerability Assessment

Attacks on CI could significantly disrupt the functioning of government and business alike and produce cascading effects far beyond the targeted sector and physical location of the incident (see Figure 22.1). Direct terrorist attacks and natural, manmade, or technological hazards could produce catastrophic losses in terms of human casualties, property destruction, and economic effects, as well as profound damage to public morale and confidence. Finally, attacks using components of the nation's CI as weapons of mass destruction could have even more devastating physical and psychological consequences.

There are different threat approaches as part of vulnerability assessment methodologies. These threat approaches can be summarized by two main approaches: asset based and scenario based. The asset-based approach examines the impact on

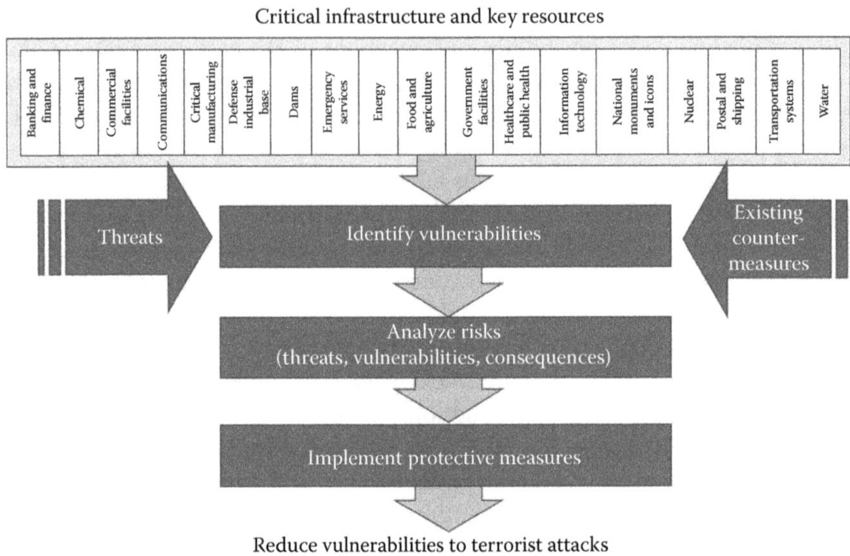

Figure 22.1 Protective security process for CIKR.

individual assets (e.g., loading dock, main lobby) if attacked, while the scenario-based approach considers multiple potential specific sequences of events (e.g., damage or destroy building) (Survey, 2002).

Vulnerability assessment methodologies are categorized as using one (or more) of the following approaches: checklist, simple rating, risk matrix, or risk equation. In some cases, a methodology may be a hybrid that incorporates elements of multiple approaches, or a range of approach options may be available to the assessor. A description of each approach category is provided below.

Checklist

Checklist-based vulnerability assessments are the simplest methodological approach. They consist of a list of questions or criteria against which the assessor compares the characteristics of the facility or asset being evaluated. The checklists may be grouped according to the various assessment elements (e.g., physical, cyber, OPSEC, and interdependencies). The questions may be answered "yes" or "no" or may require a qualitative response. Generally, if the answer to a question is "no," or if the criterion is not met, a recommended action will be requested or required. An example of a limited checklist methodology is in Table 22.1 and includes questions that might be asked at public health facilities.

On the basis of the understanding that a methodology is written documentation of a systematic process to be used by specialized teams or even normal employees

Table 22.1 Example Checklist Items for Public Health Facilities

Category	*Example Question*
Access control	Are employees required to have identification and is that required to provide access to sensitive work areas?
Barriers	Is there adequate perimeter fencing?
Monitoring and surveillance	Is there an intrusion detection system?
Communications	Is there a security awareness program for employees?
Inspection	Is there a vehicle inspection program for visitor vehicles?
Security force	Are regular and random patrols conducted and if so how often?
Cyber security	Does your water system have a procedure to deal with public information request, and to restrict distribution of sensitive information?

to conduct an assessment of the risk or vulnerabilities of an asset or facility, certain other forms of tools or guidance documents were not included in this survey. Guidance documents may present a general discussion on the objectives or scope of a vulnerability, risk, or security assessment; may provide typical security measures or criteria for various types of facilities or assets; or may outline potential mitigation measures for various types of vulnerabilities or assets. They do not, however, provide a step-by-step systematic, documented process to be followed in order to conduct a vulnerability assessment and/or do not contain necessary specific evaluation techniques (e.g., checklists, ranking scales, matrix categorization, or quantifiable equations). There are also assessment tools, which can be used to support vulnerability assessments, but they themselves are not assessment methodologies. These would include software platforms and formats that can store and manipulate data or predict impact severity.

Simple Rating

Many vulnerability assessment methodologies prioritize asset vulnerabilities for potential corrective action by defining a set of measurable criteria, rating each asset (and the associated vulnerability) on each criterion, and qualitatively or quantitatively combining the individual ratings. An example of a basic, broadly applicable rating approach is a target analysis process developed and practiced by special operations forces. This process, called CARVER analysis, has been adapted and used as part of numerous vulnerability assessment methodologies. CARVER is an acronym that stands for criticality, accessibility, recoverability, vulnerability, effect, and recognizability. Each factor in the acronym typically has an associated scale (e.g., a 10-point scale), and individual assets (i.e., potential targets) are numerically rated on each factor. A rank-order of critical assets is established on the basis of the overall CARVER score (determined by summing the points assigned to the individual factors). Other "rating and weighting" schemes also are used to provide a logical and consistent basis for prioritizing vulnerabilities for importance or potential corrective actions.

Risk Matrix

A risk matrix is often used to focus vulnerability assessment results and help categorize the assets, sites, and/or systems assessed into discrete levels of risk so that appropriate protection and mitigation measures can be applied. Figure 22.2 shows a typical risk matrix, which conveys the notion that risk is a function of event severity (i.e., the severity of consequences) and the likelihood of its occurrence. Likelihood is often determined by considering the attractiveness of the targeted assets, the degree of threat, and the degree of vulnerability.

As depicted in Figure 22.2, asset vulnerabilities that have the highest likelihood of being successfully exploited (i.e., frequent) and that result in the highest severity (i.e., catastrophic) have the highest priority for vulnerability reduction actions and protective measures to mitigate the risks. Similarly, asset vulnerabilities with the lowest likelihood of being exploited (i.e., unlikely) and that result in the lowest

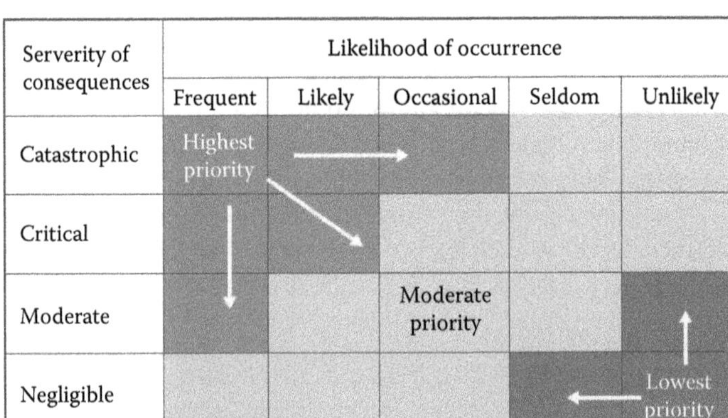

Figure 22.2 Illustrative risk matrix.

severity (i.e., negligible) have the lowest priority for mitigation. Many variations of this basic approach are used with different numbers of severity and likelihood levels, as well as definitions for those levels, to assist in focusing on the highest priority risks.

Risk Equation

Some methodologies seek a single measure that allows comparison of alternative countermeasures. Such a measure can be used to "rank order" or "prioritize" countermeasures. One approach is to calculate a risk number that is a function of probability of attack, system effectiveness, and consequence. A simple formula for risk defined in this way is

$$\text{Risk} = R = P_A * (1 - P_E) * C$$

where

P_A = likelihood of occurrence (attack)
P_E = system effectiveness [therefore, $(1 - P_E)$ = system ineffectiveness]
C = consequence value

In some approaches, a consequence value is directly addressed by assigning, for example, a low, medium, or high value. Similarly, likelihoods of occurrence and system effectiveness are combined by assigning, for example, a low, medium, or high value. Finally, to calculate a "risk value," one converts low, medium, and high assignments to numerical values (e.g., 0.1, 0.5, and 0.9), and inserts the numbers into the risk equation to calculate a risk value. Or, numerical values are determined by constructing and running models that yield likelihoods and consequences.

Another variation calls for specification of several consequences and corresponding measures (e.g., economic loss in dollars, duration of loss in hours, number of customers impacted, fatalities, and illnesses). These must then be combined in some fashion to construct a single measure of consequence value. This can be done by considering all of the measures and simply assigning a single measure that represents the set of measures (e.g., "high" for economic loss, "medium" for duration of loss, and "low" for number of customers impacted may be rated as "medium" overall).

An even more thorough approach is to carefully consider the ranges that may be obtained for each of the measures and assign "weights," which are used to construct a function that yields an overall measure of the desirability of each countermeasure (or portfolio of countermeasures). This function is sometimes called a "utility function." In this approach, the measure of risk is a "utility value" and high values are preferred over low values. Therefore, the portfolio of countermeasures that yields the largest expected utility is, by definition, the most desirable (best).

Required Expertise

To carry out a vulnerability assessment, a team of experts typically needs to visit the facility to ascertain the vulnerabilities of the critical assets and the anticipated results that would be caused by their physical destruction or impairment. Depending on the objectives and scope of the assessment, VA teams may include the following types of experts:

- Physical security experts who focus on the physical security of the facility, including access controls, barriers, locks and keys, badges and passes, intrusion detection devices and associated alarm reporting and display, closed circuit television (CCTV) assessment and surveillance, communications equipment, lighting, postings, security systems wiring, and protective force personnel.
- Explosive ordnance disposal (EOD) experts who examine and evaluate the vulnerability of critical assets to attacks that involve explosive devices of all kinds, including vulnerabilities to vehicle-delivered explosives and small charges.
- Assault planning experts who focus on terrorist strategies for the most likely method of attack, including physical security vulnerabilities (e.g., fencing or CCTV gaps) and outside surveillance/positioning vulnerabilities (e.g., areas of cover for clandestine operations and positions for using long-range weapons).
- Infrastructure systems experts who calculate the anticipated results of the loss of the asset as it pertains to the facility and the loss of the facility as it pertains to its specific infrastructure.
- Interdependencies experts who evaluate the dependence of the facility on outside infrastructures, such as electric power, water (potable and process) and wastewater, natural gas, steam, petroleum products, telecommunications, transportation (e.g., roads, railroads, and marine links), and banking and finance.
- Operation security experts who evaluate human resources security procedures, facility engineering, facility operations, administrative support organizations, telecommunications and information technologies, publicly released information, and trash and waste handling.
- Intelligence operations experts who interact with local law enforcement and intelligence/security personnel to determine if there are potential terrorist or criminal elements in the region who may have an interest in the facility.

Outline of Risk Management Steps

This section presents an outline of the risk management process which has been adapted from Department of Energy (DOE) and DHS methodologies. Table 22.2 provides an overview of representative steps in a comprehensive, asset-based vulnerability assessment methodology. This includes countermeasure (actions taken to reduce or eliminate vulnerabilities) and risk management considerations. The

Table 22.2 General Vulnerability Assessment Process

Step	*Description*	*Considerations*
1	Identify critical assets and the impacts of their loss	• Identify the critical functions of the facility. • Determine which assets perform or support the critical functions. • Evaluate the consequences or impacts to the critical functions of the facility from the disruption or loss of each of these critical assets.
2	Identify what protects and supports the critical assets	• Identify the components of the physical security system (e.g., perimeter barriers, building barriers, intrusion detection, access controls, and security forces) that protect each asset. • Identify the critical internal and external infrastructures (e.g., electric power, petroleum fuels, natural gas, telecommunications, transportation, water, emergency services, computer systems, air handling systems, fire systems, and SCADA systems) that support the critical operations of each asset (interdependencies). • Identify sensitive information about the facility and its operation that must be protected.
3	Identify and characterize the threat	• Gather threat information and identify threat categories and potential adversaries. • Identify the types of threat-related undesirable events or incidents that might be initiated by each threat or adversary. • Estimate the frequency or likelihood of each threat-related undesirable event or incident based on historical information. • Estimate the degree of threat to each critical asset for each threat-related undesirable event or incident.
4	Identify and analyze vulnerabilities	• Identify the potential vulnerabilities of each asset to each threat or adversary. • Identify the existing measures intended to protect the critical assets and estimate their levels of effectiveness in reducing the vulnerabilities of each asset to each threat or adversary. (Step 2 provides a starting point for this activity.) • Estimate the degree of vulnerability of each critical asset for each threat-related undesirable event or incident and thus each threat or adversary.

(*Continued*)

Table 22.2 (continued) General Vulnerability Assessment Process

Step	*Description*	*Considerations*
5	Assess risk and determine priorities for asset protection	• Estimate the effect on each critical asset from each threat or adversary taking into account existing protective measures and their levels of effectiveness. • Determine the relative degree of risk to the facility in terms of the expected effect on each critical asset (a function of the consequences or impacts to the critical functions of the facility from the disruption or loss of the critical asset, as evaluated in Step 1) and the likelihood of a successful attack (a function of the threat or adversary, as evaluated in Step 3, and the degree of vulnerability of the asset, as evaluated in Step 4). • Prioritize the risks based on the relative degrees of risk and the likelihoods of successful attacks using an integrated assessment.
6	Identify mitigation options, costs, and trade-offs	• Identify potential mitigation options to further reduce the vulnerabilities and thus the risks. • Identify the capabilities and effectiveness of these mitigation options. • Identify the costs of the mitigation options. • Conduct a cost–benefit and trade-off analyses for the various options. • Prioritize the alternatives for implementing the various options and prepare recommendations for decision makers.

methodologies included in this survey address, to a greater or lesser degree, some or all of these steps.

The following sections describe the steps of the risk management process in more detail. Where appropriate, the steps contain checklists of questions that could be used to guide the implementation of a risk management program (Energy Infrastructure, 2002).

Step 1. Identify Critical Assets and the Impacts of Their Loss

Estimates of the potential consequences, including economic implications, of not mitigating identified vulnerabilities or addressing security concerns are necessary to effectively apply risk management approaches to evaluate mitigation option and security recommendations. Outages because of security failures could degrade public health facility's reputation and place the community served at risk to economic losses or even losses of property and life.

In addition, the modern public health facility's telecommunication and computer network have many external connections to public and private networks. Such connections are used to communicate with customers and offer new electronic services, such as online billing and payment. Cyber security should be a primary concern, especially for utilities that operate in this interconnected environment. An information technology (IT) security architecture may need to be developed.

Possible critical assets include people, equipment, material, information, installations, and activities that have a positive value to an organization or facility. People include public health facility executives and managers, security personnel, contractors and vendors, and field personnel. Equipment includes vehicles and other transportation equipment, maintenance equipment, operational equipment, security equipment, and IT equipment (computers and servers). Material includes tools, spare parts, and specialized supplies. Information includes employee records, security plans, asset lists, intellectual property, patents, engineering drawings and specifications, system capabilities and vulnerabilities, financial data, and operating, emergency, and contingency procedures. In addition to the operational installations that make up the energy infrastructure itself, installations include headquarters offices, field offices, training centers, contractor installations, and testing, research, and development laboratories. Activities include movement of personnel and property, training programs, communications and networking, negotiations, and technology research and development.

The various public health facilities, the local government, and public health associations have roles and responsibilities for identifying assets, effects of asset loss, vulnerabilities, threats, and risk mitigation options. Coordination among public health facilities; local, state, and federal agencies; and public health associations is crucial to this process.

Public health facilities need to identify the critical functions of the facility, and determine which physical and cyber assets perform or support the critical functions. The key assets identified should be related to the criticality of overall operations of the individual facilities. Potential assets for public health facilities include emergency rooms, operating rooms, laboratories, medicines, blood supplies, manufacturing facilities, waiting areas, patient care areas, and transport vehicles. They should evaluate the consequences or impacts to the critical functions of the public health facility from the disruption or loss of each of these critical assets and prioritize the critical assets based on these.

Not all assets and activities warrant the same level of protection. The cost of reducing risk to an asset must be reasonable in relation to its overall value. The value, however, does not need to be expressed in dollars. A potential loss can be stated in terms of human lives or the impact on the local or state economy.

The first set of questions is designed to guide the process of identifying the critical functions of the public health facility and the assets that perform or support them, and evaluating the potential consequences of disruptions or loss of these critical assets.

Criticality criteria (functions and assets)

- What critical mission activities take place at the public health facility or its remote sites?
- Could these critical mission activities be conducted elsewhere if this asset or entire facility was unable to operate?
- What critical or valuable equipment is present at the facility or its remote sites?
- Where are the critical assets located?
- Have people, installations, and operations been considered when assessing assets?
- Have cyber networks and system architectures (e.g., SCADA systems, business e-mail, and e-commerce) been documented fully?

Criticality criteria (impacts of loss)

- What affects would be expected if a specific asset were compromised?
- What is the potential for immediate and significant local impacts due to the loss of the asset?
- What facility personnel, tenants, customers, and visitors could be affected by the loss of the asset?
- What would be the impact on people's lives and on national or local security due to the loss of the asset?
- What would be the financial impacts to the public health facility and the local community?

Criticality criteria (asset value)

- Is there little or no redundant capacity or capability to mitigate the loss of the asset?
- What is the potential for cascading effects (e.g., to other interdependent infrastructures or industries) due to the loss of the asset?
- Do any special situations need to be considered regarding the loss of the asset, such as the status of other public health facilities (e.g., hospitals, life support systems, or emergency services) that depend on this public health facility?
- What is the potential for catastrophic effects (weapons of mass destruction levels impact)?
- What did it cost to develop the asset?
- Would the public health facility need an extended period to make repairs to the asset?
- How does the need for protecting the asset compare with other assets also considered critical?

Once the assets critical to the operation of the public health facility have been identified and characterized, an impact assessment must be carried out to describe the consequences of losses if an undesirable event occurs. The degree of impact should be quantified by using a relative impact or criticality rating criteria and a consistent rating scale. (An example of a scale for rating criteria is presented in Step 5.) The assets are then ranked in terms of criticality.

Step 2. Identify What Protects and Supports the Critical Assets

The existing protection of critical assets provided by the physical security system and the dependence of the critical assets on both external and internal infrastructures must be known to evaluate the vulnerabilities of the assets to threats or adversaries. In addition, operating procedures and other sensitive information, which if available to adversaries might jeopardize critical assets, must be identified as their availability can also affect the vulnerabilities of assets.

Physical Security Systems

Physical security systems are used to protect public health facilities and their assets from unauthorized individuals and outside attacks. Such systems usually include perimeter barriers, building barriers, intrusion detection, access controls, and security forces.

Infrastructure Interdependencies

Today's public health facilities depend on many different infrastructures to support their critical functions and assets. These infrastructure interdependencies must be identified and the adequacy of security measures that are in place to protect and back up these infrastructures must be evaluated. Typically, these supporting infrastructures include

- Electric power supply and distribution
- Petroleum fuels supply and storage
- Natural gas supply
- Telecommunications
- Transportation (road, rail, air, and water)
- Water and wastewater
- Emergency services (fire, police, and emergency medical)
- Computers and servers
- Heating, ventilation, and air conditioning (HVAC) systems
- Fire suppression and fire fighting systems
- SCADA systems

The electric power supply and distribution infrastructure can include the local electrical distribution utility, facility-operated electric generation equipment, backup generators fueled by natural gas or petroleum fuels, uninterruptible power supplies (UPSs), and the associated switching and distribution hardware. The petroleum fuels supply and storage infrastructure include on-site storage as well as local suppliers, storage terminals, and the entire petroleum industry. The telecommunications infrastructure includes commercial telephone, fiber optic, and satellite networks and facility-owned radio, telephone, microwave, and fiber-optic pathways. Computers and servers, HVAC systems, fire suppression and fire fighting systems, and SCADA systems tend to be operated by the public health facility and, in turn, depend on the other infrastructures such as telecommunication, electric power supply and distribution, petroleum fuels supply and storage, natural gas supply, water and wastewater, and emergency services.

Sensitive Information

Protecting operating procedures and other sensitive information, the release of which might jeopardize a public health facility and its assets, is the objective of OPSEC programs. OPSEC programs utilize tools such as employee background checks, trash handling procedures, telephone policies, and IT (computer) security to protect against both industrial espionage and deliberate disruption of critical assets and functions.

The second set of questions is designed to guide the process of identifying the existing components of the physical security system that protect the critical assets, the CI systems that support the critical assets, and the operating procedures and sensitive information that must be protected to avoid jeopardizing the critical assets.

Public health facility and critical asset protection (physical assets):

- What department or person has overall responsibility for security or is that responsibility spread over many departments or people with shared responsibilities for security along with their other responsibilities?
- What perimeter barriers (e.g., fences, gates, vehicle barriers), if applicable, protect the public health facility as a whole and the individual critical assets and what levels of protection do they provide?
- What building barriers (e.g., walls, roof/ceiling, windows, doors, locks) protect each critical asset and what levels of protection do they provide?
- What is the status of the intrusion detection that protects each critical asset (e.g., intrusion sensors, alarm deployment, alarm assessment, alarm maintenance) and what level of protection does it provide?
- What is the status of the access control that is used at each critical asset (e.g., personnel access, vehicle access, contraband detection, access point illumination) and what level of protection does it provide?
- What is the nature of the security force (both the protective force and appropriate local law enforcement agencies) that protects each critical asset (e.g., number, training, armament, communications) and what level of protection does it provide?

- What types of undesirable events (e.g., surreptitious forced entry, technical implant, theft of sensitive information or materials) are protected against?
- During which hours of the day and under what conditions are the various components of the physical security system effective?
- Over what areas do the various components of the physical security system provide protection?
- What is the history of reported malfunctions of the various components of the physical security system?
- What is the correlation of the effectiveness of the various components of the physical security system to security incident reports that may indicate that the system was defeated?
- Have liaisons and working relationships been established with the local government and its departments, such as police, fire, emergency medical services, and public works?

Public health facility and critical asset protection (infrastructure interdependencies):

- Which infrastructures (both internal and external) are essential for a specific critical asset to be able to carry out its critical functions?
- What external utility or internal department and equipment is the normal provider of each essential infrastructure for each critical asset and how is each infrastructure connected to each asset (e.g., the types and pathways of power lines, pipelines, and cables)?
- What alternatives (e.g., redundant systems, alternative suppliers, backup systems established work-around plans) are available if the normal providers of the essential infrastructures are disrupted and how long can the alternatives support the critical functions of the assets?
- What is the potential for interdependency effects on external infrastructures (i.e., effects on the energy, telecommunications, transportation, water and wastewater, banking and finance, emergency services, and government services infrastructures)?

Public health facility and Critical Asset Protection (Sensitive Information):

- What types of information about the public health facility, its assets, and its operations should be considered critical or sensitive information?
- What are the methods and means by which sensitive information might fall in the wrong hands, such as via disgruntled employees; access to the facility by the public; outside construction, repair, and maintenance contractors; press contacts; briefings and presentations; public testimony; Internet information; paper and material waste; telecommunications system taps; and cyber (computer) intrusions?

- What are the existing policies and procedures that are used to protect sensitive information, such as employee background checks, disciplinary procedures, security training, trash handling procedures, paper waste handling procedures, salvage material handling procedures, dumpster control, telephone policies, and IT (computer) security.

Once public health facilities have identified their existing physical protective measures, they should coordinate with their respective local governments and law enforcement agencies to ensure that the level of protection and response that they expect will be forthcoming. They should also coordinate with the critical external infrastructure providers to ensure that the robustness and redundancy that they depend on will continue to be provided. The objective of these coordination efforts is to ensure that roles for response and recovery from a disruption are understood by all so that quick and effective measures can be taken when problems occur.

In addition, local and state governments can assist public health facilities in infrastructure restoration activities. Potential support can come in many areas, such as maintaining critical spare parts, assisting with special equipment, working with the emergency telecommunications spectrum, securing easy access to the site of the disruption for repair crews and needed equipment, working out mutual assistance programs with other energy providers, and supplying temporary staffing.

The public health facility should also check with local and state governments to ensure that critical information about their facility, its assets, and its operations will not be released to the general public in any future additions to public Internet sites, press releases, or public hearings.

Step 3. Identify and Characterize the Threat

To put the information about the critical assets of the public health facility gathered above to use in a quantitative risk assessment, the potential threats and adversaries that may be expected must be identified and quantified. The set of questions provided in this section serves as guidance for evaluating the threat environment to which the public health facility could be exposed and establishing qualitative or quantitative threat ratings for each critical asset. The goals of the threat assessment are to understand, from the adversary's point of view, the adversary's capabilities and intent to collect critical information.

The federal agencies (e.g., DOE, DHS, and the Federal Bureau of Investigation [FBI]), the Intelligence Community, state governments, and energy industry associations each collect threat information. This information should be shared among these groups and with the local public health facilities in order to have the most comprehensive and updated threat information possible. In addition, threats to public health facilities could affect state and local assets. State and local governments have access to law enforcement and intelligence data. This information should be integrated and shared, together with any information that the energy industry associations and public health facilities collect.

This third set of questions is to be used to identify and evaluate the threat environment to which a public health facility may be exposed.

Intent and capabilities of adversaries:

- What types of adversaries are expected?
- Who are the specific adversaries expected?
- What are the specific goals and objectives of each adversary?
- Which are the critical assets that each specific adversary is aware?
- Does each specific adversary know enough about the asset to plan an attack?
- What are the possible modes of attack (e.g., explosives or incendiary devices delivered by car, truck, boat, rail, mail, individuals, or standoff weapons; aircraft impacts; sabotage of equipment or operations; assaults by lightly or heavily armed individual attacker or team of attackers; theft, alteration, or release of information, materials, or equipment; contamination by chemical agents, biological agents, or radioactive material; and cyber attacks) each adversary might use?
- Are there other, less risky means for a specific adversary to attain his/her goals?
- What is the probability that an adversary will choose one method of attack over another?
- What specific events might provoke a specific adversary to act?

Information concerning potential threats and adversaries can be gathered about potential threats and adversaries by

- Joining a threat analysis working group that includes local, county, state, and federal agencies, the military, and other industry partners
- Obtaining access to the National Infrastructure Protection Center (NIPC), service provided by Analytical Services, Inc. (ANSER), FBI-sponsored InfraGuard, Carnegie Mellon University's CERT®, or other information system security warning notices
- Initiating processes to obtain real-time information from the field (e.g., on-duty offices, civilian neighborhood watch programs, local businesses, other working groups in the area)
- Arranging for threat briefings by local, state, and federal agencies
- Performing trend analyses of historical security events (both planned and actual)
- Creating possible threat scenarios based on input from the threat analysis working group and conducting related security exercises

Step 4. Identify and Analyze Vulnerabilities

In addition to identifying the critical assets of the public health facility, the impact of their disruption, the present protection provided, and the potential threats against them, the vulnerability of those assets to the potential threats must be quantified, at least to some extent, to determine the overall risk to the assets.

There are various types of vulnerabilities such as physical, technical/cyber, and operational. A public health facility, including perimeter barriers (fences, walls, gates, landscape, sewers, tunnels, parking areas, alarms), compound area surveillance (closed-circuit television, motion detectors, lighting), building perimeters (walls, roofs, windows, doors, shipping docks, locks, shielded enclosures, access control, alarms), and building interiors (doors, locks, safes, vents, intrusion sensors, motion sensors), is subject to physical vulnerabilities. Electronic equipments, such as acoustic equipment, secure telephones, computers and computer networks, and radio-frequency equipment, are subject to technical or cyber vulnerabilities. The guard force, personnel procedures, and operational procedures are subject to operational vulnerabilities.

Various characteristics of assets, including any existing protection identified in Step 2, may affect their susceptibility to attacks and must be considered when identifying susceptibilities. Such asset characteristics include building design; equipment properties; personal behavior; locations of people, equipment, and buildings; and operational and personnel practices.

Both public health facilities and local governments should be concerned with identifying and analyzing vulnerabilities. Public health facilities should analyze the vulnerabilities of their physical and cyber systems. Local governments should coordinate management of the vulnerabilities of the energy infrastructure, including individual public health facilities, which support government and community operations and assets.

This fourth set of questions is to be used to evaluate the vulnerability of the critical energy infrastructure assets to the potential threats and to establish qualitative or quantitative vulnerability ratings for each asset.

Public health facility and critical asset vulnerabilities

- How susceptible is each critical asset to physical attack if readily available weapons (guns, normal ammunition, vehicle, simple explosives) were used?
- How susceptible is each critical asset to physical attack if difficult-to-acquire weapons (assault rifles, explosive ammunition, rocket launchers, biological or chemical agents, aircraft, sophisticated explosives) were used?
- How susceptible is each critical asset to physical attack from insiders?
- Are any of the critical assets unprotected? If so, describe them.
- Are any of the critical assets minimally protected? If so, describe them.
- How susceptible is each critical asset to cyber attack?

Step 5. Assess Risk and Determine Priorities for Asset Protection

Scales for the rating criteria identified in the first four steps (asset criticality in terms of the impact of loss or disruption, threat characteristics, and asset vulnerability) must be developed. The concept of criteria development is presented below in the form of a generic example. Those that conduct an actual assessment should define rating scales that are appropriate to the specific assessment.

Using the individual rating values assigned to each combination of asset criticality, threat, and vulnerability, a relative degree of risk or a risk rating can be established for each asset for one or more postulated adverse events or consequences that could result from an attack by the identified adversary. Often a multiplicative approach involving the three rating criteria is used to obtain a risk rating:

Risk Rating = (Impact Rating) × (Threat Rating) × (Vulnerability Rating)

An additional scale must be developed to assign a qualitative overall risk level from the quantitative risk rating. The risk ratings or risk levels are used to prioritize the assets for the selection and implementation of security improvements to achieve an acceptable overall level of risk at an acceptable cost.

The following should be considered when developing and using rating criteria:

- Subject-matter expert opinions and perspectives should be documented. The team involved in the assessment should reflect a variety of different perspectives, and the team should work toward reaching a consensus regarding a set of priorities.
- Information should be presented in a usable format (e.g., table, matrix, or spreadsheet).
- Assumptions should be documented.

A generic example of possible scales for the rating criteria is presented below in the form of a set of tables to illustrate the concept. These or similar tables are used to establish qualitative or quantitative criticality, threat, and vulnerability ratings for each critical asset.

Asset Impact/Criticality Rating Criteria

Each critical asset that is identified in Step 1 of the risk management process is assigned an impact rating value that reflects the importance or criticality of a loss or disruption of that asset with regard to the continued operation of the public health facility or other organization being assessed. In Table 22.3, a quantitative criticality rating scale of 0–100% is used, which corresponds to qualitative criticality levels of critical, high, medium, and low.

Table 22.3 Asset Impact/Criticality Rating Criteria

Criticality Level	*Description*	*Rating Scale (%)*
Critical	Indicates that compromise of the asset would have grave consequences leading to loss of life or serious injury to people and disruption of the operation of the public health facility. It is also possible to assign a monetary value or some other measure of criticality.	75–100
High	Indicates that compromise of the asset would have serious consequences that could impair continued operation of the public health facility.	50–75
Medium	Indicates that compromise of the asset would have moderate consequences that would impair operation of the public health facility for a limited period.	25–50
Low	Indicates little or no impact on human life or the continuation of the operation of the public health facility.	1–25

Threat Rating Criteria

The individual potential threats against the assets of the public health facility or other organization being assessed that are identified in Step 3 are assigned a threat rating value that reflects the magnitude of the threat. In Table 22.4, a quantitative threat rating scale of 0–100% is used, which corresponds to qualitative threat levels of critical, high, medium, and low.

Vulnerability Rating Criteria

The vulnerabilities of the assets in terms of in-place measures to protect those assets that are identified in Step 2 are assigned a vulnerability rating value that reflects the extent to which the asset is protected against each threat identified in Step 3.

Table 22.4 Threat Rating Criteria

Threat Level	*Description*	*Rating Scale (%)*
Critical	Indicates that a definite threat exists against the asset and that the adversary has both the capability and intent to launch an attack, and that the subject or similar assets are targeted on a frequently recurring basis.	75–100
High	Indicates that a credible threat exists against the asset based on knowledge of the adversary's capability and intent to attack the asset and based on related incidents having taken place at similar assets or in similar situations.	50–75
Medium	Indicates that there is a possible threat to the asset based on the adversary's desire to compromise the asset and the possibility that the adversary could obtain the capability through a third party who has demonstrated the capability in related incidents.	25–50
Low	Indicates little or no credible evidence of capability or intent and no history of actual or planned threats against the asset.	1–25

In Table 22.5, a quantitative vulnerability rating scale of 0–100% is used, which corresponds to qualitative vulnerability levels of critical, high, medium, and low.

Step 6. Identify Mitigation Options, Costs, and Trade-Offs

The ultimate goal of a risk management process is to select and implement security improvements to achieve an acceptable overall risk at an acceptable cost. Step 5 of the risk management process prioritizes the combinations of assets and threats by

Table 22.5 Vulnerability Rating Criteria

Vulnerability Level	*Description*	*Rating Scale (%)*
Critical	Indicates that there are no effective protective measures currently in place and adversaries would be capable of exploiting the critical asset.	75–100
High	Indicates that although there are some protective measures in place, there are still multiple weaknesses through which adversaries would be capable of exploiting the asset.	50–75
Medium	Indicates that there are effective protective measures in place; however, one weakness does exist that adversaries would be capable of exploiting.	25–50
Low	Indicates that multiple layers of effective protective measures exist and essentially no adversary would be capable of exploiting the asset.	1–25

the risk ratings or risk levels. This, in turn, helps to identify where protective measures against risk are most needed.

In this step, potential measures to protect critical assets from recognized threats are identified, specific programs to assure that appropriate protective measures are put into place are established, and appropriate agencies and mechanisms needed to put protective measures in place are identified. Protective measures that can address more than one threat or undesirable event should be given special attention.

A variety of approaches to developing protective measures exist. Protective measures can reduce the likelihood of a failure due to an attack by adding physical security. Protective measures can also be implemented to prevent or limit the consequences of a failure or to speed the recovery following a failure, no matter what the cause of that failure.

Best practices and lessons learned from DOE and DHS programs provide some general actions, activities, and recommendations that can help identify appropriate potential mitigations measures. Some of these are listed below.

- The trend in IT until very recently has been to outsource more and more functions. If possible, cyber security should remain as an enterprise function and should not become a contractor function.
- Logging and reporting should be enabled on IT network routers and firewalls to gain a better understanding of user access and interactions with remote systems.
- Sensitive and confidential documents should not be placed on Web sites. Appropriate document review, classification, and access controls should be implemented. This practice should apply to documents and other information that is found in newsgroups, media sites, and other linked sites.
- Security measures, such as traffic filtering, authorized controls, encryption and access controls, minimizing or disabling of unnecessary services and commands, minimizing banner information, and e-mail filtering and virus control, should be implemented.
- A formal process for accessing relevant threat information and for contacting the proper government and law enforcement agencies should be instituted (if it does not already exist), and reviewed and updated on a regular basis. The public health facility may need to work with government to obtain security clearances for appropriate personnel.
- Appropriate security measures (e.g., access controls, barriers, badges, intrusion detection devices, alarm reporting and display, closed-circuit television cameras, communication equipment, lighting, and security officers) should be implemented.
- Top management support is critical in ensuring a successful security program.
- Security training programs should be formalized.
- Procedures for escorting contractors and visitors into sensitive areas should be enhanced and enforced.
- Security should be incorporated in the company goals as well as in its corporate culture.
- The foundation for security is well-informed employees acting responsibly.
- A formal review process should be established for all information released to the public, particularly through the public health facility's Web site. A periodic review of "public" information should be performed to audit the effectiveness of information protection policies.
- The public health facility should be careful about disseminating sensitive information to the press or competitors. Only minimal information should be made available about personnel (especially executives).
- Security training and awareness should be provided to all employees on a regular basis.
- At a minimum, an annual audit of overall security should be conducted.

Some illustrations specific to public health facilities are listed below as an example of specific protective measures that can be implemented. Since public health facilities vary significantly in regards to their purpose, configuration, visitor access, and so on, the protective measures below are examples that may or may not be relevant to any given site. A vulnerability assessment would help develop appropriate protective measures to implement. The following example protective measures are grouped by type:

Access Control

- Keep sensitive areas locked and not accessible to visitors.
- Prevent visitor vehicles from parking within 50 feet of buildings.
- Close and lock all nonessential gates for entry.

Measures to Limit Consequences

Improve emergency plans and procedures for continued operation during undesirable events and ensure that operators are trained to implement these contingency plans.

- Modify the physical system—improve control centers and protective devices, increase redundancy of key equipment, and increase reserve margins.

Monitoring and Surveillance

- Conduct video surveillance of buffer zone.
- Inspect interior/exterior of buildings and storage areas in regular use.
- Increase building spot checks.
- Provide CCTV video feeds to local law enforcement.
- Monitor utility supplies (e.g., electric power, natural gas, water, telecommunications).
- Check HVAC filtration, any detectors and monitors, and alarm systems.

Communications

- Enhance interface with law enforcement and safety and related emergency groups.
- Conduct additional employee briefings on security.
- Provide backup power source for communications equipment.

Security Force

- Consider guard reinforcement, and ensure guards are adequately trained in company procedures.
- Expand roving guard patrols.

Security Program

- Reduce quantities of hazardous materials on site.
- Conduct refresher course on WMD for local police.

Personnel Protection

- Update personnel when rising threats occur.
- Implement additional security measures for high-profile management.

As indicated earlier, local governments should coordinate public health facility activities related to risk management. The following questions can help guide local and state governments through the risk management process.

Local and State Government's Role in the Risk Management Process:

- Has the local or state government identified any critical issues or vulnerabilities regarding its energy infrastructure?
- If the local or state government has identified critical issues, what are they and why are they critical?
- Has the local or state government developed plans to counter these vulnerabilities?
- Has the local or state government coordinated information with other local public health facilities, local law enforcement agencies, and others concerning these vulnerabilities?

Conclusion

Vulnerability assessment methodologies generally evaluate vulnerabilities by broadly considering the threat, existing protective measures, and consequences that result if an asset is attacked (asset-based approach). Alternatively, multiple potential sequences of attack events can be considered in order to evaluate the likelihood that the current protective measures at a facility will be able to successfully deter, detect, and/or delay an attack (scenario-based approach).

Protecting and ensuring the continuity of the CI and key resources of the United States are essential to the nation's security, public health and safety, economic vitality, and way of life.

Appendix: Key Definitions and Nomenclature

Key Definitions

Adversary: An individual, group, organization, or government that conducts activities, or has the intention and capability to conduct activities, detrimental to the U.S. government or its assets. Adversaries include intelligence services of host nations, political or terrorist groups, criminals, and private interests.

Asset: Any person, equipment, material, information, installation, or activity that has a positive value to an organization or facility. The asset also may have value to an adversary, although the nature and magnitude of those values may differ.

Cost–Benefit Analysis: Part of the management decision-making process in which the costs and benefits of each alternative are compared and the most appropriate alternative is selected.

Mitigation or Protective Measure: An action taken or a physical entity used to reduce or eliminate one or more vulnerabilities. The cost of a possible mitigation measure may be monetary or nonmonetary (e.g., reduced operating efficiency, adverse publicity, unfavorable working conditions, and political consequences).

Impact: The amount of loss or damage that can be expected. The impact may be influenced by time or other factors.

Risk: The potential for damage or loss of an asset. The level of risk is a condition of two factors:

The value placed on the asset by its owner and the consequence, impact, or adverse effect of loss or change to the asset and

The likelihood that a specific vulnerability will be exploited by a particular threat.

Risk Assessment: The process of evaluating threats to the vulnerabilities of an asset to give an expert opinion on the probability of loss or damage and its impact, as a guide to taking action.

Risk Management: The process of selecting and implementing security protective measures to achieve an acceptable level of risk at an acceptable cost.

Threat: Any indication, circumstance, or incident with the potential to cause the loss of or damage to an asset. Threat can also be defined as the intention and capability of an adversary to undertake actions that would be detrimental to U.S. interests. Threat categories include insider, terrorist, intelligence service, environmental, criminal, and military.

Undesirable Event: Any incident with the potential to cause the loss of or damage to an asset. Undesirable events can be due to actions such as theft, compromise, destruction, sabotage, assault, assassination, and kidnapping or due to occurrences such as nonavailability or impaired operation of an asset.

Vulnerability: Any weakness that can be exploited by an adversary to gain access to an asset. Vulnerabilities can result from the following:

- Building characteristics
- Equipment properties
- Personal behavior
- Locations of people, equipment, and buildings
- Operational and personnel practices

Nomenclature

ANSER	Service provided by Analytical Services, Inc.
DOE	Department of Energy
EMS	Energy management system
FBI	Federal Bureau of Investigation
HVAC	Heating, ventilation, and air conditioning
IT	Information technology
NIPC	National Infrastructure Protection Center
NNSI	Nonproliferation and National Security Institute
OEA	Office of Energy Assurance
OJP	Office of Justice Programs
OPSEC	Operations security
RTU	Remote terminal unit
SCADA	Supervisory control and data acquisition
UPS	Uninterruptible power supply
VRAP	Vulnerability and Risk Analysis Program

References

Energy Infrastructure Risk Management Checklists for Small and Medium Sized Public Health Facilities, US Department of Energy, Office of Energy Assurance, August 2002.

National Infrastructure Protection Plan: Partnering to Enhance Protection and Resiliency. 2009. Retrieved August 13, 2010, from http://www.dhs.gov/xlibrary/assets/NIPP_Plan.pdf.

National Infrastructure Protection Plan. Washington, DC: US Department of Homeland Security, 2006.

Survey of Vulnerability Assessment Methodologies, US DHS Protective Security Division, October 2003.

U.S. Department of Homeland Security, http://www.dhs.gov/files/programs/gc_1189168948944.shtm, official website of the Department of Homeland Security, Critical Infrastructure, downloaded June 2013.

White House, *Homeland Security Presidential Directive/HSPD-7: Critical Infrastructure Identification, Prioritization, and Protection*, Washington, DC, December 17, 2003. http://www.whitehouse.gov/sites/default/files/omb/memoranda/fy04/m-04-15.pdf.

White House, *Presidential Policy Directive—Critical Infrastructure Security and Resilience*, Washington, DC, February 2013. http://www.whitehouse.gov/the-press-office/2013/02/12/presidential-policy-directive-critical-infrastructure-security-and-resil.

Chapter 23

Critical Infrastructures and Interdependencies

James Peerenboom and Ronald E. Fisher

Introduction

The definitions of critical infrastructure (CI), key resources, and interdependencies have steadily evolved and matured over nearly two decades. The President's Commission on Critical Infrastructure Protection (PCCIP), which was established in the mid-1990s under the Clinton administration through Executive Order 13010, states that:

> Certain national infrastructures are so vital that their incapacity or destruction would have a debilitating impact on the defense or economic security of the United States. These critical infrastructures include telecommunications, electrical power systems, gas and oil storage and transportation, banking and finance, transportation, water supply systems, emergency services (including medical, police, fire, and rescue), and continuity of government. Threats to these critical infrastructures fall into two categories: physical threats to tangible property ("physical threats"), and threats of electronic, radio-frequency, or computer-based attacks on the information or communications components that control critical infrastructures ("cyber threats"). Because many of these critical infrastructures are owned and operated by the private sector, it is essential that the government and private sector work

> together to develop a strategy for protecting them and assuring their continued operation. (Clinton, 1996, p. 37347)

This definition and the previous and subsequent presidential decision directives (PDDs), executive orders, and legislative acts have all expanded on the basic definition first used by President Roosevelt in the midst of World War II. This chapter utilizes the PCCIP definition of "infrastructure" as

> the framework of interdependent networks and systems comprising identifiable industries, institutions (including people and procedures), and distribution capabilities that provide a reliable flow of products and services essential to the defense and economic security of the United States, the smooth functioning of government at all levels, and society as a whole. (Clinton, 1996, p. 37347)

Following the passage of the U.S. Patriot Act of 2001, the president's *National Strategy for Homeland Security,* issued in July 2002, included and expanded on the definition of CI. The National Infrastructure Protection Plan (NIPP) was published in 2006 and an update in 2009, also defined and refined the definition of CI. In particular, the NIPP provides the unifying structure for integrating a wide range of efforts for the protection of CI into a single national program. Table 23.1 provides an overview of each of the 18 CI defined by the Department of Homeland Security (DHS) (DHS, 2012) and revised in 2013 to 16 CI (White House, 2013). Table 23.1 also highlights the vastness of thousands of CI facilities that comprise these sectors. For example, there are over 40,000 chemical facilities, hundreds of thousands of miles of pipelines, and thousands of substations. The vastness of CI facilities increases the complexity of maintaining and protecting such a large number of facilities.

Protecting and ensuring the continuity of the CI of the United States is essential to the nation's security, public health and safety, economic vitality, and way of life. As defined in the 2009 NIPP, CI is the systems and assets, whether physical or virtual, so vital that their incapacity or destruction may have a debilitating impact on the security, economy, public health or safety, environment, or any combination of these, across any federal, state, regional, territorial, or local jurisdiction. The key resources are publicly or privately controlled resources essential to the minimal operations of the economy and government.

CI needs to be understood and assessed not only to apply a further layer of defense, but also to create an organizational understanding of the complex infrastructure network we live in today and to generate a more resilient and robust nation in the face of uncertain threats (Auerswald et al., 2005). This chapter explores further the dependencies and interdependencies among and between the CIs to assist emergency managers in understanding better the complexities and interconnectedness of CIs as they play a critical role in supporting CIs. This chapter also

Table 23.1 Critical Infrastructure and Key Resources Sectors

Sector	*Sector Description*
Chemical	The chemical sector is an integral component of the U.S. economy, employing nearly 1 million people and earning revenues of more than $637 billion per year. This sector can be divided into five main segments, based on the end product produced: basic chemicals, specialty chemicals, agricultural chemicals, pharmaceuticals, and consumer products. Each of these segments has distinct characteristics, growth dynamics, markets, new developments, and issues. The majority of chemical sector facilities are privately owned, requiring the DHS to work closely with the private sector and its industry associations to identify and prioritize assets, assess risks, develop and implement protective programs, and measure the effectiveness of the program. The chemical sector is dependent on, depended on by, and overlaps with a wide range of other sectors, including transportation systems, energy, water, agriculture and food, information technology, and communications.
Commercial facilities	Facilities associated with the commercial facilities sector operate on the principle of open public access, meaning that the general public can move freely throughout these facilities without the deterrent of highly visible security barriers. The majority of the facilities in this sector are privately owned and operated, with minimal interaction with the federal government and other regulatory entities. The commercial facilities sector consists of the following eight subsectors: 1. Public assembly (e.g., arenas, stadiums, aquariums, zoos, museums, and convention centers) 2. Sports leagues (e.g., professional sports leagues and federations) 3. Gaming (e.g., casinos) 4. Lodging (e.g., hotels, motels, and conference centers)

(Continued)

Table 23.1 (continued) Critical Infrastructure and Key Resources Sectors

Sector	*Sector Description*
	5. Outdoor events (e.g., theme and amusement parks, fairs, campgrounds, and parades) 6. Entertainment and media (e.g., motion picture studios, broadcast media) 7. Real estate (e.g., office/apartment buildings, condominiums, mixed-use facilities, and self-storage), and 8. Retail (e.g., retail centers and districts, shopping malls).
Communications	The communications sector is an integral component of the U.S. economy as it underlies the operations of all businesses, public safety organizations, and government. Over 25 years, the sector has predominantly evolved from a provider of voice services into a diverse, competitive, and interconnected industry using terrestrial, satellite, and wireless transmission systems. The transmission of these services has become interconnected; satellite, wireless, and wireline providers depend on each other to carry and terminate their traffic and companies routinely share facilities and technology to ensure interoperability. A majority of the communications sector is privately owned, requiring the department to work closely with the private sector and its industry associations to identify infrastructure, assess and prioritize risks, develop protective programs, and measure the effectiveness of the program. The communications sector is closely linked to other sectors: • The energy sector provides power to run cellular towers, central offices, and other critical communication facilities. • The information technology sector provides critical control systems and services, physical architecture, and Internet infrastructure. • The banking and finance sector relies on telecommunications for the transmission of transactions and operations of financial markets.

	• The emergency services sector depends on telecommunications for directing resources, coordinating response, alerting the public, and receiving emergency 911 calls, and • The postal and shipping sector uses telecommunications for its control systems, tracking shipments, and regular communication requirements.
Critical manufacturing (CM)	The critical manufacturing sector is crucial to the economic prosperity and continuity of the United States. U.S. manufacturers design, produce, and distribute products that provide more than one of every 8$ of the U.S. gross domestic product and employ more than 10% of the nation's workforce. A direct attack on or disruption of certain elements of the manufacturing industry could disrupt essential functions at the national level and across multiple other CI. On the basis of guidance provided by HSPD-7, the following nine industries currently meet the critical infrastructure and key resources (CIKR) criteria of the CM sector and are not included within an existing sector: *Primary metal manufacturing* 1. Iron and steel mills and ferro alloy manufacturing 2. Alumina and aluminum production and processing 3. Nonferrous metal (except aluminum) production and processing *Machinery manufacturing* 4. Engine, turbine, and power transmission equipment manufacturing *Electrical equipment, appliance, and component manufacturing* 5. Electrical equipment manufacturing

(*Continued*)

Table 23.1 (continued) Critical Infrastructure and Key Resources Sectors

Sector	*Sector Description*
	Transportation equipment manufacturing 6. Motor vehicle manufacturing 7. Aerospace product and parts manufacturing 8. Railroad rolling stock manufacturing 9. Other transportation equipment manufacturing The products made by these manufacturing industries are essential in varying capacities to many other CIKR sectors. The CM sector focuses on the identification, assessment, prioritization, and protection of nationally significant manufacturing industries that may be susceptible to terrorist attacks.
Dams	The dams sector comprises the assets, systems, networks, and functions related to dam projects, navigation locks, levees, hurricane barriers, mine tailings impoundments, or other similar water retention and/or control facilities. There are over 82,000 dams in the United States; approximately 65% are privately owned and more than 85% are regulated by State Dam Safety Offices. The dams sector is a vital and beneficial part of the nation's infrastructure and continuously provides a wide range of economic, environmental, and social benefits, including hydroelectric power, river navigation, water supply, wildlife habitat, waste management, flood control, and recreation. The dams sector has dependencies and interdependencies with a wide range of other sectors, including • The agriculture and food sector, as a continued source of water for irrigation and water management • The transportation systems sector uses dams and locks to manage navigable waters throughout inland waterways

	• The water sector supplies potable water to concentrated populations and commercial facilities in the United States • The energy sector provides approximately 8–12% of the nation's power needs with hydropower dams • The emergency services sector, which relies on dams sector assets for firefighting water supply, emergency water supply, and waterborne access in the event of a significant disaster
Defense industrial base	The defense industrial base (DIB) sector includes Department of Defense (DoD), government, and the private sector worldwide industrial complex with the capabilities of performing research and development, design, production, delivery, and maintenance of military weapons systems, subsystems, components, or parts to meet military requirements. The DIB sector includes more than 100,000 DIB companies and their subcontractors. DIB companies include domestic and foreign entities, with production assets located in many countries. The DIB sector provides products and services that are essential to mobilize, deploy, and sustain military operations. The DIB sector does not include the commercial infrastructure of providers of services such as power, communications, transportation, or utilities that DoD uses to meet military operational requirements. These requirements, including cyber, are addressed in DoD's broader defense critical infrastructure program (DCIP), where they are integrated in all DIB sector activities.
Emergency services	The emergency services sector (ESS) is a system of response and recovery elements that forms the nation's first line of defense, prevention, and reduction of consequences from any terrorist attack. It is a sector of trained and tested personnel, plans, redundant systems, agreements, and pacts that provide life safety and security services across the nation via the first-responder community composed of federal, state, local, tribal, and private partners. The ESS is representative of the following first-responder disciplines: emergency management, emergency medical services, fire, hazardous material, law enforcement, bomb squads, tactical operations/special weapons, assault teams, and search and rescue. All first responders within the ESS are individuals possessing specialized training from one or more of these disciplines.

(*Continued*)

Table 23.1 (continued) Critical Infrastructure and Key Resources Sectors

Sector	*Sector Description*
	The ESS has numerous interdependencies with all CIKR sectors. Most significantly, it is the primary protector for all other CIKR, including nuclear reactors, chemical plants, and dams. All other CIKR facilities depend on the ESS to assist with planning, prevention, and mitigation activities, as well as respond to day-to-day incidents and catastrophic situations.
Energy	The U.S. energy infrastructure fuels the economy of the twenty-first century. Without a stable energy supply, health and welfare is threatened and the U.S. economy cannot function. More than 80% of the country's energy infrastructure is owned by the private sector. The energy infrastructure is divided into three interrelated segments: electricity, petroleum, and natural gas. The U.S. electricity segment contains more than 6413 power plants with approximately 1075 gigawatts of installed generation. Approximately 48% of electricity is produced by combusting coal (primarily transported by rail), 20% in nuclear power plants, and 22% by combusting natural gas. The remaining generation is provided by hydroelectric plants, oil, and renewables. The sector's reliance on pipelines highlights the interdependency with the transportation sector and the reliance on the energy sector for power means that virtually all sectors have dependencies on this sector. The energy sector is well aware of its vulnerabilities and is leading a significant voluntary effort to increase its planning and preparedness. Cooperation through industry groups has resulted in substantial information sharing of effective and best practices across the sector. Many sector owners and operators have extensive experience abroad with infrastructure protection and have more recently focused their attention on cybersecurity.
Food and agriculture	The agriculture and food sector has the capacity to feed and clothe people well beyond the boundaries of the nation. The sector is almost entirely under private ownership and is composed of an estimated 2.1 million farms, approximately 900,000 restaurants and more than 400,000 registered food manufacturing, processing, and storage facilities. This sector accounts for roughly one-fifth of the nation's economic activity and is overseen at the federal level by the U.S. Department of Agriculture and the Department of Health and

	Human Services' Food and Drug Administration. The agriculture and food sector has critical dependencies with water (for clean irrigation and processed water), transportation systems (for movements of products), energy (to power the equipment needed for agriculture production and food processing), banking and finance, chemical, dams, and other sectors as well.

Financial services sector	The financial services sector, the backbone of the world economy, is a large and diverse sector primarily owned and operated by private entities. In 2007, the sector accounted for more than 8.0% of the U.S. gross domestic product.
	This sector consists of over 29,000 financial firms, including
	• Depository financial institutions
	  • Banks
	  • Thrifts
	  • Credit unions
	• Insurers
	• Securities brokers/dealers
	• Investment companies
	• Certain financial utilities
	Financial services firms provide a broad array of products to their customers. These products
	• Allow customers to deposit funds and make payments to other parties,
	• Provide credit and liquidity to customers,
	• Allow customers to invest funds for both the long and short term, and
	• Transfer financial risks between customers.

(Continued)

Table 23.1 (continued) Critical Infrastructure and Key Resources Sectors

Sector	*Sector Description*
	The financial institutions that provide these services are all somewhat different, each within a specific part or parts of the financial services marketplace. Financial institutions operate to provide customers the financial products that they want, ensure the institution's financial integrity, protect customers' assets, and guarantee the integrity of the financial system. As such, financial institutions and the financial markets that they organize manage a wide variety of financial and certain nonfinancial risks. In addition to the actions of financial institutions, direct financial regulation applies to many, but not all, financial services providers. The U.S. system of financial regulation is complex and exists at both the federal and state levels. The regulatory systems for financial services firms manage and regulate various forms of risk and guard against prohibited practices.
Government facilities	The government facilities sector includes a wide variety of buildings, owned or leased by federal, state, territorial, local, or tribal governments, located domestically and overseas. Many government facilities are open to the public for business activities, commercial transactions, or recreational activities. Others that are not open to the public contain highly sensitive information, materials, processes, and equipment. This includes general-use office buildings and special-use military installations, embassies, courthouses, national laboratories, and structures that may house critical equipment and systems, networks, and functions. In addition to physical structures, the sector considers cyber elements that contribute to the protection of the assets of the sector (e.g., access control systems and closed-circuit television systems) as well as the protection of individuals who possess tactical, operational, or strategic knowledge or perform essential functions.
Health care and public health	The healthcare and public health sector constitutes approximately 15% of the gross national product with roughly 85% of the sector's assets privately owned and operated. Operating in all U.S. states, territories, and tribal areas, the healthcare and public health sector plays a significant role in response and recovery across all other sectors in the event of a natural or man-made disaster.

	While health care tends to be delivered and managed locally, the public health component of the sector, primarily focused on population health, is managed across all levels of government—local, tribal, territorial, state, regional, and national. The healthcare and public health sector is highly dependent on fellow sectors for continuity of operations and service delivery including transportation systems, agriculture and food, energy, water, emergency services, information technology, and communications.
Information technology	The IT sector is central to the nation's security, economy, and public health and safety. Businesses, governments, academia, and private citizens are increasingly dependent on IT sector functions. These virtual and distributed functions produce and provide hardware, software, and IT systems and services, and—in collaboration with the communications sector—the Internet. The IT sector functions are operated by a combination of entities—often owners and operators and their respective associations—that maintain and reconstitute the network, including the Internet. The Internet encompasses the global infrastructure of packet-based networks and databases that use a common set of protocols to communicate. The networks are connected by various transports and the availability of these networks and services is the collective responsibility of the IT and communication sectors. The DHS is the sector-specific agency for the IT sector.
Nuclear reactors, materials, and waste	Nuclear power accounts for approximately 20% of the nation's electrical use, provided by 104 commercial nuclear reactors licensed to operate in the United States. The nuclear reactors, materials, and waste (nuclear) sector includes nuclear power plants; nonpower nuclear reactors used for research, testing, and training; nuclear materials used in medical, industrial, and academic settings; nuclear fuel fabrication facilities; decommissioning reactors; and the transportation, storage, and disposal of nuclear material and waste. The nuclear sector has identified interdependencies with other sectors, including • Energy as a supplier to the nation's electrical grid • Transportation systems through the movement of radioactive material

(Continued)

Table 23.1 (continued) Critical Infrastructure and Key Resources Sectors

Sector	*Sector Description*
	• Chemical as related to hazardous chemicals at fuel cycle facilities • Health care and public health through nuclear medicine, radiopharmaceuticals, and sterilization of surgical supplies, and • Government facilities through federal and state facilities that use radioactive material for various purposes.
Transportation systems	The nation's transportation system quickly, safely, and securely moves people and goods through the country and overseas. The transportation systems sector consists of six key subsectors or modes: *Aviation* includes aircraft, air traffic control systems, and approximately 450 commercial airports and 19,000 additional airfields. This mode includes civil and joint use military airports, heliports, short takeoff and landing ports, and seaplane bases. *Highway* encompasses more than 4 million miles of roadways and supporting infrastructure. Vehicles include automobiles, buses, motorcycles, and all types of trucks. *Maritime transportation system* consists of about 95,000 miles of coastline, 361 ports, over 10,000 miles of navigable waterways, 3.4 million square miles of exclusive economic zone to secure, and intermodal landside connections, which allow the various modes of transportation to move people and goods to, from, and on water. *Mass transit* includes multiple occupancy vehicles, such as transit buses, trolleybuses, vanpools, ferryboats, monorails, heavy (subway) and light rail, automated guideway transit, inclined planes, and cable cars designed to transport customers to local and regional routes.

	Pipeline systems include vast networks of pipeline that traverse hundreds of thousands of miles throughout the country, carrying nearly all of the nation's natural gas and about 65% of hazardous liquids, as well as various chemicals. *Rail* consists of hundreds of railroads, more than 143,000 route miles of track, more than 1.3 million freight cars, and roughly 20,000 locomotives. *Postal and Shipping* moves over 574 million messages, products, and financial transactions each day. Postal and shipping activity is differentiated from general cargo operations by its focus on letter or flat mail, publications, or small- and medium-size packages and by service from millions of senders to nearly 152 million destinations.
Water and wastewater systems	Homeland Security Presidential Directive-7 (HSPD-7) designates the Environmental Protection Agency (EPA) as the federal lead for the water sector's CI protection activities. All activities are carried out in consultation with the department and the EPA's water sector partners. There are approximately 160,000 public drinking water systems and more than 16,000 publicly owned wastewater treatment systems in the United States. Approximately 84% of the U.S. population receives their potable water from these drinking water systems, and more than 75% of the U.S. population has its sanitary sewerage treated by these wastewater systems. The water sector is vulnerable to a variety of attacks through contamination with deadly agents, physical attacks such as the release of toxic gaseous chemicals, and cyber attacks. If these attacks were realized, the result could be large numbers of illnesses or casualties and/or a denial of service that would also impact public health and economic vitality. Critical services such as firefighting or health care and other dependent and interdependent sectors, such as energy, transportation systems, agriculture, and food would be negatively impacted by a denial of service from the water sector.

Source: Adapted from DHS, 2012, http://www.dhs.gov/files/programs/gc_1189168948944.shtm, official website of the U.S. Department of Homeland Security, "Critical Infrastructure" (downloaded June 2013).

provides a building block for the next chapter on vulnerability assessments so that infrastructure interdependencies are considered in risk analysis.

Concepts and Terminology

A variety of concepts and definitions can be used to describe interdependencies among the 18 CI (Peerenboom 2001, Peerenboom et al. 2002). The NIPP defines interdependency as the "multi- or bi-directional reliance of an asset, system, network, or collection thereof, within or across sectors, on input, interaction, or other requirement from other sources in order to function properly" (DHS, 2005, p. 104). Figure 23.1 provides a list of the dimensions of infrastructure interdependencies along with each of the 18 CI. They are represented by gears that are all interconnected. The threats in the middle highlight that both natural (e.g., hurricane, earthquake) and man-made (intentional and unintentional) threats can directly or indirectly lead to infrastructure failures that can impact

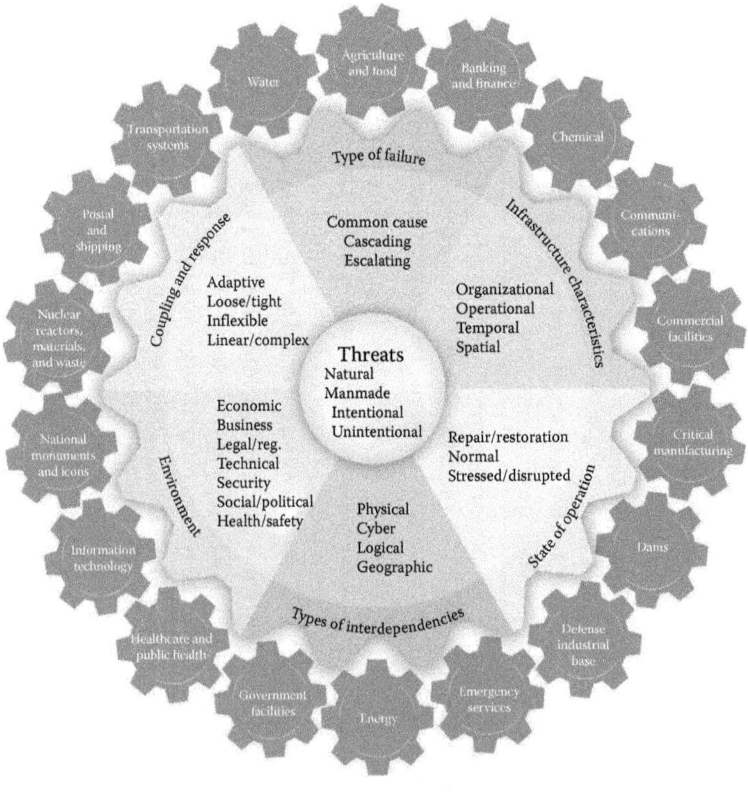

Figure 23.1 Infrastructure interdependencies dimensions.

multiple infrastructures. For example, a lightning strike that disrupts an electric substation has immediate impacts on the energy sector. These impacts may quickly cascade to some or all the other CI. The infrastructure interdependencies relationships are explained through the six dimensions in Figure 23.1. The six infrastructure interdependencies dimensions include infrastructure characteristics, state of operation, types of interdependencies, environment, coupling and response behavior, and type of failure. It is important for emergency managers to be familiar with the infrastructure interdependencies' dimensions to increase their understanding of the interconnectedness of CI.

For example, Figure 23.1 shows four types of infrastructure interdependencies:

- Physical (e.g., the material output of one infrastructure is used by another),
- Cyber (e.g., infrastructures utilize electronic information and control systems),
- Geographic (e.g., infrastructures are colocated in a common corridor), and
- Logical (e.g., infrastructures are linked through financial markets) (Rinaldi, 2001).

The proliferation of information technology (IT), along with the increased use of automated monitoring and control systems and the increased reliance on the open marketplace for purchasing and selling infrastructure commodities and services (e.g., electric power), has increased the prevalence and importance of cyber and logical interdependencies. Physical, cyber, geographic, and logical infrastructure interdependencies transcend individual infrastructure sectors (by definition) and generally transcend individual public and private-sector companies. Further, they vary significantly in scale and complexity, ranging from local linkages (e.g., municipal water supply systems and local emergency services), regional linkages (e.g., electric power coordinating councils), national linkages (e.g., interstate natural gas and transportation systems), and to international linkages (e.g., telecommunications, banking, and finance systems). These scale and complexity differences create a variety of spatial, temporal, and system representation complexities that are difficult to identify and analyze.

From an analytic perspective, infrastructure interdependencies must be viewed from a "system of systems," or holistic perspective. Figure 23.2 illustrates the complexities of infrastructure interdependencies with electric power. Other energy subsectors (i.e., natural gas, oil, and petroleum) rely on electric power and electric power relies on them as well. For example, natural gas compressors require electric power to operate and some electric power plants require natural gas. Electric power is also interdependent with each of the other CIs (e.g., transportation, telecommunications, and water). Even in this simplification of interdependencies, the figure starts to become populated with interrelated lines showing the dependencies and interdependencies.

Failures affecting interdependent infrastructures can be described in terms of three types of failures:

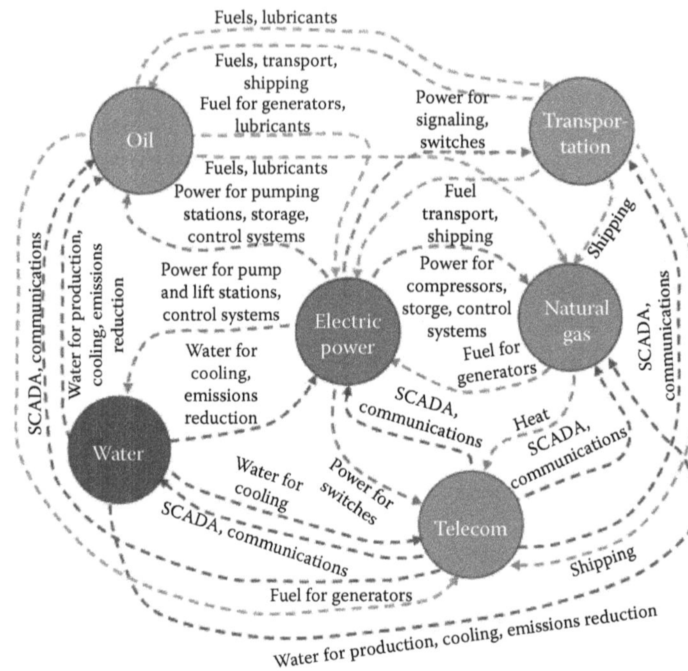

Figure 23.2 Example of infrastructure interdependencies.

- *Cascading failure.* A disruption in one infrastructure causes a disruption in a second infrastructure (e.g., the August 2003 blackout led to communications and water supply outages, air traffic disruptions, chemical plant shutdowns, and other interdependency-related impacts).
- *Escalating failure.* A disruption in one infrastructure exacerbates an independent disruption of a second infrastructure (e.g., the time for recovery or restoration of an infrastructure increases because another infrastructure is not available).
- *Common cause failure.* A disruption of two or more infrastructures at the same time is the result of a common cause (e.g., Hurricane Katrina simultaneously impacted electric power, natural gas, petroleum, water supply, emergency services, telecommunications, and other infrastructures).

As an illustration of cascading and escalating failures, consider the disruption of a microwave communications network that is used for the supervisory control and data acquisition (SCADA) system in an electric power network (Figure 23.3). The lack of monitoring and control capabilities could cause generating units to be taken offline, an event that, in turn, could cause a loss of power at a distribution substation (designated with 1 and 2 in the figure). This loss could then lead to

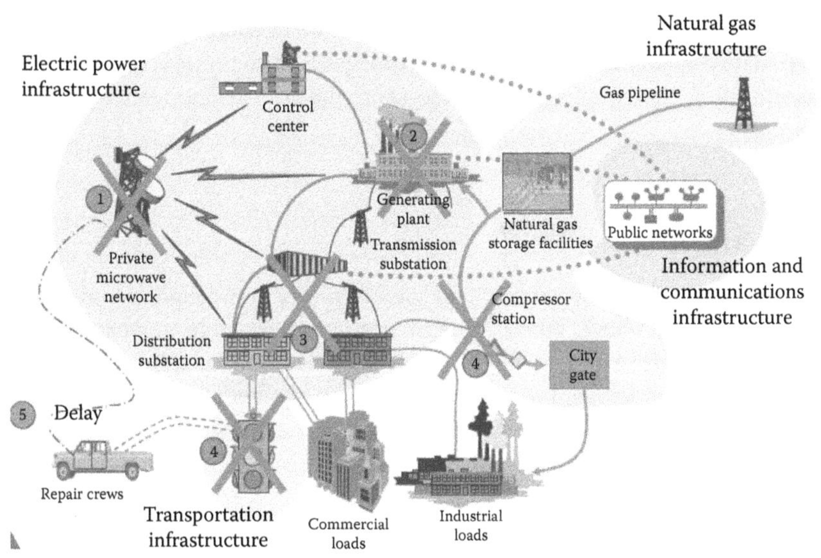

Figure 23.3 Example of infrastructure failure.

blackouts in the area served by the substation (item 3). Such an outage could affect multiple dependent infrastructures (subject to the availability of backup systems), such as transportation and water systems, commercial office buildings, schools, chemical facilities, banking and financial institutions, and many others. These disruptions could lead to delays in repair and restoration activities (i.e., an escalating failure) because of logistics, communications, business services, and other interdependency-related problems (item 4). This simplified example reinforces the notion that understanding and analyzing cascading and escalating failures requires a systems perspective and a broad set of interdisciplinary skills. The state of operation of an infrastructure, which can range from normal operation to various levels of stress, disruption, or repair and restoration, also must be considered in examining interdependencies. Further, it is necessary to understand backup systems, other mitigation mechanisms that reduce interdependency-related problems, and the change in interdependencies as a function of outage duration and frequency. Such considerations add complexity to the calculus of quantifying infrastructure interdependencies.

The coupling and response behavior is another important interdependencies dimension for emergency managers. Understanding how tightly or loosely coupled infrastructures are and their response behaviors can help emergency managers prioritize resources and reduce consequences from infrastructure failures. An example of tightly and loosely coupled systems is the energy sector and its subsectors (electricity, petroleum, and natural gas). When the electricity goes out, there is an instantaneous outage. In contrast, natural gas has a line pack in the system such that an outage at a pipeline may not be impacted for minutes, hours, or even days depending on how much natural gas is in the system and in the load. Understanding the coupling dimension can help prioritize emergency response resources.

Application

Although types of interdependencies, type of failure, and coupling and response behavior are noted above, all six dimensions are important in understanding the relationship of infrastructure interdependencies. Emergency managers can take these concepts and principles and apply them to their respective jurisdiction(s). Talking with utility providers, emergency managers can identify CI and begin to understand the interconnectedness that is characteristic of infrastructure interdependencies. Also, talking with key users of CI helps to identify consequences that could result from infrastructure interdependencies and mitigation strategies to reduce these consequences.

Identifying, understanding, and analyzing the interdependencies among infrastructures have taken an increasing importance. The key technological, economic, and regulatory changes have dramatically altered the relationships among infrastructures and the IT revolution has substantially led to more interconnected and complex infrastructures with generally greater centralization of control. Globalization also continues to drive interconnectedness and complexity. Indeed, the trend toward greater infrastructure interdependency has accelerated and shows little sign of abating. The six "dimensions" of the taxonomy frame major aspects of interdependencies: types of interdependencies, infrastructure environment, coupling and response behavior, infrastructure characteristics, types of failures, and state of operations. A public–private partnership with close coordination with emergency managers is needed to support resilient and robust critical infrastructures.

Emergency managers can play a vital role in assisting in the management of critical infrastructures in their respective areas. For example, in the SCADA outage example (Figure 23.3), emergency managers could work with the utility companies and communities in prioritizing resources to assist in restoration efforts, minimizing the escalating and cascading effects of the outage. Specifically, having an increased understanding of the dependencies and interdependencies could lead emergency managers to assist the transportation crew to reach the disrupted communication tower quicker. This could lead to a reduced outage time, thereby reducing the escalating outage time and possibly eliminating the cascading outage. If the communication tower was fixed quickly enough, the electric power sector might not be disrupted.

Emergency managers, especially those working in emergency operations centers (EOCs), have insights into the consequences from infrastructure incidents (e.g., natural disasters, terrorist attacks, and human error). Planners do not always consider or fully understand the complexities and interconnectedness of interdependencies. Shortly, after August 2003 East Coast power failure, water systems began shutting down due to their reliance on electric power. This dependence was not well understood throughout many communities.

Emergency managers should take these infrastructure dimensions and apply them to their respective areas. Working with the owners and operators of CI along with government organizations (city, state, and federal) can help foster a collaborative environment to understand and address the key interdependencies of common concern. Two great examples of collaboration are:

- The Pacific Northwest Economic Region (PNWER)—Founded in 1991, PNWER is the only statutory, nonpartisan, binational, and public/private partnership in North America. PNWER is the forum for collaborative binational planning involving both the public and private sectors and offers leadership at the state/provincial level in Salem, Olympia, Boise, Helena, Juneau, Edmonton, Regina, Victoria, Yellowknife, and Whitehorse, and at the national level in Washington, DC and Ottawa. PNWER facilitates working groups consisting of public and private leaders to address specific issues impacting our regional economy (PNWER, 2012).
- Channel Industries Mutual Aid (CIMA)—A nonprofit organization combining the firefighting, rescue, hazardous material handling, and emergency medical capabilities of the refining and petrochemical industry in the Houston Ship Channel area. Since 1955, this organization has been providing cooperative assistance and expertise for all kinds of emergencies—both natural and man made (CIMA, 2012).

A combined public–private partnership that involves emergency managers will help increase the resiliency of CI. Table-top exercises, face-to-face meetings, Internet data sharing, and emails are all great ways to help increase the partnership. Some of the actions needed to understand better interdependencies are low cost. For example, getting utilities and communities together to go through "what if" scenarios for different types of events that could occur is useful to help in prioritizing resources when an event does occur. A better understanding of interdependencies can assist in supporting CI and preventing cascading, escalating, and common cause failures.

References

Auerswald, P., Branscomb, L., La Porte, T., and Michel-Kerjan, E. 2005. The challenge of protecting critical infrastructures, *Issues in Science and Technology*, University of Texas at Dallas, Fall 2005, 77–83.

Channel Industries Mutual Aid (CIMA), 2012, http://www.cimatexas.org/cima/, official website, Diana Becerra, Deere Park, TX (downloaded June 2013).

Clinton, W. J. 1996. Executive Order 13010—Critical Infrastructure Protection, Presidential Document, *Federal Register*, July 17, 1996, Vol. 61, No. 138, pp. 37345–37350.

DHS, 2005. *U.S. Department of Homeland Security, Interim National Infrastructure Protection Plan*, Washington, DC, February 2005.

DHS, 2012, http://www.dhs.gov/files/programs/gc_1189168948944.shtm, official website of the U.S. Department of Homeland Security, "Critical Infrastructure," (downloaded June 2013).

Peerenboom, J. P. 2001. Infrastructure interdependencies: Overview of concepts and terminology, invited paper, *National Science Foundation/Office of Science and Technology Policy Workshop on Critical Infrastructure: Needs in Interdisciplinary Research and Graduate Training*; Washington, DC, June 14–15, 2001.

Peerenboom, J. P., Fisher, R. E., Rinaldi, S., and Kelly, T. 2002. Studying the chain reaction, *Electric Perspectives*, pp. 22–35, Washington, DC, January/February 2002.

Global Challenges Regional Solutions, Pacific Northwest Economic Region, 2012 Annual Report, Lyle Stewart Hon, President, Seattle, Washington, www.pnwer.org.

Rinaldi, S., Peerenboom, J. P., and Kelly, T. 2001. Identifying, understanding, and analyzing critical infrastructure interdependencies, feature article, *IEEE Control Systems Magazine*, 11–25, December 2001.

White House. 2013. *Presidential Policy Directive—Critical Infrastructure Security and Resilience*, Washington, DC, February 2013.

SPECIAL CONSIDERATIONS

VII

Chapter 24

Nuclear and Radiological Incidents

Andrew M. Bramnik

Introduction

When the average person hears the word radiation, their immediate thoughts often turn to disasters like Chernobyl and Hollywood movies about radiation exposure or nuclear war. In fact, nuclear materials benefit our society: We use them to power our homes and businesses, and to diagnose and treat illnesses. When presented out of context, the hazards of radiation are enough to scare any emergency manager: an invisible, odorless force that can only be detected by specialized equipment and that has the potential to harm humans, affect food and water, and contaminate the environment. Many local officials and emergency responders are not aware of radioactive materials (RAM) being transported, used, or stored within their jurisdiction. Additionally, radiation is not well understood by members of the public, creating the potential for confusion and fear of the subject. RAM are used in many industries across the world, however, and provide a tremendous benefit to our way of life. When examined as a whole, RAM are highly regulated by federal and state governments. Rigorous security and safety requirements are inspected by trained individuals who have unrestricted access to licensed facilities and can employ strong enforcement actions to address noncompliance. The United States' arsenal of nuclear weapons and weapons-grade materials are similarly kept safe and secured.

Although incidents involving public citizens and RAM are very rare, and there is no imminent threat of a nuclear detonation within the United States, any such incident would affect large areas and whole communities. Whether intentional or accidental, some significant incidents can expand into mass casualty and mass confusion events. Media coverage of any nuclear or radiological event combined with perceptions about radiation from popular culture will almost certainly influence response and recovery actions. In addition, much of the nation's RAM are manufactured or possessed by private companies; outside of military applications, the government's role is mostly regulatory with respect to radioactive sources. Therefore, nongovernment resources and experience in radiation protection may need to be leveraged for the most effective response to a significant event.

Entire books have been written about radiation safety in emergency management. This chapter will focus on three primary objectives: to introduce readers to basic information about radiation; to describe a range of potential incidents involving radiological and/or nuclear materials; and to discuss methods to protect the public during any event. Emergency managers can utilize this information to evaluate and enhance preparedness in their individual communities. Students and other readers can use this chapter to better understand a topic that is often misunderstood and learn how radiation benefits society in diverse fields. Hopefully, every reader can use this chapter to recognize and dispel some fears about radiation that have been generated by popular culture. A summary of abbreviations in this chapter has been provided after the conclusion to help readers navigate the large list of terms and agencies.

Section I: Background

Radiation Basics

A basic foundation is important to prepare for nuclear or radiological incidents, including understanding some key concepts. Although it is used differently, radiation is actually a general term for any energy that comes from a source and travels through some material or through space. There are many types of natural and man-made radiation, such as heat, light, microwaves, or sound. Many people only think of radiation as something that can damage matter. This is known as ionizing radiation: a type of radiation that can produce charged particles—particles with a net positive or negative charge—in matter.[1] This chapter will focus on ionizing radiation, which is produced from unstable or "radioactive" atoms that have too much mass, too much energy, or both. These atoms give off their excess energy or mass, creating radiation.

The excess energy emitted by unstable atoms can be either a particle or an electromagnetic wave. Emergency managers and responders should be familiar with three basic types of radiation (Figure 24.1):

- *Alpha:* A particle consisting of two protons and two neutrons. Alpha particles can only travel short distances in air, and are not strong enough to

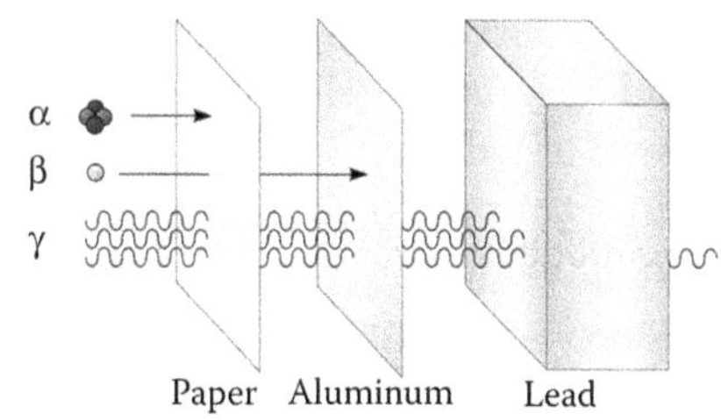

Figure 24.1 Alpha, beta, and gamma radiation penetration. (From Wikimedia Commons; licensed under the Creative Commons Attribution 2.5 Generic license and the Creative Commons Attribution-Share Alike 3.0 Unported license, added December 27, 2009: http://en.wikipedia.org/wiki/File:Alfa_beta_gamma_radiation_penetration.svg.)

penetrate human skin. Dry clothing is sufficient to protect the skin from alpha radiation. Particles that have been inhaled, absorbed through open wounds, or swallowed can cause internal damage. Because of their difficulty in penetrating objects, alpha particles generally require special instruments to be detected. Most common radiation survey meters found at licensed facilities or emergency services offices cannot accurately detect alpha radiation. There is very low risk from alpha radiation unless introduced into the human body.

- *Beta:* A particle consisting of one electron. Beta particles can travel moderate distances in air, and may be able to affect human skin in higher concentrations. The U.S. Nuclear Regulatory Commission's (NRC's) website states that a negatively charged beta particle is identical to an electron, while a positively charged beta particle is called a positron. Large amounts of beta radiation may cause skin burns, and beta emitters are harmful if they enter the body. Beta particles may be stopped by thin sheets of metal or plastic. Without reasonable protection, beta radiation is both an internal and external hazard. RAM that emit high-energy beta particles can be detected by most common radiation survey meters; lower-energy beta particles may not be detected by these same instruments.[2]
- *Gamma:* A wave of energy similar to an x-ray; gamma radiation can travel several feet in the air and can easily penetrate most materials. Without reasonable protection, gamma rays also represent both an external and internal hazard to humans. Gamma rays are similar to light and radio waves; only they give off more energy. Because of this higher energy level, gamma radiation requires dense materials such as lead or tungsten to provide reasonable shielding. The ability to penetrate materials and the higher energy level also make gamma radiation detectable by many survey instruments, including legacy Civil Defense Force survey meters.[3]

Exposure to ionizing radiation may have short-term or long-term health effects, depending on the type and amount of radiation. Suffering an acute exposure—being exposed to a high level of radiation in a short time—may cause symptoms such as skin reddening, burns, or cataracts (if the lens of the eye is exposed). Similar to being outside in the sun for too long, these effects can worsen with more exposure to radiation. This is known as a "deterministic" effect. Being exposed to moderate levels of radiation over a long period of time will lead to an increased chance of developing cancer. Using the same example as before, this "nondeterministic" or "stochastic" effect might be seen in individuals who are regularly outside in the sun for very long periods: over time, their risk of developing skin cancer increases. In all cases, however, no effects of cancer are made better or worse because of the radiation dose.

Common Uses of Radioactive Materials

Radiation is always around us. For example, the Earth is constantly being irradiated by the Sun, and uranium ore is mined out of the ground before being processed for use in nuclear reactors. This does not mean that we are in any danger from naturally occurring RAM. In fact, RAM are used to benefit society in several ways across a range of fields, including power generation, medical, industrial, research, and even home applications.

Prior to the 1950s, nuclear materials were the exclusive domain of governments attempting to develop and test nuclear weapons. During a speech to the United Nations General Assembly in December 1953, President Dwight Eisenhower presented his vision of "Atoms for Peace":

> The atomic energy agency could be made responsible for the impounding, storage and protection of the contributed fissionable and other materials. The ingenuity of our scientists will provide special safe conditions under which such a bank of fissionable material can be made essentially immune to surprise seizure. The more important responsibility of this atomic energy agency would be to devise methods whereby this fissionable material would be allocated to serve the peaceful pursuits of mankind. Experts would be mobilized to apply atomic energy to the needs of agriculture, medicine and other peaceful activities. A special purpose would be to provide abundant electrical energy in the power-starved areas of the world. Thus the contributing Powers would be dedicating some of their strength to serve the needs rather than the fears of mankind.[4]

The industries for commercial nuclear power and other nonmilitary applications grew out of President Eisenhower's vision.

According to the U.S. Energy Information Agency, as of 2012, there were 104 operable commercial nuclear reactors at 65 nuclear power plants. Since 1990, about 20% of the nation's total electricity supply has been provided by nuclear power.[5] The basic method by which nuclear power plants produce electricity is similar to other power plants: the plant uses heat to create steam, which spins turbine generators to produce electricity. The biggest difference at nuclear power plants is how that heat is generated: fission. When a uranium fuel atom is struck by a neutron, it breaks apart—creating a small amount of heat and one or two more neutrons. This breaking of atomic bonds is called fission. The newly created neutrons then strike different uranium atoms and repeat that process. Nuclear power plants create a chain reaction of fission, creating more neutrons and more heat. The United States utilizes two types of reactors: boiling water reactors (BWRs) that use the heat to boil water directly into steam and pressurized water reactors (PWRs) that take superheated water (under pressure) from a primary loop and create steam inside a steam generator. Diagrams of both types of plants are shown in Figures 24.2 and 24.3. Both types of reactors have redundant safety systems that receive power from the electrical grid, and can be powered by onsite diesel generators if necessary.[6]

Aside from nuclear power, perhaps no application has become more accepted in American society than medical uses of RAM. While almost everybody is familiar with x-ray machines, those are not the same as RAM: in hospitals and private medical offices, x-rays are produced by a machine—when the machine is off, there is no radiation risk from that device. RAM cannot be "turned off": they are always emitting energy or matter. Therefore, nuclear medicine departments use sources of

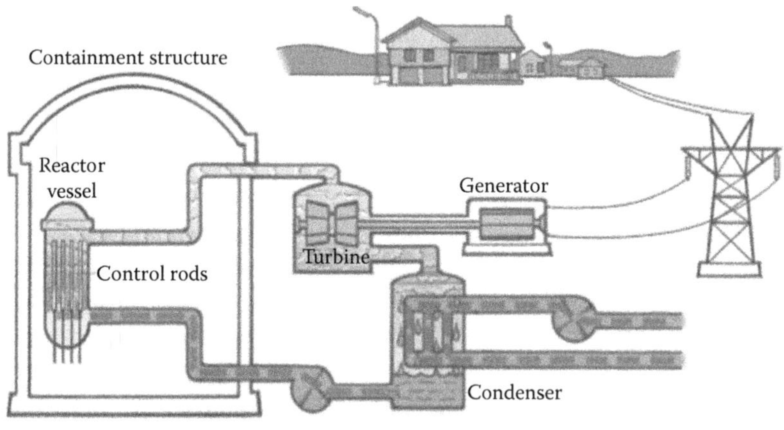

Figure 24.2 Diagram of boiling water reactor. (From the U.S. Nuclear Regulatory Commission's "Students Corner" website, last updated March 29, 2012; accessed online at: http://www.nrc.gov/reading-rm/basic-ref/students/animated-bwr.html and http://www.nrc.gov/reading-rm/basic-ref/students/animated-pwr.html.)

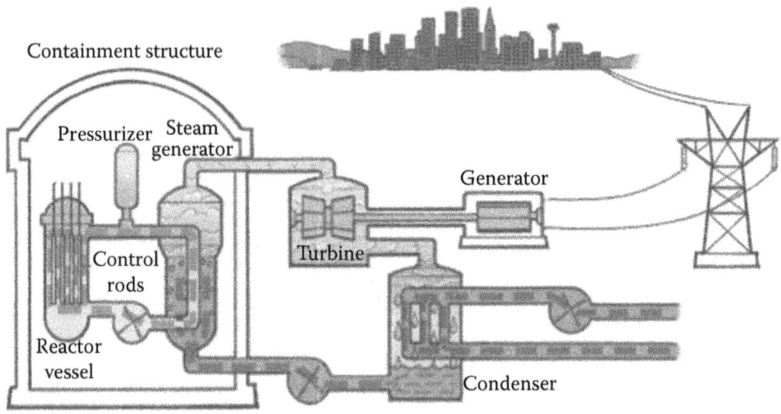

Figure 24.3 Diagram of pressurized water reactor. (From the U.S. Nuclear Regulatory Commission's "Students Corner" website, last updated March 29, 2012; accessed online at: http://www.nrc.gov/reading-rm/basic-ref/students/animated-bwr.html and http://www.nrc.gov/reading-rm/basic-ref/students/animated-pwr.html.)

licensed materials that are introduced into the human body for diagnostic or therapeutic purposes. Nuclear materials may be injected for use as tracers for heart and bone scans; swallowed as capsules to treat thyroid cancer, or implanted into tumors to directly kill cancerous cells.

While medical applications of radiation have become widely known, RAM are also used for industrial purposes. Small sealed sources of RAM are used in gauges that measure the moisture or the density of materials. Additionally, different sources of RAM are employed to examine welds for cracks or defects by a process known as industrial radiography. In this process, crews tape a piece of film on one side of the weld and crank the radiation source out to the other side, obtaining a picture of the weld—similar to a medical x-ray. Because the gauges and devices used for these operations are portable, they require multiple layers of security. Some of the same principles are also used at manufacturing or production facilities that use fixed gauges containing RAM to measure the thickness of materials or the level of product in a container.

As described in the earlier sections, radiation can be detected using calibrated instruments. RAM are often used as tracers for scientific research and development because of that property. Laboratories can chemically attach a radioactive isotope to a new drug for life sciences research, and then use survey readings to measure the presence or absence of radiation in the target organ. Separately, other research facilities use small-scale research and test reactors to expose materials to a reactor core using specially designed chambers; these facilities can then conduct experiments on these items, such as to study whether electronic components for unmanned space probes will work normally when exposed to radiation, and to test whether materials will become more brittle after being irradiated for different lengths of time.

Lastly, RAM have been used in our homes and offices with tremendous benefits. Most residential smoke detectors contain a low-activity alpha source called americium-241. Alpha particles emitted by the americium ionize the air, making the air conductive. Any smoke particles that enter the unit reduce the current and set off an alarm.[7] Additionally, small quantities of radioactive hydrogen can be used to make objects glow faintly in the dark, even without a power source. Exit signs, watches, and military weapon sights sometimes employ RAM for this purpose. These are just some examples of how radiation can provide beneficial uses to society in our everyday lives. Products for home or commercial use are manufactured to strict requirements, so that these materials are safe to be around every day.

Individual Protection

Regardless of the level of inherent safety built into a product, personal safety is always a priority. As occupational workers who work near or use RAM know, the industry-wide mantra for individual radiation safety is: time, distance, and shielding. These three basic principles apply to any and all sources of radiation for individual protection. Implementing any one tenet by itself will reduce a person's exposure to RAM, but implementing all three can help keep a person's dose as low as reasonably achievable—a standard known as ALARA.

Time

The less time an individual spends near a source of radiation, the less dose the individual will receive. Occupational radiation workers employ this principle by studying and practicing work procedures before going into areas with high radiation doses, so they can work more quickly and then leave the area. First responders can apply this principle if they discover a radioactive source by observing and/or documenting any identifiable information about the source and then retreating to report that data.

Distance

The greater the distance between an individual and a source of radiation, the less dose the individual will receive. That is because the further away a person gets from the source (whether it is ionizing radiation, light, heat, etc.), the amount of radiation over a given area will decrease proportional to the distance. This concept is called the *inverse square law*: doubling the distance from a source reduces its intensity to 1/2; tripling the distance reduces the intensity to 1/9; and quadrupling the distance reduces the intensity to 1/16 (Figure 24.4). The U.S. Environmental Protection Agency (EPA) website makes the analogy of a radiation source to a bare light bulb. Because the bulb gives off light equally in all directions (like a circle),

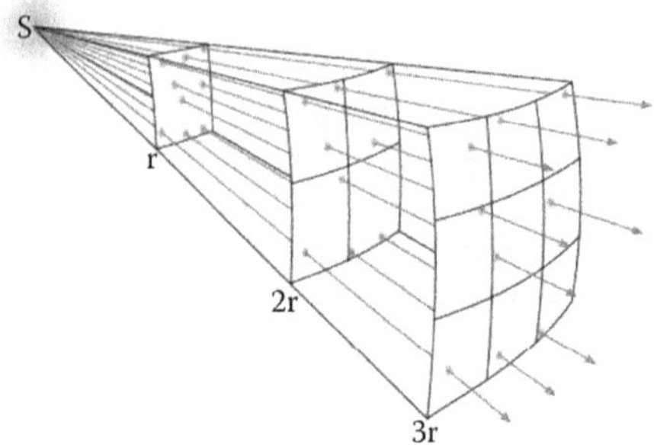

Figure 24.4 Diagram of inverse square law. (From Wikimedia Commons; licensed under the Creative Commons Attribution-Share Alike 3.0 Unported license, added April 2, 2008: http://en.wikipedia.org/wiki/File:Inverse_square_law.svg.)

the energy from that light is distributed evenly over that circle. Therefore, moving further away from the light bulb means that the energy must be evenly distributed over a larger area.[8]

Occupational radiation workers employ this principle by using long-handled tools when manipulating certain sources. Visitors and inspectors at licensed facilities may apply this principle by observing activities such as research experiments from a distance. First responders can apply this principle by keeping themselves—and others—further away from suspected sources of RAM.

Shielding

The greater the amount of shielding between an individual and a source of radiation, the less dose the individual will receive. Different types of shielding are better for the different types of radiation discussed above. For example, a thin sheet of paper is sufficient to shield alpha particles, but will not protect against gamma radiation. Heavy clothing or plastics (such as Lexan) provide adequate shielding for beta particles. Dense, heavy materials such as lead and tungsten provide better shielding for gamma radiation than other substances. Occupational radiation workers employ this principle by surrounding RAM storage locations with lead or concrete, and working behind clear plastic "L-blocks" when handling certain isotopes. First responders can apply this principle by putting shielding materials between themselves and a suspected source of radiation: a lead blanket draped over the source, moving to the opposite side of a car or truck, and relocating inside a building are three examples.

As previously discussed, radiation is either present or not. Similar to procedures for chemical and biological safety, individuals should observe hazard labels, heed warning signs, and apply these three principles to keep radiation exposure ALARA. Because radiation is used in so many different applications, several government organizations are responsible for regulating the safe use of RAM.

Roles and Responsibilities

The National Response Framework (NRF) is the guiding document that describes the proposed national response to an incident on U.S. soil. The framework is an all-hazards plan, and information about the types of events discussed in this chapter is presented in the Nuclear/Radiological Incident Annex (N/RIA) to the NRF. The N/RIA defines the roles and responsibilities of federal agencies in responding to nuclear and/or radiological incidents; discusses the specific authorities, capabilities, and assets the federal government has for responding to nuclear and/or radiological incidents; discusses the integration of the concept of operations with other elements of the NRF, including unique organization, notification, and activation processes and specialized incident-related actions; and provides guidelines for notification, coordination, and leadership of federal activities.[9]

As a general rule, state and local officials are the primary drivers in response to an incident or event under the NRF. The federal government should only become involved when those resources are exhausted, or the emergency will require more advanced capabilities. For low- and midlevel events involving RAM, the entire response may be limited exclusively to one agency at either the state or local level. For example, events at some individuals or companies may only warrant a response from the NRC or another federal agency with no state or local involvement. Significant events that are likely to impact members of the public will likely overwhelm local responders, and may trigger activations across numerous federal agencies without being specifically requested.

The N/RIA is intended to provide specific guidance to federal agencies during a response to an incident involving release of RAM. The annex designates coordinating agencies and cooperating agencies for different types of incidents. The lead agencies depend on factors such as what agency regulates the company or material involved, and whether or not the incident was deliberately caused. As described by the N/RIA, "coordinating agencies are those federal agencies that own, have custody of, authorize, regulate, or are otherwise assigned responsibility for the nuclear/RAM, facility, or activity involved in the incident. Cooperating agencies include other federal agencies that provide additional technical and resource support specific to nuclear/radiological incidents to U.S. Department of Homeland Security (DHS) and the coordinating agencies."[9] Some of the departments involved in potential radiological or nuclear event response include the following.

State and Local Agencies

As discussed above, incidents should be handled at the jurisdictional level closest to the event. Local police, fire, and emergency medical service (EMS) responders know their communities best, and have critical knowledge of the environment as well as public–private resources that may be valuable to the overall response. Some state governments are "Home Rule" states, meaning that implementation of protective actions recommendations to protect the public rests with county authorities. Additionally, some states have entered into formal agreements with the NRC to license and inspect individuals or companies that possess and utilize certain RAM. As of 2012, approximately 19,000 nuclear material users were regulated by 37 states under such agreements with the NRC.[10]

U.S. Department of Homeland Security

The DHS is the primary department responsible for the federal government's overall response to a significant event involving nuclear or radiological materials. When DHS is coordinating the federal response to such an event, it will also coordinate—in concert with applicable other governments—the overall federal recovery pursuant to the NRF.[9]

Federal Emergency Management Agency

Federal Emergency Management Agency (FEMA) is responsible for consequence management after a disaster occurs. In this capacity, FEMA is the conduit between state and local authorities and the federal government. FEMA reviews and evaluates emergency plans for counties that have a nuclear power plant. Those counties are also required to participate in annual exercises with the nuclear plant in their jurisdiction; the state and county's participation is evaluated every other year.

U.S. Nuclear Regulatory Commission

The NRC is an independent agency that licenses and inspects the civilian use of RAM. Generally, the agency's mission focuses on nuclear reactors, materials, and waste. As mentioned above, the NRC may enter into agreements with state governments to transfer regulatory oversight for nuclear materials users within a given state; however, the NRC is the primary regulator for commercial power reactors.

U.S. Environmental Protection Agency

The EPA's mission is to protect human health and the environment. The agency is best known for setting limits on the amount of materials that can be emitted into the air, water, and soil, including RAM. The EPA also conducts independent research on how radiation affects human health, and works with other federal agencies during emergencies involving RAM. EPA operates a national network of radiation monitoring stations known as RadNet, which can provide environmental readings in nearly real time. Lastly, the EPA is the coordinating agency for responses to international events involving RAM.[11]

U.S. Department of Health and Human Services, the Centers for Disease Control and Prevention, the Food and Drug Administration, and the U.S. Department of Agriculture

Collectively known as the Advisory Team for Environment, Food, and Health—or, the "A-Team" for short—representatives from these agencies provide advice to federal, state, and local governments during radiation emergencies. Each agency also has some responsibility for radiation protection within their areas of regulatory oversight.

U.S. Department of Energy

The U.S. Department of Energy (DOE) has several roles and responsibilities in this field. The department has primary oversight for research and production of nuclear weapons, in conjunction with the Department of Defense. These applications include nonproliferation efforts such as new detection technologies and tracking materials from the former Soviet Union. DOE also sponsors or conducts research and development for nuclear power sources, fuel studies, and decommissioning. Lastly, DOE has several resources that can be activated during an event through the National Nuclear Security Administration, including the Radiological Assistance Program Team for first-response assistance; the Federal Radiological Monitoring and Assessment Center for advanced dose modeling; and the Nuclear Emergency Support Team for terrorist threats.[12]

Types of Incidents

There are several different types of incidents or events that may involve RAM. Not all of these events warrant a response by federal, state, or local officials; even fewer events require any actions by members of the public. Like other hazards, nuclear and radiological incidents can be assessed in terms of risk. Risk is the product of an event's likelihood of occurring and the consequences of that event. Weighing risk is not always straightforward and should be assessed for individual communities. Accidents involving RAM happen infrequently but regularly across the country: the NRC reported over 100 "medical events" in 2010 and 2011; however, the effect of these incidents on the public is very small.[13] Weighing which type of event has the highest risk to a given population is critical for emergency managers and local officials to determine to create effective plans and procedures.

The Idaho National Laboratory (INL) maintains a database of reportable events that have occurred across the country, known as the Nuclear Materials Events Database (NMED). Although access to NMED is restricted to authorized users, the INL publishes an annual report of its data on the public website nmed.inl.gov. When available, the laboratory will also publish reports with trend data. These reports are an excellent source of information for local officials or emergency responders wishing to learn more about types of events concerning RAM. One take-away from these reports is that the low- and midlevel events described below happen infrequently but regularly, with minimal or no effect on public health and safety.

For purposes of this chapter, incidents or events involving RAM can be grouped into several different categories by relative risk. Furthermore, the words "incident" and "event" are used interchangeably here, but may have specific definitions for different government agencies. The following discussion will provide some basic information about the different groups. To prepare for the risks of incidents involving RAM, emergency managers may be best served by creating similar "tiers" of risk based on the sources in their area.

Low-Level, Contained Event

An event of this type is contained entirely by the company that possessed, stored, or utilized the RAM. Depending on the type and quantity of material and the applicable regulations, events in this category may not need to be reported at all. These events include certain medical, shielding, or disposal incidents.

Administrations of the wrong type or amount of RAM to patients for diagnostic imaging are generally considered to be low-risk. That is because the materials involved in diagnostic imaging decay quickly and have minimal effect on the body in low doses. Depending on what level of government has oversight of the hospital or clinic where the event occurred, no report may be required when the wrong dose or patient is injected. Similarly, contamination events stemming from single doses of RAM are small, well-contained, and usually involve isotopes that decay very quickly. Of course, during any such event, the responsible party should evaluate what happened, understand the root cause, and document and implement corrective actions to prevent recurrence.

Another low-level event may include shielding of RAM. Commercial, industrial, and research facilities sometimes use gauges containing sealed sources of RAMs to measure the fill level, moisture, or density of objects. These devices are often built with dense shielding that will protect the source unless opened. To enhance safety, gauges often use pneumatic systems to open the shielding that will return to a safe position if power or air flow is disrupted. Occasionally, these devices may become stuck in the unshielded position, creating a potential hazard to employees. While these devices require a response from a licensed individual or company, engineering or process controls may be sufficient to protect workers until repairs have been completed.

Lastly, some licensed companies have lost, misplaced, or even disposed of devices containing small sources of RAM. These companies are required to report the lost material to the appropriate agency, but—depending on the sources involved—no additional follow-up may be available. The reason is that some devices contain little enough material and in forms that are inherently safe. Some "generally licensed" exit signs may fall into this category because they contain a small amount of RAM that cannot be detectable by a radiation survey meter. Therefore, if the licensee and the regulator determine that one of those items has been improperly disposed of—such as to a landfill—the item may not be detectable or retrievable. These events present little or no danger to any member of the public. The NMED Annual Report for Fiscal Year 2010 indicated that there were almost 2500 reports of lost, abandoned, or stolen material between 2001 and 2010; however, this value does not reflect the actual number of lost sources, and the last few years of the decade showed decreases in the number of events reported to the NRC.[14]

Mid-Level, Localized Event

An event of this type is usually contained within the company or organization where it occurred, but likely requires some follow-up by a local, state, or federal agency. Depending on the type and quantity of RAM involved, an event of this type may have a small potential to impact members of the public. These midlevel events rarely require notification to local officials or emergency responders unless they involve lost or stolen materials. In those instances, local law enforcement response may focus more on stolen property procedures than reacting to missing RAM. As with low-level events, these incidents span a range of disciplines, including medical, commercial, and industrial facilities.

NRC regulations require that hospitals and medical providers have procedures to provide high confidence that administrations of RAM are in accordance with the doctor's written treatment plan. An unintended deviation from the medical professionals' planned administration may be a "medical event." As stated earlier in this chapter, a medical event does not necessarily mean that any harm was done to an individual; rather, the report is to inform the federal or state regulator so that they can examine the procedures and personnel involved with the event.

The NMED Annual Report for 2010 stated that 399 medical events were reported between 2001 and 2010. Many different types of incidents may qualify as a medical event and require reporting to the appropriate regulatory agency, including delivering treatment doses to the wrong site within the body; delivering an incorrect treatment dose; administering an incorrect isotope or radiopharmaceutical to a patient; and administering a treatment dose to the wrong individual. These types of incidents are more significant than the low-level events discussed previously because they have the potential to negatively affect the patient—either through the effects of unintended radiation dose on the body (such as a higher dose to an internal organ), or due to insufficient treatment (such as an underdose to a cancerous tumor). Lastly, because some medical treatments require implanting patients with RAM, there is some potential that nearby individuals could potentially be exposed. Medical facilities are required to determine whether or not a patient can be released from their facility after such a procedure. Under these rules, the facility must provide instructions to keep the probability of others being exposed as low as possible. However, owing to the possibility of people other than the patient being exposed, medical events are scrutinized very closely by regulatory agencies. Local agencies and EMSs are not usually informed about medical events unless the medical facility is concerned about other individuals being exposed to undue amounts of radiation.

Releases of licensed materials and/or contamination represent another midlevel event that may be contained entirely within a facility, or may have impacts that extend offsite. The NMED Annual Report for Fiscal Year 2010 reported 137 such events between 2001 and 2010, including several different types of incidents: contamination on personnel or equipment, accidental disposal of RAM to the normal trash, and leaking sources of RAM. Licensed facilities are required to have a radiation safety officer (RSO) who is responsible for implementing a radiation safety program. This individual is primarily responsible for providing information to regulatory agencies during an incident. Additionally, the RSO should be trained and qualified to respond to contamination. In the event that contamination or other sources of radiation spread outside of a licensed facility, the RSO should conduct surveys, wipe tests, and decontamination procedures. Local agencies may or may not be notified, either by the RSO or by the regulator, after such an event.

Another midlevel event that will likely involve the local emergency response community is physical damage to sources of radiation. Just like homes and high-rises, commercial, industrial, and research facilities are vulnerable to unexpected floods, fires, and even explosions. In the event of physical damage, responding police, fire, and EMS may not know beforehand that the facility possesses RAM. The RSO is especially important during these incidents, as both a liaison and source of information for emergency personnel. Well-prepared licensees should have emergency and operating procedures that include appropriate notifications to first responders. Hazard indicators such as warning and/or caution signs, Department of Transportation (DOT) labels, and shielded areas can help inform responders where licensed materials are located during an emergency. Unless there is a large fire or explosion that spreads licensed material indiscriminately, there should be no action required by members of the public during this type of event.

Section II: Significant Events

Types of Significant Events

The final category is significant events, or those that require a larger and more coordinated response. Here, it is important to understand the difference between "nuclear" and "radiological" events: nuclear events are those that involve nuclear weapons, such as a bomb or improvised nuclear device (IND); radiological events are those that use, spread, divert, or sabotage RAM. These events include a radiological dispersal device (RDD), a radiological exposure device (RED), a transportation incident, or a significant release of RAM that forms a plume. These events may be accidental or deliberate, and it is extremely unlikely for officials to know which during the initial response phase. One commonality shared by these events is that they are not contained to a licensed facility or location: the presence of RAM somewhere it should not be may require protective actions for nearby members of the public.

Nuclear Detonation

The U.S. government, joined by countries around the world, has expressed its concern about nuclear terrorism. In March 2012, President Barack Obama stated during a Nuclear Security Summit that "nuclear terrorism is one of the most urgent and serious threats to global security."[15] Nuclear weapons are assessed by the amount of energy that can be released in their detonation, measured in kilotons (kT). For example, a one kT yield nuclear device has roughly the same amount of explosive force as one thousand tons of TNT. As described by the *Journal of Nuclear Medicine*, even a very low-yield weapon, "0.01 kT, would have an explosive impact greater than the Oklahoma City bombing in 1995. The nuclear weapons detonated at Hiroshima and Nagasaki had a yield of 15 and 21 kT, respectively."[16]

Besides the explosive force of a nuclear detonation, such an event will also produce a delayed effect: radioactive fallout. While the initial impact will damage or destroy objects within a radius depending on its yield, fallout will spread outward from the impact site. A report from Lawrence Livermore National Laboratory stated that fallout "is generated when the dust and debris excavated by the explosion are combined with radioactive fission products produced in the nuclear explosion and drawn upward by the heat of the event, often forming a 'mushroom cloud' from which highly radioactive particles drop back down to earth as they cool."[17] Radioactive fallout will not be limited to the impact area of the nuclear detonation; it will be spread by the wind. This creates additional safety challenges to members of the public as well as first responders attempting to operate near the scene, and generates critical decision points for responding officials.

A nuclear detonation is a very low probability event but would have a very high impact. Although emergency management principles state that incidents should be handled at the level of jurisdiction that is closest to the event, a domestic nuclear detonation would immediately trigger a large-scale federal response. In that scenario, state and local resources may be called upon to support a federal mobilization rather than the other way around.

One very important distinction between a nuclear detonation and the other types of significant events discussed in this chapter is that RDDs, REDs, and nuclear power plants physically cannot create a nuclear explosion. Specific types and quantities of highly enriched RAM are required to construct a bomb capable of the devastation caused by a nuclear detonation. Other events can spread material around and/or expose individuals, but no other incident can cause such an explosion.

Radiological Dispersal Device

RDD are designed to intentionally spread RAM over an area. Although these devices are commonly presumed to be explosives-based "dirty bombs," an explosion is not required for an RDD to distribute RAM. Similar to biological or chemical agents, RAM can be dispersed through nonexplosive means; however, modern definitions of RDDs generally consider the most significant threat to be from an explosive device. An RDD spreads material over an area, whether by an explosive force or another means. Most injuries or fatalities after an explosive RDD will likely be caused by the explosion itself, and not any radioactivity spread by the event.

An RDD is a disruptive device: by itself, an explosion affects a small area; however, an explosion that spreads RAM could contaminate nearby people and the surrounding environment. Depending on the type and quantity of RAM being used, as well as the dispersion pattern following the explosion, individuals located near an RDD detonation are not likely to be contaminated to a level that requires medical attention. The buildings and areas surrounding the explosion, however, will likely require some kind of decontamination. That effort may take a significant amount of time. The uncertainty of detecting, assessing, and responding to a dispersal device introduce challenges in dealing with an RDD.[18]

First, emergency responders called to an explosion may not have any indication of the presence of RAM. As described previously, radiation cannot be seen or smelled, and requires survey instruments to detect. To the extent possible, first responders and emergency managers should be mindful that RAM may not be detected until after the initial response to the bomb detonation is completed. As with any explosive device, hazards such as fire, shrapnel, or chemicals may affect the response effort. Second, contamination caused by an RDD may require buildings or outdoor areas to be closed for decontamination. This may affect shipping corridors, traffic patterns, and the local economy. A 2012 study published in the journal *Risk Assessment* estimated that a hypothetical dirty bomb attack on downtown Los Angeles' financial district could affect the region's economy at a cost nearly $16 billion, fueled primarily by psychological effects that could persist for a decade. Of that total, only about $1 billion included the immediate costs such as injuries, cleanup, and business closures. The rest would be from lasting psychological aversion to shopping or traveling through the affected area.[19]

Radiological Exposure Device

Although many people have heard of a "dirty bomb" through movies or television shows, the concept of a radiological exposure device is not so well known (Figure 24.5). In simple terms, an RED consists of a radiation source that is hidden so that it exposes individuals without their knowledge. The potential harm from an

150 Ci Iridium-192 source under seat

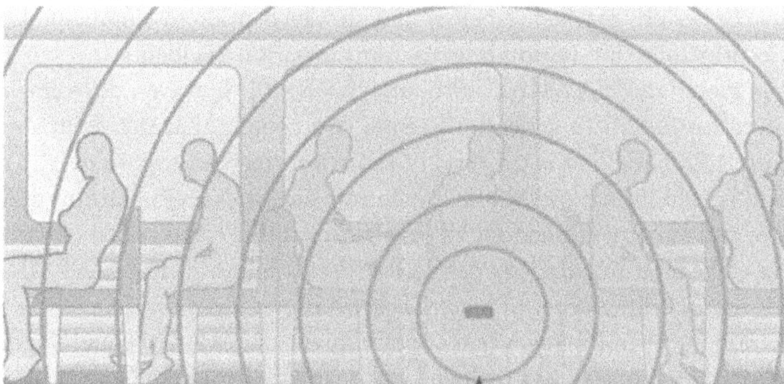

Figure 24.5 Diagram of a radiological exposure device in train car. (Adapted from U.S. Department of Health & Human Services, Radiological Emergency Medical Management, Radiological Exposure Devices (REDs). Last updated December 28, 2011, copied from REMM website; original image online at: http://www.remm.nlm.gov/red.htm.)

RED can be evaluated using the basic principles of time, distance, and shielding: individuals will be exposed to more radiation if they are closer to the source, if they remain near the source longer than others, and they have less shielding between them and the source.[20]

For purposes of this discussion, an RED is considered to be a sealed source of RAM. Unsealed RAM (in a liquid or powder form, for example) can spread and cause contamination; this would more accurately be described as an RDD. One unique challenge with an exposure device is that there is no telltale explosion or event to reveal its presence; it is only known about when discovered. Therefore, it may be difficult to determine when the device was placed and when the first person was exposed. An attack of this type has a low likelihood of causing mass casualties because the device remains in a fixed location while people come and go around it. In a public setting, it could be very difficult to discover who was exposed to low doses of radiation. Officials responding to a suspected RED should exercise caution and employ the principles of time, distance, and shielding for personal protection.

Transportation Incident

Emergencies and incidents during transportation can happen involving any type of material—including chemical, biological, explosive, or radiological. As with any of the events discussed in this section, such an incident may be deliberate or accidental. Because of this, the U.S. DOT has regulations that specify how packages must be tested, marked, labeled, and surveyed prior to shipment. Many of the same principles for responding to transportation incidents involving other hazardous materials (HAZMAT) also apply to RAM; however, different hazards are involved for RAM. In general, there are two primary hazards from a package containing RAM: the first is the level of radiation that can expose nearby people; the second is contamination. While it is in transit, packages containing RAM are covered by the aforementioned DOT regulations. DOT regulations in Title 49 of the Code of Federal Regulations define RAM as a Class 7 HAZMAT. That does not mean that six other types of materials are more or less hazardous—it is just an identification number. Packages that contain quantities of RAM equal or greater to set thresholds are required to be labeled and marked (Figure 24.6).

Packages with a white RADIOACTIVE-I label have practically no radiation levels at their surface; packages with a yellow RADIOACTIVE-II label have some detectable radiation levels at their surface; and packages with a yellow RADIOACTIVE-III label have the highest radiation levels of the three. Packages with a very small quantity of RAM are not required to have any of these labels. A carrier transporting one or more RADIOACTIVE-III packages must also use placards on the outside of his

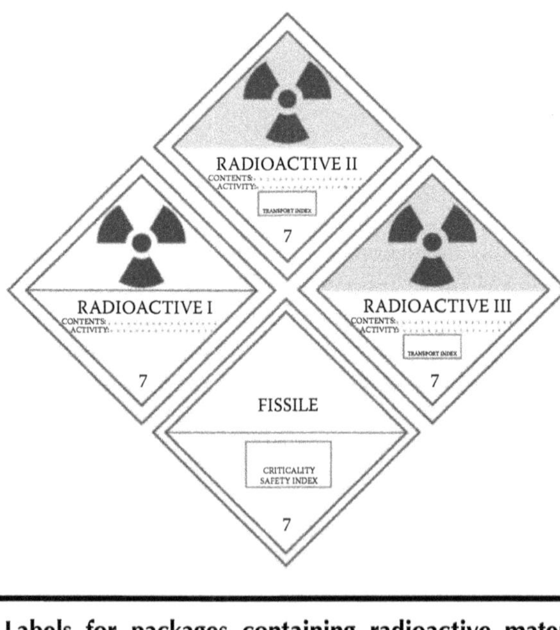

Figure 24.6 Labels for packages containing radioactive materials. (Adapted from U.S. Department of Transportation, Pipeline and Hazardous Materials Safety Administration. *How to Handle Radioactive Materials Packages: A Guide for Cargo Handlers*. Washington, DC. Revised September 2006; obtained online at: http://phmsa.dot.gov/staticfiles/PHMSA/DownloadableFiles/Files/Handle_Radioactive.pdf.)

or her vehicle. This should be an immediate indicator to first responders arriving onto a scene. All packages, regardless of their labels, may also have an amount of removable (nonfixed) radioactive contamination on their surface. Shippers and receivers of these packages are required to test for contamination under DOT, NRC, and state regulations. Although the presence of contamination does not necessarily represent a threat to public health and safety, detecting contamination using wipe tests or survey instruments may indicate a nearby source of RAM.

As mentioned, responding to a transportation incident shares many elements with responding to other types of HAZMAT. Responders should use the principles of time, distance, and shielding to minimize potential doses. In addition to the RADIOACTIVE labels that display relative levels of radiation, packages should be marked with a four-digit ID number, such as "UN 3332." The DOT releases an "Emergency Response Guidebook" every four years that first responders can utilize for transportation incidents. The book has guides that correspond to the ID number on package exteriors. The ID number in the example above corresponds to Guide No. 164 in the 2008 Emergency Response Guidebook: "Radioactive Materials (Special Form/Low to High Level External Radiation)." Guide No. 164 then describes the potential hazards, public safety concerns, and emergency response procedures for different emergency scenarios.[21]

Release of Material

Local, state, and federal responders must work together to form a coordinated response for plumes of RAM outside a licensed facility. Local and state responders are best equipped to identify populations that may be affected by a release, communicate protective action recommendations (such as to evacuate or to shelter in place, for example), and conduct life-saving operations if needed. Federal and state resources will support local responders as appropriate, while also calculating plume models and assessing event escalation pathways. There may be similarities in the responses for RDDs and nuclear detonations, not only in the event response phase but also in the investigation of exactly what happened and why.

Emergency managers from communities that are concerned about these types of incidents may wish to develop standard operating procedures or prearranged plans with state emergency management agencies. Another resource may be counties that contain or neighbor a county with a nuclear power plant. FEMA requires that each of these counties participate in an annual emergency preparedness exercise with the nuclear power plant, and evaluates the counties every other year. County emergency plans—including mutual aid compacts, evacuation routes, procedures for obtaining resource allocation, and command and control protocols—are regularly tested and updated.

When a significant incident requires a law enforcement investigation—such as during suspected terrorist attacks—it can add additional complexity to a potentially large-scale response. As mentioned above, officials will likely not know whether the event was accidental or deliberate during the initial phase of the response. During a significant event, the Federal Bureau of Investigation (FBI) is the lead agency in charge of the federal government's investigation. The FBI may investigate whether the act was terrorism, industrial sabotage, an accident, or something else. While lifesaving operations and protecting the public will be first responders' primary responsibility, the impact site may be considered an active crime scene. As such, the FBI and/or local law enforcement agencies may need access to the site to gather information for rapid assessment and comparison to available intelligence.

Events at Commercial Power Reactors

Emergencies at commercial nuclear power plants are not listed in the event categories above. That is because power plants have their own scale for assessing incidents involving the reactor, power-generating equipment, or other areas of the plant property. After stories about the events at Three Mile Island, Chernobyl, and more recently Fukushima Dai-ichi, many members of the public may believe that any emergency at a power reactor will result in a large-scale, mass-casualty event. That is not correct. Each nuclear power plant in the United States is required to develop, implement, and maintain an emergency plan. These emergency plans describe specific conditions or occurrences that will trigger the activation of their plan, called emergency action levels or EALs.

The existence of an EAL at a nuclear power plant does not mean that a threat to public health and safety exists. EALs can range from disruption of communications systems and temporary equipment unavailability all the way up to the potential for a release of materials in a plume. Based on its EALs, nuclear power plants can declare one of four emergency classifications to the NRC. The following definitions are copied directly from NRC's website[22]:

Notification of Unusual Event

Events are in process or have occurred that indicate a potential degradation of the level of safety of the plant or indicate a security threat to facility protection. No releases of RAM requiring offsite response or monitoring are expected unless further degradation of safety systems occurs.

Alert

Events are in process or have occurred that involve an actual or potential substantial degradation of the level of safety of the plant or a security event that involves probable life-threatening risk to site personnel or damage to site equipment because of intentional malicious dedicated efforts of a hostile act. Any releases are expected to be limited to small fractions of the EPA Protective Action Guideline exposure levels.

Site Area Emergency

Events are in process or have occurred that involve an actual or likely major failure of plant functions needed for the protection of the public or security events that result in intentional damage or malicious acts (1) toward site personnel or equipment that could lead to the likely failure of or (2) prevents effective access to equipment needed for the protection of the public. Any releases are not expected to result in exposure levels that exceed EPA Protective Action Guideline exposure levels beyond the site boundary.

General Emergency

Events are in process or have occurred that involve actual or imminent substantial core degradation or melting with the potential for loss of containment integrity or security events that result in an actual loss of physical control of the facility. Releases can be reasonably expected to exceed EPA Protective Action Guideline exposure levels offsite for more than the immediate site area.

Nuclear power plants are required to notify state and local officials within 15 min of an emergency classification, and the NRC as soon as possible thereafter. The appropriate response to an emergency classification depends on several factors, including the type of event and the amount of uncertainty in the event details. Based on these factors, the appropriate reaction may range from a simple notification receipt by government officials up to response actions by a range of agencies at the local, state, and federal levels.

Radiation Response Myths and Myth-Busting

Myths, urban legends, and public perceptions are difficult to overcome under the best of circumstances. It is understandable for people to readily accept their fears of the "worst-case scenario" during an event involving RAM. Public officials, emergency responders, technical experts, members of the media, and many more will likely present lots of information to concerned citizens all across the world. Even if that information is meant to inform and allay peoples' fears, in some people, it will have the opposite effect. Another chapter in this textbook addresses crisis communications using traditional as well as social media tools, so that will not be discussed here. Below are some response-related myths and facts regarding nuclear and radiological events:

1. Because of their potential health effects, securing RAM is the highest priority during an emergency response.
   - *Fact*: Under the Incident Command System, priorities and objectives are determined by the first responders to any emergency scene. For most situations, life safety, responder safety, and minimizing property damage are three top priorities. In May 2011, an EF5 tornado with 200 mph winds severely damaged Joplin, Missouri, including the St. John's Mercy Hospital. The top priority during that disaster was patient evacuation followed by triage. Recovery of hazardous materials such as RAM, biohazards, and chemicals occurred over the first 2–3 days, with a focus on risk mitigation. The presence of RAM should not by itself drastically alter first responders' missions.
2. During a nuclear or radiological event, the federal government will "take over" whatever it wants, including local resources and private assets.
   - *Fact*: Federal agencies' first priority during any significant event will be to make the situation safe and work with state and local officials to protect the public. The federal agencies and departments listed in this chapter do not have the regulatory or statutory right to "take over" privately held resources. This is unlike a 2006 episode of the NBC television show *The West Wing*, which featured Martin Sheen as the President of the United States who directed the Chairman of the NRC to send engineers to take control of a nuclear plant suffering from a catastrophic accident. In reality, federal, state, and local officials work in collaboration with each other to support private companies—including nuclear power plants—to protect people and the environment.
3. Nuclear detonations and/or RDD create fallout that makes items radioactive.
   - *Fact*: Radioactive fallout is caused by dust and debris being caught-up with the RAM from the explosion's origin. RAM are different from radioactive contamination: RAM emit radiation until they have fully decayed; contamination, on the other hand, can be cleaned and removed. If an object such as clothing or a building

becomes contaminated, it is not made radioactive—the clothing or building itself does not give off energy. Therefore, if an individual believes that he or she has become contaminated by fallout, the best thing to do is to remove the affected clothing (if facilities are available), place them in a sealed plastic bag (to allow for testing later), and take a shower to wash off dust and dirt (shampoo hair, but no need for conditioner). These simple steps can reduce up to 95% of the contamination on a person's body.

4. Government agencies tell us that any amount of radiation has some risk; therefore, the only truly safe level of radiation is zero.
   - *Fact*: There is no such thing as "zero levels" of radiation ... radiation is always around us. Even when completely isolated from any identifiable RAM, a background level of radiation surrounds us every day. This background radiation is caused by natural and man-made sources: naturally occurring radon contributes to this background, as do the structures we live and work in. Some elements and materials are naturally radioactive: rocks, minerals, and ores often contain a very small percentage of radioactive isotopes. Because of this, some building materials such as marble and granite are ever-so-slightly radioactive. Anything grown in the ground also has a tiny amount of radiation. Take bananas, for example: bananas contain potassium, which human beings need to survive. The natural potassium in bananas and other foods contains a minuscule amount of radioactive potassium-40. It would take a comically large number of bananas or granite countertops or other sources to cause any kind of health effects from radiation.
5. Radiation from a nuclear detonation or a dirty bomb is worse than naturally occurring radiation because it is man-made.
   - *Fact*: There is no difference between man-made and naturally occurring radiation. The health effect of these materials depends on the type of radiation they emit: alpha, beta, or gamma, for example. Individuals can protect themselves from all RAM, whether they occur naturally or not, by using the basic principles of time, distance, and shielding.

Section III: Protective Actions

Protective Action Recommendations

After a nuclear or radiological incident, it is reasonable to assume that confusion and even panic may be prevalent. Historical evidence has shown that media portrayals may not have fully vetted their facts in the rush to provide "breaking news." Additionally, even peer-reviewed articles or websites may present facts using terminology that sound very similar, but actually have different meanings.

Regardless of the incident's representation in the media, emergency responders have identified protective actions and recommendations that can be implemented to protect public health and safety. These recommendations may come from various sources: federal or state agencies, the utility owning the affected nuclear power plant, the academic community, and even the military. Responsibility for approving and implementing the recommended course of action usually falls to the state government. Some states are designated as "Home Rule" states in their constitution, meaning that the responsibility for approving protective action recommendations has been delegated to the county governments.

Protective actions are designed to reduce or eliminate the public's exposure to radiation or other hazards following a significant incident. These actions can have two goals: primary means to protect the public in the short term and secondary means that focus on both long-term health and safety in addition to specific procedures to respond to the event. Generally, these actions should focus on managing the scene or managing the victims.[23]

Primary Protective Actions

These are specific actions that can be implemented by members of the public or responders to have a significant, short-term effect on health and safety. Sheltering, evacuating, relocating, and intercepting food or water represent four primary protective actions that should be evaluated following a significant event. It is important to realize, however, that there may not be sufficient information in the immediate aftermath of a significant event to determine radiation levels or make adequate dose projections.[20] Models and assumptions can be helpful when determining what primary protective actions to take. Any assumptions and models should be documented in detail so that they can be compared against data taken from the scene at the earliest opportunity.

Sheltering in place may sound simple, but it is one of the most effective ways to minimize exposure to RAM. When considered against the basic principles discussed at the beginning of this chapter, sheltering provides both shielding and some distance from radioactive fallout, contamination, or a plume. An effective shelter is especially vital after a nuclear detonation. A study conducted by Lawrence Livermore National Laboratory determined that the largest potential for reducing

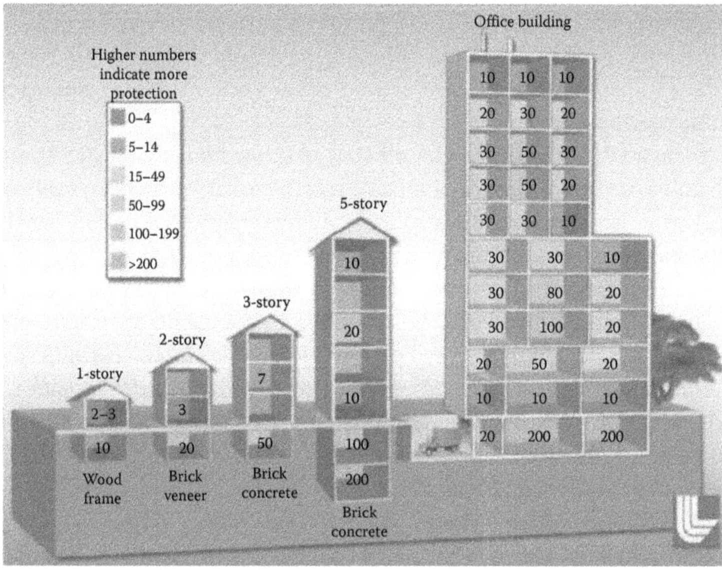

Figure 24.7 Protection factors for sample shelters. (Adapted from B.R. Buddemeier and M.B. Dillon. *Key Response Planning Factors for the Aftermath of Nuclear Terrorism,* Lawrence Livermore National Laboratory LLNL-TR-410067, August 2009, and adapted from the Armed Forces Radiobiology Research Institute Training and Reports.)

injuries and fatalities during a postdetonation response phase comes from reducing exposure to radioactive fallout. The report categorized adequate shelters as "locations that have as much earth, building materials, or distance between the occupants and exposed horizontal surfaces as possible. Exposed horizontal surfaces accumulate fallout. Buildings do not have to be air-tight. Broken windows do not greatly reduce the protection offered by a shelter." Basements, multistory brick or concrete structures, multistory shopping mall interior spaces, and subway tunnels are examples of adequate shelters. Individuals in an adequate shelter should remain there, while individuals in cars, single-story structures, or above ground should move to an adequate shelter as soon as possible.[19] A graphic showing the relative protection for various shelters is presented in Figure 24.7.

The greater the protection factor, the more protected an individual in that space is from the effects of radioactive fallout. The report concluded that a person on the top floor or an outer room on the ground level of the office building pictured would have a protection factor of 10: that individual would receive 1/10 (or 10%) of the exposure that someone outside would receive. In contrast, someone in the core of the building halfway up would have a protection factor of 100 and receive only 1/100 (or 1%) of the outdoor exposure. In fallout areas, knowing locations with adequate protection factors could prevent a potentially lethal exposure. After a nuclear detonation, individuals in fallout areas should shelter for at least the first hour. After a transportation incident or an event at a commercial power reactor, the shelter time may be different.

The discussion above shows why officials may recommend sheltering in place, although an immediate and understandable reaction to an emergency may be to run away. Sheltering is not a permanent solution, however, and evacuation may become necessary. In some communities—such as those located near commercial power reactors—evacuation plans may have been developed that utilize specific, labeled escape routes. Listening to emergency responders and local officials, traveling along prescribed routes, and adjusting based on radiation levels and travel times is referred to as an informed evacuation. Sheltering followed by an informed evacuation is considered the best way to reduce exposure to radioactive fallout. Conversely, a radial or uninformed evacuation may be recommended by emergency responders. In this scenario, individuals within certain distances from the event may be advised to evacuate in directions away from the incident. For example, a state government may recommend that everyone located within 5 miles from a power plant as well as everyone located within 10 miles in the downwind direction should evacuate. This recommendation may be appropriate as a precautionary action, or as a reaction to events that do not have a high probability of immediate fallout.

For incidents involving RAM, evacuation is different from relocation. While evacuation is an immediate action, relocation is a long-term—or even permanent—removal from an area. The purpose of relocation is to protect individuals from contamination or fallout by allowing it to decay and creating space for workers to decontaminate, if possible. The EPA has developed protective action guides that specify what dose levels would require individuals to relocate for 1, 2, or 50 years. Although considered a primary protective action, relocation would be considered in the intermediate to late phase of an incident. It is mentioned here as a reminder for emergency responders and officials to consider potential long-term actions when responding to a significant event. Failing to do consider this and subsequently adjusting the relocation area will confuse the public and could cause people to distrust the responders.

Another strategy to protect the public during the early to intermediate phase of an event is to implement controls on food and drinking sources. Interdicting food and water that may have been contaminated can minimize doses in both the short and long terms. In the short term, recommending the avoidance of certain contaminated items can prevent individuals from ingesting RAM. After the March 2011 disaster at the Fukushima Dai-ichi nuclear power plant, Japanese officials found elevated levels of radioactive iodine-131 that were up to five times the allowable levels in milk supplies at nearby dairy farms. Although this level did not pose any immediate threat to the public's health, government officials asked all dairy farms within 18 miles of the plant to halt all milk shipments.[24] For long-term health and safety, interdiction efforts for significant incidents may involve moving livestock indoors, using covered feed, or delaying or halting shipments of produce.

Every protective action recommendation has benefits and drawbacks. While sheltering may provide the best protection from radioactive fallout, it also limits individuals to the food and supplies they have at their homes. Officials may recommend that citizens evacuate from an area as a precaution, but with evacuation comes traffic, accidents, and potential hoarding of supplies. Even when the best recommendation is to take no protective action, some people may self-evacuate or buy supplies for an extended shelter. Emergency managers and officials facing a significant incident need to weigh all of the potential pros and cons to determine how to best protect the public. There are a number of resources that can help communities develop emergency plans and standard operating procedures to aid these deliberations.

Secondary Protective Actions

Secondary protective actions are more specific to the incident type. These steps may focus on either specialized response to the event or long-term health and safety. Accordingly, these activities may be taken during the intermediate or late phase of the response. A sample secondary action includes medical countermeasures. Countermeasures may include issuing potassium iodide, a substance that saturates the thyroid and minimizes the risk from accumulating radioactive iodine in the body. Further medical responses may focus on treating victims to either radiation exposure or contamination. Decontamination methods can differ based on the type and quantity of contamination: contaminated clothes should be removed, while tepid water and soap are sufficient for most personnel decontamination. Do *not* vigorously scrub individuals who have contamination on their skin with rough fabrics. This practice was previously thought to be the best way to remove contamination, but has the potential to damage the skin and therefore allow radioactive particles to enter the body. Based on dose projections and radiation levels, responding officials may implement additional secondary protective actions such as area access controls and further restrictions on food and water.

Ongoing Protective Actions

Individuals and first responders alike can take actions to protect themselves during or after a significant event involving RAM. Anyone without specific training or radiation detection equipment should follow responder instructions. Individuals can minimize their exposure to radioactive fallout or a plume by utilizing the basic principles of time, distance, and shielding: for members of the public, that could mean remaining in an adequate shelter until further instructions become available; for response personnel, that could mean reducing time near the affected area.

First responders should be aware of equipment that may help to minimize their exposure. Personal protective equipment (PPE) is frequently used by fire fighters, but may not be as helpful after one of the significant events described in this chapter. That is because gamma radiation from fallout can penetrate even thick clothing, including anticontamination suits, self-contained breathing apparatus (SCBAs), and HAZMAT suits. Additionally, wearing bulky suits and elaborate respiratory protection may increase responders' exposure because they reduce workers' speed, efficiency, and impact the ability to communicate.[25] Personal dosimeters are passive devices that do not require electrical power, and may be distributed for responders to assess their overall dose. Alternatively, staff members may be given electronic and/or alarming dosimeters that produce an audible warning when the ambient dose rate exceeds a set level. If provided, these devices should be worn and checked regularly.

During a response to a significant event, one of the most important tools for police, fire, EMS, and other responders to utilize is guidance on how long to stay in the affected area. The desire to work extra hours during lifesaving operations or victim triage may result in undue exposure to responders. Following prescribed guidance on how long to work or remain in affected areas and other steps for personal safety is especially vital after any of these significant events. In April 2011, HHS developed a "State & Local Planners Playbook for Medical Response to a Nuclear Detonation" that may serve as an additional resource to assist organizations in developing appropriate guidance.

Conclusion

RAM provide a number of benefits to American society across a variety of fields, from electricity to medicine, and from industry to research. Although events can occur involving these materials, they are infrequent and usually contained within licensed facilities. By reviewing basic information about radiation, assessing potential incidents, and summarizing protective actions, hopefully you agree that the stories told by Hollywood and Science Fiction authors are just that: fiction. Although many community leaders and emergency managers may not know where RAM is located within their jurisdiction, in most cases, they may not need to. Regulatory and safety protocols require RAM to be secured and individuals to be protected.

Despite how concerning the elements of this chapter may sound for individuals who are only just beginning to learn about radiation, there is no imminent or expected threat from any of these events. Even if an event were to occur, the organizations that would be involved have developed and practiced detailed plans to protect public health and safety. Every concerned citizen can assess his or her own preparedness using the DHS website ready.gov. Individuals living near a nuclear power plant or a large licensed facility should be familiar with information distributed by the owner of that facility. This information provides details about emergency preparedness, warning sirens, and evacuation plans. Lastly, many communities offer automated telephone and text message systems to alert residents of emergency conditions; these represent two tools to stay informed before, during, and after an incident.

As with all hazards, concerned officials, responders, and communities should not wait until an event has occurred to develop an adequate response plan. Emergency workers and officials should know what to look for when responding to an incident that potentially involves nuclear and/or radiological materials. A variety of resources from the federal government and private companies exist to help state and local communities prepare for incidents involving RAM. Many of the citations and agencies referenced in this chapter maintain comprehensive websites that can aide communities, including the HHS "Radiation Emergency Medical Management" (REMM) website, the CDC "Radiation Emergencies" website, and the NRC "How Can I Prepare for a Radiological Emergency" website. By staying informed, using common sense, and employing the basic principles of time, distance, and shielding, everyone can minimize their exposure to radiation.

Abbreviations

ALARA	As low as reasonably achievable
BWR	Boiling water reactor
CDC	U.S. Centers for Disease Control and Prevention
DHS	U.S. Department of Homeland Security
DOE	U.S. Department of Energy
DOT	U.S. Department of Transportation
EAL	Emergency action level
EMS	Emergency medical service
FBI	Federal Bureau of Investigation
FDA	U.S. Food and Drug Administration
FEMA	Federal Emergency Management Agency
EPA	U.S. Environmental Protection Agency
HAZMAT	Hazardous materials
HHS	U.S. Department of Health and Human Services
IND	Improvised nuclear device
kT	Kiloton
NMED	Nuclear Materials Events Database
N/RIA	Nuclear/Radiological Incident Annex
NRC	U.S. Nuclear Regulatory Commission
NRF	National Response Framework
ORAU	Oak Ridge Associated Universities
ORISE	Oak Ridge Institute for Science and Education
PPE	Personal protective equipment

PWR	Pressurized water reactor
RAM	Radioactive materials
RDD	Radiological dispersal device
REAC/TS	Radiation emergency assistance center/training site
RED	Radiological exposure device
REMM	Radiation Emergency Medical Management
RSO	Radiation safety officer
USDA	U.S. Department of Agriculture

References

1. Radiation Emergency Assistance Training Center/Training Site (REAC/TS) at the Oak Ridge Institute for Science and Education (ORISE), managed by Oak Ridge Associated Universities (ORAU) for the U.S. Department of Energy. http://orise.orau.gov/reacts/guide/define.htm.
2. U.S. Nuclear Regulatory Commission. *Glossary*. Last updated March 29, 2012. http://www. nrc.gov/reading-rm/basic-ref/glossary.html.
3. Radiation Emergency Assistance Training Center/Training Site. http://orise.orau.gov/reacts/guide.
4. President Dwight Eisenhower, presentation to the 470th Plenary Meeting of the United Nations General Assembly, December 8, 1953. http://www.iaea.org/About/history_speech.html.
5. U.S. Energy Information Agency. *Energy in Brief: What Is the Status of the U.S. Nuclear Industry?* Last updated April 22, 2011. http://www.eia.gov/energy_in_brief/nuclear_industry.cfm.
6. U.S. Nuclear Regulatory Commission. *Power Reactors*. Last updated March 29, 2012. http://www.nrc.gov/reactors/power.html.
7. Health Physics Society website. *Is Anything We Use in Everyday Life Radioactive?* Last updated August 17, 2011. http://hps.org/publicinformation/ate/faqs/consumerproducts.html.
8. U.S. Environmental Protection Agency website. *Radiation Protection Basics*. Last updated July 8, 2011. http://www.epa.gov/radiation/understand/protection_basics.html.
9. Nuclear/Radiological Incident Annex, June 2008. http://www.fema.gov/pdf/emergency/nrf/nrf_nuclearradiologicalincidentannex.pdf.
10. The Honorable Gregory B. Jaczko, Chairman, U.S. *Nuclear Regulatory Commission. Nuclear Security in the New Threat Environment*. At GovSec Conference, Washington, DC, April 4, 2012.http://www.nrc.gov/reading-rm/doc-collections/commission/speeches/2012/s-12-007.pdf.
11. U.S. Environmental Protection Agency. *Radiation Protection: Basic Information*. Last updated March 13, 2012. http://www.epa.gov/rpdweb00/basic/index.html.
12. U.S. National Nuclear Security Administration, part of the U.S. Department of Energy. *Responding to Emergencies*. http://nnsa.energy.gov/aboutus/ourprograms/emergencyoperationscounterterrorism/respondingtoemergencies.

13. U.S. Nuclear Regulatory Commission (NRC) presentation. *Status of Medical Events FY2011, Donna-Beth Howe, PhD, September 23, 2011.* http://pbadupws.nrc.gov/docs/ML1126/ML11264A176.pdf.
14. Nuclear Materials Events Database Annual Report, Fiscal Year 2010. Idaho National Laboratory. Dated February 2011. http://nmed.inl.gov/AnnualReports/NMEDFY10%20Annual.pdf.
15. President Barack Obama. *Remarks by President Obama at Opening Plenary Session of the Nuclear Security Summit*, Seoul, South Korea, March 26, 2012. http://www.whitehouse.gov/the-pressoffice/2012/03/26/remarks-president-obama-opening-plenary-session-nuclear-security-summit.
16. D.J. Barnett et al. Understanding radiologic and nuclear terrorism as public health threats: Preparedness and response perspectives, *Journal of Nuclear Medicine*, 47(10):1653–1661, 2006.
17. B.R. Buddemeier and M.B. Dillon. *Key Response Planning Factors for the Aftermath of Nuclear Terrorism*. Lawrence Livermore National Laboratory LLNL-TR-410067, August 2009.
18. U.S. Department of Health & Human Services, Radiation Emergency Medical Management, Radiological Dispersal Devices (RDDs). Last updated December 28, 2011. http://www.remm.nlm.gov/rdd.htm.
19. J.A. Giesecke et al. Assessment of the regional economic impacts of catastrophic events: CGE analysis of resource loss and behavioral effects of an RDD attack scenario, *Risk Analysis* 32(4):583–600, April 2012.
20. U.S. Department of Health and Human Services, Radiation Emergency Medical Management. Last modified on March 6, 2013. http://www.remm.nlm.gov/.
21. U.S. Department of Transportation, Pipeline and Hazardous Materials Safety Administration. *2008 Emergency Response Guidebook*. http://phmsa.dot.gov/staticfiles/PHMSA/DownloadableFiles/Files/erg2008_eng.pdf.
22. NRC Bulletin 2005-02. *Emergency Preparedness and Response actions For Security-Based Events*. July 18, 2005. http://www.nrc.gov/reading-rm/doc-collections/gen-comm/bulletins/2005/bl200502.pdf.
23. U.S. Department of Health and Human Services, Radiation Emergency Medical Management, *Explaining Protective Actions and Protective Action Guides (PAGs)*. Last updated December 28, 2011. http://www.remm.nlm.gov/pag_terms.htm.
24. K. Belson, and H. Tabuchi, Japan finds tainted food up to 90 miles from nuclear sites, *The New York Times* (online), Published March 19, 2011. http://www.nytimes.com/2011/03/20/world/asia/20japan.html?pagewanted=all.
25. B.R. Buddemeier and M.B. Dillon, *Key Response Planning Factors for the Aftermath of Nuclear Terrorism*, Lawrence Livermore National Laboratory LLNL-TR-410067, August 2009.
26. U.S. Department of Health & Human Services, Radiological Emergency Medical Management, Radiological Exposure Devices (REDs). Last updated December 28, 2011. http://www.remm. nlm.gov/red.htm.
27. U.S. Department of Transportation, Pipeline and Hazardous Materials Safety Administration. *How to Handle Radioactive Materials Packages: A Guide for Cargo Handlers*. Washington, DC. Revised September 2006; http://phmsa.dot.gov/staticfiles/PHMSA/DownloadableFiles/Files/Handle_Radioactive.pdf.

Chapter 25

Agroterrorism

Michael J. Fagel and Kelly Hamilton

While sometimes described as a low probability, yet very high consequence event, the potential of terrorist attacks against agricultural targets (agroterrorism) is increasingly recognized as a national security threat, especially after the events of September 11, 2001. In this context, agroterrorism is defined as the deliberate introduction of an animal disease (Figure 25.1) or plant disease with the goal of generating fear, causing economic losses, and/or undermining stability. Additionally, other WMD (weapons of mass destruction) agents could be used in an attack against the agriculture and food sectors of the United States.

Agriculture as a Target: Overview of Terrorist Threat

Agroterrorism is a specific subset of terrorism and bioterrorism and can be appealing to both domestic and international terrorists. People generally associate bioterrorism with outbreaks of human illness (such as from anthrax or smallpox), rather than diseases first affecting animals or plants. Agriculture has several characteristics that pose unique problems for managing the threat:

- Agricultural production is geographically disbursed in unsecured environments (e.g., open fields and pastures throughout the country). Although some livestock are housed in secure facilities, agriculture in general requires large expanses of land that are difficult to secure from intruders.
- Production livestock are frequently concentrated in confined locations (e.g., feedlots with thousands of cattle in open-air pens, farms with tens of thousands of pigs, or barns with hundreds of thousands of poultry). Concentration

Figure 25.1 Cattle and other livestock could be targeted through intentional disease-caused outbreaks both by commingling and through feed and water.

in transportation, slaughter, processing, and distribution also makes large-scale contamination more likely.

- Live animals, grain, and processed food products are routinely transported and commingled in the production and processing system; these factors circumvent natural barriers that could slow pathogenic dissemination.
- The presence (or rumor) of certain pests or diseases in a country can quickly stop all exports of a commodity, and can take months or years to resume.
- The past success of eradicating and keeping many diseases out of the United States means that many veterinarians and scientists lack direct experience with foreign diseases. This may delay recognition of symptoms in case of an outbreak.
- The number of lethal and contagious biological agents is greater for animals and plants than for humans. Most of these diseases are environmentally resilient, endemic in foreign countries, and not harmful to humans—making it easier for terrorists to acquire, handle, and deploy the pathogens with little or no danger to the perpetrator.

The general susceptibility of the agriculture and food industry to bioterrorism is difficult to address in a systematic manner because of the highly dispersed, yet concentrated nature of the industry and the inherent biology of raising animals and growing plants.

The results of an agroterrorist attack may include major economic crises in the agricultural and food industries, loss of confidence in government, and possibly human casualties. Humans could be at risk in terms of food safety or public health, especially if the chosen disease is transmissible to humans (zoonotic). But an agroterrorist attack need not cause human casualties for it to be effective or to cause large-scale economic consequences.

The production agriculture sector (Figure 25.2) would suffer economically in terms of plant and animal health, and the supply of food and fiber may be reduced, especially in certain regions. The demand for certain types of food may decline based on the products targeted in the attack (e.g., dairy, beef, pork, poultry, grains, fruit, or vegetables), whereas demand for other types of food may rise because of food substitutions.

An agroterrorism event would cause economic losses to individuals, businesses, and governments through costs to contain and eradicate the disease and to dispose of contaminated products. Economic losses would accumulate throughout the farm-to-table continuum as the supply chain is disrupted, especially if domestic markets for food become unstable or if trade sanctions are imposed by other countries on the U.S. exports. The economic impact can spread to farmers, input suppliers, food processors, transportation, retailers, and food service providers.

Public opinion may be particularly sensitive to a deliberate attack affecting the food supply. Public confidence in government could be eroded if authorities appear unable to prevent or effectively respond to such an attack or to protect the population's food supply. As the United States evolved away from an agrarian society during the twentieth century, food and the fear of inadequate food supplies moved further from the minds of most U.S. residents. However, because food remains an important part of everyone's daily routine and survival, significant threats to the currently held notion of food protection in the United States could cause a reordering of people's priorities. The nation has undergone a consolidation and clearer definition of what food protection encompasses. There are generally three recognized elements of food

Figure 25.2 Food chain supply could be dramatically affected by targeted terrorist attacks.

protection: Food safety which is protecting the food supply from unintentional acts, food defense which is protecting the food supply from intentional acts and food security, ensuring the nation has an adequate and wholesome supply of food, particularly in the time of catastrophic events such as Hurricane Sandy in 2012.

Because an agroterrorist attack may not necessarily cause human casualties, be immediately detected, or have the "shock factor" of an attack against the more visible public infrastructure or human populations, agriculture may not be a terrorist's first choice of targets. Nonetheless, some types of agroterrorism could be relatively easily achieved and have significant economic impacts. Thus, the possibilities are treated seriously, especially in the post-September 11 world.

Importance of Agriculture in the United States

The agriculture industry and food industry are intricately interwoven and both are very important to the social, economic, and psychological well-being of our nation. Each of us typically eats on a daily basis and relies on a protected food supply assured by many laws and regulations; a safe and wholesome food supply also helps maintain the political stability of the United States. Although farming employs less than 2% of the country's workforce, 16% of the workforce is employed in the food and fiber sector, ranging from farmers and input suppliers, to processors, shippers, grocers, and restaurateurs.

While in a slight decline over the past three decades, the food and fiber sector contributes upwards of 15%, to the gross domestic product (GDP), even though the farm sector itself contributed less than 1–2%. Gross farm sales can exceed $200 billion yearly, and are relatively concentrated throughout the Midwest, parts of the East Coast, and California. Production is split nearly evenly between crops and livestock.

Agriculture in the United States is highly advanced and productive. This productivity allows Americans to spend less than 11% of their disposable income on food, compared with a global average of 20–30%.[1]

The import and export of agriculture and food products account for billions of dollars in trade each year; because of agriculture produce the trade balance for the nation is positive.

Cattle are the most widely distributed given the prevalence of small cow–calf herds throughout the country and pockets of dairy on the West Coast, upper Midwest, and Northeast. However, beef cattle feedlots are particularly concentrated from northern Texas through Kansas, Nebraska, eastern Colorado, and western Iowa. In rural states such as Montana and Wyoming, cattle roam freely on government-owned grazing allotments.

Hog inventories are concentrated in the Midwest, especially in Iowa, southern Minnesota, and in North Carolina. The production of broilers for poultry meat is concentrated throughout the Southeast, ranging from the Oklahoma–Arkansas border up to the Delmarva peninsula (Delaware–Maryland–Virginia).

The U.S. Department of Agriculture produces a census of agriculture every five years. For information from the most current census and past censuses dating back to 1840, browse Google™ USDA Census of Agriculture.

A Brief History of Agricultural Bioweapons

Attacks against agriculture are not new, and have been conducted both by nation-states and by sub-state organizations throughout history. At least nine countries had documented agricultural bioweapons programs during some part of the twentieth century (Canada, France, Germany, Iraq, Japan, South Africa, the United Kingdom, the United States, and the former USSR). Four other countries are believed to have or have had agricultural bioweapons programs (Egypt, North Korea, Rhodesia, and Syria).

Despite extensive research on the issue, however, biological weapons have been rarely used against crops or livestock, especially by state actors. Thus, in recent decades, using biological weapons against agricultural targets has remained mostly theoretical consideration. With the ratification of the Biological and Toxin Weapons Convention in 1972, many countries, including the United States, stopped military development of biological weapons and destroyed their stockpiles.[2]

Although individuals or sub-state groups have used bioweapons against agricultural or food targets, only a few can be considered terrorist in nature. In 1952, the Mau Mau (an insurgent organization in Kenya) killed 33 head of cattle at a mission station using African milk bush (a local plant toxin). In 1984, the Rajneeshee cult spread salmonella in salad bars at Oregon restaurants to ultimately influence a local election.[3]

Chemical weapons have been used somewhat more commonly against agricultural targets. During the Vietnam War, the United States used Agent Orange to destroy foliage so bombing planes could see their targets, affecting some crops. Among possible terrorist events, chemical attacks against agricultural targets include a 1997 attack by Israeli settlers who sprayed pesticides on grapevines in two Palestinian villages, destroying up to 17,000 metric tons of grapes. In 1978, the Arab Revolutionary Council poisoned Israeli oranges with mercury, injuring at least 12 people and reducing orange exports by 40%.

Economic Consequences

Economic losses from an agroterrorist incident could be large and widespread.

- First, losses would include the value of lost production, the cost of destroying diseased or potentially diseased products, and the cost of containment (drugs, diagnostics, pesticides, and veterinary services). The development of new vaccines and herbicides to combat an emerging pathogen can take years and cost millions of dollars.

- Second, export markets would be lost as importing countries place restrictions on U.S. products to prevent possibilities of the disease spreading.
- Third, multiplier effects would ripple through the economy due to decreased sales by agriculturally dependent businesses (farm input suppliers, food manufacturing, transportation, retail grocery, and food service) and tourism.
- Fourth, the government could bear significant costs, including eradication and containment costs, and compensation to producers for destroyed animals, crops, or processed food.

Depending on the erosion of consumer confidence and export sales, market prices of the affected commodities may drop. This would affect producers whose herds or crops were not directly infected, making the event national in scale even if the disease itself were confined to a small region.

For food types or product lines that are not contaminated, however, demand may become stronger, and market prices could rise for those products. Such goods may include substitutes for the food that was the target of the attack (e.g., chicken instead of beef), or product that can be certified not to come from regions affected by the attack (e.g., beef from another region of the country or imported beef). When Canada announced the discovery of bovine spongiform encephalopathy (BSE)] in May 2003, farm-level prices of beef in Canada dropped by nearly half, whereas beef prices in the United States remained very strong at record or near record levels.[4]

Consumer confidence in government may also be tested depending on the scale of the eradication effort and means of destroying animals or crops. The need to slaughter perhaps hundreds of thousands of cattle (or tens of millions of poultry) could generate public criticism if depopulation methods are considered inhumane or the destruction of carcasses is questioned environmentally. Dealing with these concerns can add to the cost for both government and industry.

Depending on the disease and means of transmission, the potential for economic damage depends on a number of factors, such as the disease agent, location of the attack, rate of transmission, geographical dispersion, how long it remains undetected, availability of countermeasures or quarantines, and incident response plans. Potential costs are difficult to estimate and can vary widely based on compounding assumptions.

The ability of farm commodity programs to compensate for losses due to agroterrorism is limited. Government income support programs subsidize about 25 agricultural commodities (such as corn, wheat, soybeans, rice, and cotton). These supported commodities represent about one-third of gross farm sales. The list of commodities that normally do not receive direct support includes meats, poultry, fruits, vegetables, nuts, hay, and nursery products. These nonsupported commodities account for about two-thirds of gross farm sales.

The food products more vulnerable to attack (meats, fruits, and vegetables) do not have existing federal farm income support programs, nor are there income support programs beyond the farm gate for food processors or retailers. Thus, any federal assistance to producers or processors stemming from an agroterrorist attack would likely come in the form of ad hoc disaster assistance. Making disaster payments to producers who do not normally receive government payments is technically more difficult than supplementing regular program payments due to drought or flood and can take months or years to reach those who need it. Agriculture and food producers typically do not have funds to survive extended periods of time waiting for subsidy.

Federal Recognition of Agroterrorism Threats

Agriculture and food production generally have received relatively less attention, or sometimes were overlooked, in counterterrorism and homeland security. After what many observers claim to be a slow start after September 11, 2001, agriculture now is garnering more attention in the expanding field of terrorism studies and policies (Figure 25.3).

Congress has held hearings on agroterrorism and, while addressing terrorism more broadly, has implemented laws and appropriations with provisions important to agriculture. The Government Accountability Office has studied aspects of food safety, border inspections, and physical security with respect to agroterrorism. The executive branch has responded by implementing the new laws, issuing several Presidential directives, and creating terrorism and agroterrorism task forces. Two initiatives that have closely studied agriculture production vulnerabilities are the Strategic Partnership Agriculture Alliance and most recently the Regionally Resiliency Assessment Program.

In its report, the 9/11 Commission (National Commission on Terrorist Attacks upon the United States) does not make any direct references to agroterrorism or terrorist attacks on the food supply. However, agriculture obviously would be affected, along with other sectors of the economy, by some of the commission's recommendations regarding coordination of intelligence, information sharing, and first responders.

Figure 25.3 Agriculture is garnering more attention in policy discussions and terrorism studies.

Congressional Hearings and Laws

On November 19, 2003, the Senate Committee on Governmental Affairs held a hearing titled, "Agroterrorism: The Threat to America's Breadbasket," including witnesses from the administration, state governments, and a private think tank.

This was the first congressional hearing devoted entirely to agroterrorism since October 27, 1999. At that time, the Subcommittee on Emerging Threats of the Senate Committee on Armed Services held a hearing titled, "Agricultural Biological Weapons Threat to the United States." During the 4 years between these hearings, a few individual panelists at more general hearings on food safety, homeland security, or terrorism discussed agroterrorism in reference to other topics.

Bioterrorism Preparedness Act

The Public Health Security and Bioterrorism Preparedness and Response Act (P.L.107-188, June 12, 2002) contained several provisions important to agriculture. These provisions accomplish the following:

- Expand Food and Drug Administration (FDA) authority over food manufacturing and imports (particularly in Sections 303–307).
- Tighten control of biological agents and toxins ("select agents" as discussed in Sections 211–213, the "Agricultural Bioterrorism Protection Act of 2002") through rules issued by the Animal and Plant Health Inspection Service (APHIS) and the Centers for Disease Control and Prevention (CDC).
- Authorize expanded agricultural security activities and security upgrades at USDA facilities (Sections 331–335).
- Address criminal penalties for terrorism against enterprises raising animals (Section 336) and violation of the select agent rules (Section 231).

New FDA Rules on Food Processors and Importers

The Bioterrorism Preparedness Act responded to long-standing concerns about whether the FDA in the Department of Health and Human Services (HHS) had the authority to assure food safety. FDA was instructed to implement new rules for

- Registration of food processors
- Prior notice of food imports
- Administrative detention of imports
- Recordkeeping

Proposed rules were issued in the spring 2003 followed by a comment period. On October 10, 2003, FDA published two interim final rules for registration of food facilities and prior notice of imports. Those rules were implemented on December 12, 2003, but FDA allowed flexible enforcement during a transition period. The rule on administrative detention of imports was effective upon enactment, with FDA procedures announced on May 27, 2004.

Registration of Food Processors

The act required FDA to establish a one-time registration system for any domestic or foreign facility that manufactures, processes, packs, and handles food. For the first time, all food facilities supplying food for the United States were required to register with the FDA.

Registering involved providing information about the food products (brand names and general food categories), facility addresses, and contact information. Restaurants, certain retail stores, farms, nonprofit food and feeding establishments, fishing vessels, and trucks and other motor carriers were exempt from registration requirements. However, many farms had a difficult time determining whether they needed to register based on the amount of handling or processing they performed.

Registration documents are protected from public disclosure under the Freedom of Information Act. The registry provides a complete list of companies subject to FDA authority, and enhances the agency's capability to trace contaminated food. Critics argued that registration created a recordkeeping burden without proof that facilities will be able to respond in an emergency.

Prior Notice of Imports

As of December 12, 2003, importers are now required to give advance notice to FDA before importing food. Electronic notice must be provided by the importer within a specified period before arrival at the border (within 2 h by road, 4 h by air or rail, and 8 h by water). With prior notice, FDA can assess whether a shipment meets criteria that can trigger an inspection. If notice is not given, the food will be refused entry and held at the port or in secure storage. Some critics are concerned that the administrative cost of compliance may raise the price of food and this was once again raised during discussion and ultimately passage of the Food Safety Modernization Act of 2012. Others have argued that perishable imports are subject to increased spoilage if delays arise, or that certain perishables (especially from Mexico) are not harvested or loaded onto trucks before the 2-h notification period. However, implementation of the new system generally has not caused delays and most shippers have been accommodated.

To facilitate compliance, FDA and the Department of Homeland Security's (DHS) Customs and Border Protection (CBP) integrated their information systems to allow food importers to provide the required information using CBP's existing system for imports. In December 2003, the two agencies agreed to allow CBP officers to inspect imported foods on FDA's behalf, particularly at ports where FDA has no inspectors. Today, it is not uncommon for FDA to have an office and import compliance officer physically at ports of entry, ports of arrival, and/or ports of unlading. Additionally, USDA import compliance employees have either integrated into CBP or are working alongside CBP officials at exit and entry points. Philosophically, federal agencies have strived to keep unapproved product out of our nation by working closely with other countries and keeping the unapproved product from even being shipped to the United States.

Administrative Detention

FDA has the authority to detain food imports under certain conditions. FDA procedures for making detention were issued on June 4, 2004. To use the authority, the agency must show credible evidence that a shipment presents a serious health threat. Food may be detained for 20 days and up to 30 days, if necessary. The owners must pay the expense of moving any detained food to secure storage. Perishable foods (e.g., fruits, vegetables, and seafood) are to receive expedited review.

Maintenance of Records

In the event of a suspected food safety problem, FDA has access to records including the facility's immediate supplier, and the immediate customer. However, trade secrets can be kept confidential.

Security for Biological Agents and Toxins

In December 2002, the USDA APHIS issued regulations to reduce the threat that certain biological agents and toxins could be used in domestic or international terrorism. APHIS determined that the "select agents" on the list have the potential to pose a severe threat to agricultural production or food products.

The select agent regulations (9 CFR 121 for animals, 7 CFR 331 for plants) establish the requirements for possession, use, and transfer of the listed pathogens. The rules affect many research institutions, including federal, state, university, and private laboratories, as well as firms that transport such materials. The laboratories have had to assess security vulnerabilities and upgrade physical security, often without additional financial resources. Some have been concerned that certain research programs may be discontinued or avoided because of regulatory difficulties in handling the select agents.

Homeland Security Act

The main purpose of the Homeland Security Act of 2002 (P.L. 107–296, November 25, 2002) was to create the DHS, primarily by transferring parts or in a few cases all of many agencies throughout the federal government into the new cabinet-level department. In doing so, the law made two major changes to the facilities and functions of the Department of Agriculture.

The Homeland Security Act transferred:

- Agricultural border inspections from APHIS to DHS
- Possession of the Plum Island Animal Disease Center from USDA to DHS

Agricultural Border Inspections

Section 421 of the Homeland Security Act authorized the transfer of up to 3200 APHIS border inspection personnel to DHS. As of March 1, 2003, approximately 2680 APHIS inspectors became employees of DHS in the Bureau of Customs and Border Inspection (CBP). Because of its scientific expertise, USDA retains a significant presence in border inspection.

Historically, the APHIS Agricultural Quarantine Inspection (AQI) program was considered the most significant and prominent of agricultural and food inspections at the border. Because of this prominence, AQI was one of the many programs selected for inclusion when DHS was created. Some drafts of the bill creating the new department would have transferred all of APHIS (including, for example, animal welfare and disease eradication) to DHS. Concerns from many farm interest groups about the impact this might have on diagnosis and treatment of natural plant and animal diseases prompted a legislative compromise that transferred only the border inspection function and left other activities under USDA.

DHS–CBP personnel now inspect international conveyances and the baggage of passengers for plant, animal, and related products that could harbor pests or disease organisms. They also inspect ship and air cargo, rail and truck freight, and package mail from foreign countries.

Although the border inspection functions were transferred to DHS, the USDA retains a significant presence in border activities. APHIS employees who were not transferred continue to pre-clear certain commodities, inspect all plant propagative materials, and check animals in quarantine. APHIS personnel continue to set agricultural inspection policies to be carried out by DHS border inspectors, and negotiate memoranda of understanding to assure that necessary inspections are conducted. APHIS manages the data collected during the inspection process, and monitors smuggling and trade compliance. USDA is also statutorily charged in Section 421 (e)(2)(A) of the act to "supervise" the training of CBP inspectors in consultation with DHS.

This separation of duties is designed to allow for consolidated border inspections for intelligence and security goals, but preserve USDA's expertise and historical mission to set agricultural import policies.

Adding Agricultural Specialists

Under the CBP cross-training initiative in 2003 (also known as "one face at the border"), most CBP inspectors are trained to perform inspections in all three areas of customs, immigration, and agriculture. However, due to criticism from USDA, inspection unions, and the agricultural industry, DHS created another class of inspectors called agricultural specialists. Agricultural specialists will staff, primarily, secondary inspection stations. These specialists will include former APHIS inspectors who decided not to convert to CBP generalist inspectors and new agricultural specialist trainees.

Before DHS was created, APHIS trained its inspectors in a 9-week course that had science prerequisites. The initial DHS cross-training program announced in 2003 had only 12–16 h for agriculture in a 71-day course covering customs, immigration, and agriculture. With the creation of the agricultural specialist position, DHS created a 43-day training program for agricultural specialists.

Although DHS is training new agricultural specialists, the future size of the agricultural specialist corps is not certain, given the eventual attrition of former APHIS inspectors. Also, details are not available as to how these inspectors will be deployed and how many ports of entry will be staffed with agricultural specialists (compared with the APHIS deployment prior to DHS). Without agricultural specialists, primary agricultural inspections—the first line of defense for agricultural security—may be conducted by cross-trained inspectors with limited agricultural training. Additionally, APHIS budgets have been cut significantly in the past few years.

Executive Branch Actions

Shortly after September 11, 2001, USDA created a Homeland Security Staff in the Office of the Secretary to develop a department-wide plan to coordinate agroterrorism preparedness plans among all USDA agencies and offices. Efforts have been focused on three areas: food supply and agricultural production, USDA facilities, and USDA staff and emergency preparedness. The Homeland Security Staff also has become the department's liaison with Congress, the DHS, and other governmental agencies on terrorism issues.

The White House's National Security Council Weapons of Mass Destruction preparedness group, formed by Presidential Decision Directive 62 (PDD-62) in 1998, included agriculture, especially in terms of combating terrorism. Many observers note that, as a latecomer to the national security table, USDA has been invariably overshadowed by other agencies.

Homeland Security Presidential Directive 7

In terms of protecting critical infrastructure, agriculture was added to the list in December 2003 by Homeland Security Presidential Directive 7 (HSPD-7), "Critical Infrastructure Identification, Prioritization, and Protection." This directive replaces the 1998 Presidential Decision Directive 63 (PDD-63) that omitted agriculture and food. Both of these critical infrastructure directives designate the physical systems that are vulnerable to terrorist attack and are essential for the minimal operation of the economy and the government.

These directives instruct agencies to develop plans to prepare for and counter the terrorist threat. HSPD-7 mentions the following industries: agriculture and food; banking and finance; transportation (air, sea, and land, including mass transit, rail, and pipelines); energy (electricity, oil, and gas); telecommunications; public health; emergency services; drinking water; and water treatment.

Homeland Security Presidential Directive 9

More significant recognition came on January 30, 2004, when the White House released Homeland Security Presidential Directive 9 (HSPD-9), "Defense of United States Agriculture and Food." This directive establishes a national policy to protect against terrorist attacks on agriculture and food systems.

HSPD-9 generally instructed the Secretaries of Homeland Security (DHS), Agriculture (USDA), and HHS, the Administrator of the Environmental Protection Agency (EPA), the Attorney General, and the Director of Central Intelligence to coordinate their efforts to prepare for, protect against, respond to, and recover from an agroterrorist attack. In some cases, one department is assigned primary responsibility, particularly when the intelligence community is involved. In other cases, only USDA, HHS, and/or EPA are involved regarding industry or scientific expertise.

The directive instructs agencies to develop awareness and warning systems to monitor plant and animal diseases, food quality, and public health through an integrated diagnostic system. Animal and commodity tracking systems are included, as is gathering and analyzing international intelligence. Vulnerability assessments and subsequently risk calculations throughout the sector help prioritize mitigation strategies at critical stages of production or processing, including inspection of imported agricultural products.

Response and recovery plans are increasingly being coordinated across the federal, state, and local levels. A National Veterinary Stockpiles (NVS) of vaccine, antiviral, and therapeutic products has been developed for deployment within 24 h of an attack.

HSPD-9 encourages USDA and HHS to promote higher education programs that specifically address the protection of animal, plant, and public health. It suggests capacity-building grants for universities, and internships, fellowships, and postgraduate opportunities. HSPD-9 also formally incorporates USDA and agriculture into the ongoing DHS research program of university-based "centers of excellence."

As a presidential directive, HSPD-9 addresses the internal management of the executive branch and does not create enforceable laws. Moreover, it is subject to change without Congressional consent. Although Congress has oversight authority of federal agencies and may ask questions about implementation of the directive, a public law outlining an agroterrorism preparedness plan would establish the statutory parameters for such a plan, and, as a practical matter, might result in enhanced oversight by specifically identifying executive branch entities responsible for carrying out particular components of such a plan.

In implementing HSPD-9, the USDA Homeland Security Staff and other agencies are drawing upon HSPD-5 (regarding the national response plan) and the original HSPD-8 (regarding preparedness; HSPD 8 has since been replaced by Presidential Policy Directive 8). Implementing many of the HSPD-9 directives depends on the executive branch having sufficient appropriations for those activities.

Federal Appropriations

The President's annual budget request to Congress now includes a cross-cutting budget analysis of homeland security issues, as mandated by the Homeland Security Act of 2002 (P.L. 107–296, Section 889). In USDA, six agencies and three offices receive (or have requested) funding related to homeland security:

- Agricultural Research Service (ARS)
- Animal and Plant Health Inspection (APHIS)
- Cooperative State Research, Education, and Extension Service (CSREES)
- Food Safety and Inspection Service (FSIS)
- Economic Research Service (ERS)
- Agricultural Marketing Service (AMS)
- Departmental Administration (including Office of the Secretary, Homeland Security Staff [HSS], and Office of Chief Information Officer [OCIO])

Classifying spending on agroterrorism and homeland security requires judgments about which programs are relevant, especially when some have dual purposes. For example, animal health and plant health programs would be needed at some level due to natural and accidental outbreaks, regardless of the need for agroterrorism preparedness. For budgets, all or part of such dual-use activities may be counted as homeland security spending, especially when those functions are expanded due to agroterrorism concerns. As the nation moves further past the tragic events of September 11, 2001, budgets for terrorism are becoming more restrictive. But, money is also being spent on preparedness based on the discovery of vulnerabilities and the calculation of actual risk.

Possible Pathogens in an Agroterrorist Attack

Of the hundreds of animal pathogens, plant pathogens, and pests available to an agroterrorist, it is likely that less than two dozen represent significant economic threats. Determinants of this level of threat are the agent's virulence, contagiousness, and potential for rapid spread, and its international status as a "reportable" pest or disease (i.e., subject to international quarantine) under rules of the World Organization for Animal Health [also commonly known as the Office International des Epizooties (OIE)]. Additionally, the potential host must be particularly susceptible to the pathogen and other factors, even climatic conditions, must be favorable.

A widely accepted view among scientists is that livestock herds are much more susceptible to agroterrorism, rather than crops. Much of this has to do with the success of efforts to systematically eliminate animal diseases from U.S. herds, which leaves the current herds either unvaccinated or relatively unmonitored for such diseases by farmers and veterinarians. Once infected, livestock can often act as the vector for continuing to transmit the disease, facilitating an outbreak's spread, especially when live animals are transported. Certain animal diseases may be more attractive to terrorists because they can be zoonotic, or transmissible to humans. (Source: *Small Scale Terrorist Attacks Using Chemical and Biological Agents: An Assessment Framework and Preliminary Comparisons*, by Dana Shea and Frank Gottron.)

In contrast, a number of plant pathogens continue to exist in small areas of the United States and continue to infect limited areas of plants each year, making outbreaks and control efforts more routine. Moreover, plant pathogens are generally more technically difficult to manipulate. Some plant pathogens may require particular environmental conditions of humidity, temperature, or wind to take hold or spread. Other plant diseases may take a longer time than an animal disease to become established or achieve destruction on the scale that a terrorist may desire. The wildlife and companion animals may spread animal disease and plant pathogens once introduced into an environment become very difficult to bring about containment and eradication.

Animal Pathogens

The Agricultural Bioterrorism Protection Act of 2002 (Subtitle B of P.L. 107–188, the Public Health Security and Bioterrorism Preparedness and Response Act) created the current, official list of animal pathogens that are of greatest concern for agroterrorism. The list is specified in the select agent rules implemented by USDA APHIS and the CDC. The act requires that these lists be reviewed at least every 2 years.

The select agent list for animal pathogens draws heavily from the enduring and highly respected OIE lists of high-concern pathogens. The select agent list is composed of an APHIS-only list (of concern to animals) and an overlap list of agents selected both by APHIS and CDC (of concern to both animals and humans).[5]

OIE List

Before the Agricultural Bioterrorism Protection Act, the commonly accepted animal diseases of concern were all of the OIE's "List A" diseases and some of the "List B" diseases. In 2004, the OIE replaced its Lists A and B with a single list that is more compatible with the Sanitary and Phytosanitary Agreement of the World Trade Organization. The new OIE list available classifies diseases equally; giving each the same degree of importance in international trade. Many of these OIE-listed diseases are included in the select agent list.[6]

The OIE's List A diseases were transmissible animal diseases that had the potential for very serious and rapid spread, irrespective of national borders. List A diseases had serious socioeconomic or public health consequences and were of major importance in international trade. List B diseases were transmissible diseases considered to be of socioeconomic or public health importance within countries and significant in international trade. In creating the new list, OIE reviewed its criteria for including a disease, and the disease or epidemiological events that require member countries to file reports.

Select Agents List

The regulations establishing the select agent list for animals (9 CFR 121.3) set forth the requirements for possession, use, and transfer of these biological agents or toxins. They are intended to ensure safe handling and for security to protect the agents from use in domestic or international terrorism. APHIS and CDC determined that the biological agents and toxins on the list have the potential to pose a severe threat to agricultural production or food products.

The 23 animal diseases listed exclusively by APHIS in 9 CFR 121.3(d) include 20 of the OIE-listed diseases and three other disease agents (Akabane, Camel pox, and Menangle) considered to be emerging animal health risks for terrorism. The much larger OIE list includes other diseases that are not listed as "select agents." However, the select agent list was created to account for the additional risks perceived to be posed by terrorism.

The 21 diseases and overlap agents/toxins included by both APHIS and CDC in 9 CFR 121.3(b) pose a risk to both human and animal health. The most updated select agent list can be found at: http://www.selectagents.gov/.

Agent Analysis

It is important to note that the select agent list designates and regulates pathogens, not diseases. Thus, the overlap list between APHIS and CDC is somewhat more comprehensive than a disease-only list, particularly because certain pathogens may not cause a disease, *per se*, but may cause symptoms such as food poisoning or central nervous system responses.

Some of the pathogens in the select agent list receive more attention than others in discussions about agroterrorism. One reason is that the select agent list was designed to regulate access to and handling of high-consequence pathogens, not the diseases directly.

For example, the causative agent of BSE is considered dangerous enough to be a select agent, even though it is less likely to be a terrorist's choice than other diseases. With BSE, infection is not certain, symptoms take years to manifest, and the disease may not be detected—all making credit for an attack more doubtful.

On the other hand, foot and mouth disease (FMD) is probably the most frequently mentioned disease when agroterrorism is discussed, at least in cloven hoofed animals, because of its ease of use, ability to spread rapidly, and potential for great economic damage. In testimony before the Senate Governmental Affairs Committee on November 19, 2003, Dr. Thomas McGinn of the North Carolina Department of Agriculture described a simulation of an FMD attack by a terrorist at a single location. Only after the fifth day of the attack would the disease have been detected, by which time it may have spread to 23 states. By the eighth day, 23 million animals may need to be destroyed in 29 states.

Widespread animal diseases such as influenza, or tuberculosis receive relatively less attention than FMD, hog cholera, or Newcastle disease. However, emerging diseases such as Nipah virus, Hendra virus, and the H5N1 strain of avian influenza (zoonotic diseases that have infected people, mostly in Asia) can be lethal since vaccines are elusive or have not been developed.

Plant Pathogens

The Agricultural Bioterrorism Protection Act of 2002 (Subtitle B of P.L. 107–188) also instructed APHIS and CDC to create the current official list of potential plant pathogens. The Federal government lists biological agents and toxins for plants in 7 CFR 331.3 (Table 16.4). The act requires that these lists be reviewed at least every 2 years, and revised as necessary.

Before the act, there was no commonly recognized list of the most dangerous plant pathogens, although several diseases were usually mentioned and are now included in the APHIS select agent list.

The list of nine biological agents and toxins in 7 CFR 331.3 was compiled by the Plant Protection and Quarantine (PPQ) program in APHIS, in consultation with USDA's Agricultural Research Service; Forest Service; Cooperative State Research, Education, and Extension Service; and the American Phytopathological Society.

Countering the Threat

The goal of the U.S. animal and plant health safeguarding system is to prevent the introduction and establishment of exotic pests and diseases, to mitigate their effects when present, and to eradicate them when feasible. In the past, introductions of pests and pathogens were presumed to be unintentional and occurred through natural migration across borders, weather events, or accidental movement by international commerce (passengers, conveyance, or cargo). However, a system designed for accidental or natural outbreaks is not sufficient for defending against intentional attack.

Consequently, the U.S. system is being upgraded to address the reality of agroterrorism.

The National Research Council, which dates back to 1916, outlines a three-pronged strategy for countering the threat of agroterrorism[7]:

- Deterrence and prevention
- Detection and response
- Recovery and management

Even though no foreign terrorist attacks on crops or livestock have occurred in the United States that we are aware of, government agencies and private businesses have not taken the threat lightly. Because of the importance of brand names in marketing, many agribusinesses have prepared response plans or added security measures to protect their product line, looking at threats ranging from the source of their inputs to their retail distribution network. Since the terrorist attacks of 2001, biosecurity is an increasing priority among food manufacturers, merchandisers, retailers, and commercial farmers nationwide.

Deterrence and Prevention

Primary prevention and deterrence interventions for foreign pests and diseases include international treaties and standards (such as the International Plant Protection Convention, and those of the OIE/World Organization for Animal Health), bilateral and multilateral cooperative efforts, offshore activities in host countries, port-of-entry inspections, quarantine, treatment, and post-import tracking of animals, plants, and their products.

DHS and USDA already conduct such inspection and quarantine practices, but continued oversight is necessary to determine which preparedness activities and threats require more attention. Offshore activities include pre-clearance inspection by APHIS of U.S. imports before products leave their port of origin. APHIS has personnel in a number of host countries. Although many of these inspection programs were built to target unintentional threats, they are being augmented with personnel and technology to look for intentional threats.

Various U.S. intelligence, fusion centers, and law enforcement/investigative agencies collect information about biological weapons that could be used against U.S. agriculture. Building and maintaining a climate of information sharing between FDA, USDA, DHS, and the intelligence community is necessary, especially so that agriculture is not overlooked compared to other infrastructure and human targets.

Once inside the United States, many parts of the food production chain may be susceptible to attack with a biological weapon. For example, terrorists may have unmonitored access to geographically remote crop fields and livestock feedlots. Diseases may infect herds more rapidly in modern concentrated confinement livestock operations than in open pastures. An undetected disease may spread rapidly because livestock are transported more frequently and over greater distances between farms, and to processing plants. Processing plants and shipping containers need to be secured and/or tracked to prevent tampering.

An important line of defense is biosecurity, or the use of preventive security measures. On the farm, biosecurity is the use of farm management practices that both protect animals and crops from the introduction of infectious agents and contain a disease to prevent its rapid spread within a herd or to other farms. Biosecurity practices include structural enclosures to limit outside exposure to people and wild animals, and the cleaning and disinfection of people, clothing, vehicles, equipment, and supplies entering the farm.

Most farm specialists agree that livestock farmers are increasingly aware of the importance of biosecurity measures, particularly since the FMD outbreaks in European cattle and the avian flu and Exotic Newcastle infections in U.S. poultry populations and wild ranging birds. More farm operators are requiring visitors to wear boot covers to guard against bringing in disease through cross contamination. Regardless of the reason for following biosecurity measures (terrorism or accidents), these precautions can help prepare farms against intentional events (agroterrorism) or unintentional events (all hazards).

Detection and Response

Biological attacks on production animals and food crops may not be immediately apparent; for example, some detrimental pathogens can live in the flora of animals with no outward signs. Therefore, existing frameworks for detecting, identifying, reporting, tracking, and managing natural and accidental disease outbreaks are being applied to combating agroterrorism. Appropriate responses are being developed based on specific pathogens, targets, and other circumstances that may surround an attack; vetted information that has been turned into intelligence can also help in a shorter detection epidemiological curve.

DHS, FDA, and USDA have responded with more detailed, focused, and coordinated plan to secure the food supply. The departments are cooperating on research funding, detection technology, certified trainings, surveillance, partnerships with private industry, and state and local response coordination.

Within private industry, which owns over 80% of the agriculture and food critical infrastructure, the Food and Agriculture Information Sharing and Analysis Center (ISAC) shares information with government intelligence bureaus through the DHS. The Food and Agriculture ISAC includes more than 40 of the primary trade associations representing food and agriculture. Such ISAC centers exist in several industries and are one of the primary partnerships between government and industry for counterterrorism cooperation. By combining information among members in the same industry, security problems or attacks may become apparent more quickly than observations within individual companies. In the event of a terrorist incident, the ISAC would facilitate communication within the industry and coordinate response efforts with government officials. The Food and Agriculture ISAC was created in February 2002 and is administered by the Food Marketing Institute. In 2003, three sub-ISACs were created to cover more specific threats and information sharing for[8]

- Agriculture
- Food manufacturing and processing
- Retail

In addition to the ISAC, DHS created the Food and Agriculture Sector Coordinating Council, which oversees food security and incident management. The exact methods for control and eradication operations are difficult to predict. Past experience and simulations have shown that day-to-day decisions would be made using "decision trees" that include factors such as the geographical spread, rates of infestation, available personnel, public sentiment, and industry cooperation. All of the response would be coordinated through the National Incident Management System (NIMS) with guidance from the National Response Framework (NRF) in a large-scale event. For USDA, response procedures are outlined in the *USDA Emergency Response Manual.*

In an agriculture outbreak, damage is proportional to the time it takes to first detect the pathogen. If a foreign disease is introduced, responsibility for recognizing initial symptoms rests with farmers, producers, veterinarians, plant pathologists, and entomologists. Cooperative Extension Service agents at state universities are receiving additional training on recognizing the likely symptoms of an agroterrorism attack.

Effective detection depends on a heightened sense of awareness, and on the ability to rapidly determine the level of threat (e.g., developing and deploying rapid disease diagnostic tools). Lessons from disease outbreaks, including the FMD outbreaks in Europe and avian flu in Asia and the United States, show that the speed of detection, diagnosis, and control spell the difference between an isolated incident and an economic and public health disaster.

However, in recent years, the number of veterinarians with experience to recognize many foreign animal diseases has declined. This is because the United States has been successful in eradicating many animal diseases. Also, the number of veterinarians available across the country with large animal experience and within APHIS has declined. In light of this trend, APHIS has initiated efforts to increase training for foreign animal diseases and create registries of veterinarians with appropriate experience.

Most of the initial response to the diagnosis of a foreign animal disease is at the state and local level. If an outbreak spreads across state lines or if state and local efforts are unable to control the outbreak, federal involvement quickly follows and the U.S. Secretary of Agriculture has the authority to assume control of the response. Numerous simulation exercises have been conducted by federal, state and local authorities to test the response and coordination efforts to an agroterrorism attack.

The last line of defense, and the costliest, is the isolation, control, and eradication of an epidemic. The more geographically widespread a disease outbreak, the costlier and more drastic the control measures become. Officials gained valuable experience from recent agricultural disease outbreaks such as avian influenza in the United States, Canada, and Asia; FMD in the United Kingdom; and citrus canker in Florida. Each one of these epidemics has required the depopulation and destruction of livestock and crops in quarantine areas, indemnity payments to farmers, and immediate suspension of trade.

Of all lines of defense, mass eradication is the most politically sensitive and difficult. Actions taken in each of these outbreaks have met with varying degrees of resistance from groups opposed to mass slaughter of animals, citizens concerned about environmental impacts of destroying carcasses and large fields of crops, or from farmers who fear the loss of their livelihood. During the 2001 outbreak of FMD in the United Kingdom, the public was clearly opposed to the large piles of burning carcasses. The disposal of millions of chicken carcasses in British Columbia, Canada, during 2004 also caused a significant public debate. Thus, scientific alternatives are needed for mass slaughter and carcass disposal. Citrus canker eradication efforts in Florida's residential neighborhoods illustrate how science-based measures have been challenged and delayed in the courts, or how farmers may be reluctant to voluntarily test crops or livestock.

Laboratories and Research

Since September 11, 2001, the United States has expanded its agricultural laboratory and diagnostic infrastructure, and created networks to share information and process samples; this has created a system of "surge capacity." The USDA-funded National Plant Diagnostic Network (NPDN) which divides the United States into five regions and its sister group, the National Animal Health Laboratory Network a collection of animal health laboratories networked across the nation, has greatly reduced the detection time of animal diseases and plant pathogens. A main goal of each is to improve the diagnostic and detection system in the event of a deliberate or accidental disease outbreak.

The effectiveness of these networks requires coordinated outreach and cooperative extension services have taken on new prominence in their role of providing information about diseases such as soybean rust to farmers and others who have regular contact with farms. New Mexico has networked their cooperative extension agents into a state and local food alliance for the information sharing and recall notification.

Within the USDA, several agencies have upgraded their facilities to respond better to the threat of agroterrorism by expanding laboratory capacity and adding physical security. These programs include the ARS research on foreign animal diseases at the Plum Island Animal Disease Center in New York (the physical facility is now managed and operated by DHS and being moved to Manhattan, KS) and the ARS Southeast Poultry Research Laboratory in Athens, Georgia.

Also at USDA, three major laboratories have consolidated operations in a new BSL-3 facility in Ames, Iowa. These include the ARS National Animal Disease Center, the APHIS National Veterinary Services Laboratories (NVSL), and the APHIS Center for Veterinary Biologics. The complex is currently USDA's largest animal health center for research, diagnosis, and product evaluation. The NVSL is especially visible because it makes the final determination of most animal diseases when samples are submitted for testing.

USDA also cooperates with other federal agencies on counterterrorism research and preparedness, including the ARS and APHIS partnership with the U.S. Army Medical Research Institute for Infectious Diseases at Ft. Dietrick , Maryland. The Ft. Dietrick site offers USDA access to additional high-level biosecurity laboratories. In the recent past, USDA has conducted research on soybean rust at Ft. Dietrick.

Federal Authorities

When a foreign animal disease is discovered, whether accidentally or intentionally introduced, the Secretary of Agriculture has broad authority to eradicate it or prevent it from entering the country. The use of these authorities is fairly common, as shown by the import restrictions imposed during the 2004 outbreak of avian influenza in Asia. Federal quarantines and restrictions on interstate movement within the United States are also common for certain pest and disease outbreaks, such as for sudden oak death in California and citrus canker in Florida. In addition to federal authorities, most states have similar authorities, at least for quarantine and import restrictions.

For example, if an animal disease outbreak is found in the United States, the Secretary of Agriculture is authorized, among other things, to

- Stop imports of animals and animal products into the United States from suspected countries (7 U.S.C. 8303).
- Stop animal exports (7 U.S.C. 8304) and interstate transport of diseased or suspected animals (7 U.S.C. 8305).

- Seize, quarantine, and dispose of infected livestock to prevent dissemination of the disease (7 U.S.C. 8306).
- Compensate owners for the fair market value of animals destroyed by the Secretary's orders (7 U.S.C. 8306(d)).
- Transfer the necessary funding from USDA's Commodity Credit Corporation to cover costs of eradication, quarantine, and compensation programs (7 U.S.C. 8316).

Similar authorities cover plant pests and diseases (7 U.S.C. 7701–7772).

Recovery Management

Several activities such as confinement and eradication start in the response phase but continue throughout the management and recovery phase. Long-term economic recovery includes resuming the husbandry of animals and plants in the affected areas, introducing new genetic traits that may be necessary in response to the pest or disease, rebuilding confidence in domestic markets, and regaining international market share.

Confidence in food markets, by both domestic and international customers, depends on continuing surveillance after the threat is controlled or eradicated. Communication and education programs would need to inform growers directly affected by the outbreak, and inform consumers about the source and safety of their food. The social sciences and public health institutions play a complementary role to the agricultural sciences in responding to and recovering from agroterrorism.

If eradication of the pest or disease is not possible, an endemic infestation would result in a lower equilibrium level of production or quality. Resources would be devoted to acquiring plant varieties with resistance characteristics and breeds of animals more suitable to the new environment.

Summary

In the wake of the events of September 11, 2001, it must be understood that the terrorist threat exists in the nation at all times, and it is certainly possible that some form of agroterrorism, perhaps in conjunction with biological, chemical, radiological or even nuclear threats, could happen and therefore, preparation is necessary. This is especially true in the realm of agroterrorism, where such an incident, even one that is considered a low probability, but high consequence event, could have severe effects on consumer confidence, the supply-and-demand economy, and the various associated businesses that would be affected by some form of terrorism-caused outbreak related to agriculture and food. Agroterrorism will not carry the shock value that bombings and hijackings do, and the effect on human life may not be as severe or immediate, but significant economic issues can arise that affect many facets of the population beyond the farmer. These issues could expand all the way to U.S. import and export markets, to the federal government itself, which could incur significant costs to contain and eradicate the threat, as well as potentially compensating farmers for destroyed animals and crops. Such destruction, because of the discovery of a terrorist threat, could also raise environmental and other health issues that must be addressed, using more time and resources at all levels.

After September 11, 2001, there was a significant increase in the attention paid to a variety of terrorist threats, both large and small. But for various reasons, there was less attention paid to agroterrorism. This has somewhat been addressed, through acts such as HSPD-7, which added agriculture to the list of critical infrastructure that must be protected. HSPD-9 took this a step further by establishing a national policy to protect against terrorist attacks on agriculture and food systems.

Maladies that could potentially strike in large scales against herds of cattle and other types of animals have been thought to be dealt with through vaccination and other programs to educate farmers on their potential dangers. However, scientists now believe that livestock herds are much more susceptible to agroterrorism than crops, because current herds either are not vaccinated against threats or are relatively unmonitored against such threats, because they may have been thought to be eradicated previously. Certain animal diseases may be more attractive to terrorists because they can be zoonotic or transmissible to humans.

The Agricultural Bioterrorism Protection Act of 2002 created the current, official list of animal pathogens that are of greatest concern for agroterrorism. The act requires that these lists be reviewed at least every 2 years. In addition, there is overlap between the CDC and APHIS because some pathogens on the list may not cause a disease, but may cause symptoms such as food poisoning or responses in the central nervous system. One pathogen, FMD, is mentioned often when agroterrorism is discussed, because of its ease of use, ability to spread quickly, and potential for tremendous economic damage. There is also a similar list of plant pathogens, as required by the Agricultural Bioterrorism Protection Act of 2002. The goal of the U.S. animal and plant health safeguarding system is to prevent the introduction and establishment of exotic plants and diseases to mitigate their effects and eradicate them where necessary/possible. Part of this effort requires coordination and cooperation between federal agencies to not only safeguard domestic products and resources, but also those that may be imported in from foreign countries. Through inspection of cargo and the requirement of importers to report specific types of cargo that could fall into an agroterrorism issue, agencies are increasingly obtaining the legal authority to fight the entry of foreign diseases or agents. However, should a foreign animal disease be discovered, whether accidentally or intentionally introduced, the U.S. Secretary of Agriculture has broad authority to eradicate it. The use of these authorities is fairly common because the agriculture and food industry is so dynamic. Federal quarantines and restrictions on interstate movement within the United States are also common for certain pest and disease outbreaks.

The federal government, state governments, and local authorities have, through the involvement of many agencies and offices, taken steps to prevent agroterrorism wherever possible, and to effectively respond to an incident should an outbreak occur. The fact remains, however, that agroterrorism while considered a low probability event could have far-reaching and devastating effects to our nation in many ways.

References

1. H. S. Parker, Agricultural bioterrorism: A federal strategy to meet the threat, McNair Paper 65, National Defense University, March 2002.
2. Monterey Institute of International Studies.
3. P. Chalk, *Hitting America's Soft Underbelly: The Potential Threat of Deliberate Biological Attacks Against US Agriculture and Food Industry*, Rand National Defense Research Institute, January 2004.
4. C. Hanrahan and G. Beckier, Mad cow disease and US beef trade, CRS Report RS21709, August 4, 2004.
5. *United States Animal Health Association's "Gray Book,"* http:www.bt.cdc.gov/agent/agentlist-category.asp, accessed September 13, 2013.
6. *Terrestrial Animal Health Code*, 13th Edition, May 2004.
7. J. Graham, J. Hutton, S. Coa, M. Fagel, and W. Wright, National Cooperative Highway Research Program, *A Guide to Traffic Control of Rural Roads in an Agricultural Emergency*. Midwest Research Institute, Kansas City, MO, 2008.
8. Food Marketing Institute, "Food and Agriculture ISAC," (Information Sharing and Analysis Center), at: http://www.fmi.org/news-room/news-archive/view/2002/02/15/fmi-establishesfood-industry-security-information-and-analysis-center-in-cooperation-with-fbiand-nipc

Index

B

C

D

F

G

H

J

K

L

M

N

O

R

S

T

U

V

Milton Keynes UK
Ingram Content Group UK Ltd.
UKHW022143181223
434627UK00003B/7

9 781466 555051